CORROSION PREVENTION AND CONTROL IN WATER TREATMENT AND SUPPLY SYSTEMS

CORROSION PREVENTION AND CONTROL IN WATER TREATMENT AND SUPPLY SYSTEMS

by

J.E. Singley, B.A. Beaudet, P.H. Markey
Environmental Science and Engineering, Inc.
Gainesville, Florida

D.W. DeBerry, J.R. Kidwell, D.A. Malish
SumX Corporation
Austin, Texas

np
NOYES PUBLICATIONS
Park Ridge, New Jersey, U.S.A.

Library of Congress Catalog Card Number 85-4915
ISBN: 0-8155-1031-4
ISSN: 0090-516X
Printed in the United States

Published in the United States of America by
Noyes Publications
Mill Road, Park Ridge, New Jersey 07656

10 9 8 7 6 5 4 3 2 1

Library of Congress Cataloging in Publication Data
Main entry under title:

Corrosion prevention and control in water treatment
and supply systems.

(Pollution technology review, ISSN 0090-516X ;
no. 122)
Includes bibliographies and index.
1. Waterworks--Corrosion. 2. Corrosion and anti-
corrosives--Handbooks, manuals, etc. I. Singley, J.E.
II. Series.
TD487.C67 1985 628.1 85-4915
ISBN 0-8155-1031-4

Foreword

Corrosion prevention and control methodology for water treatment and supply systems is detailed in this book. The information supplied will provide water treatment managers and operators with an understanding of the causes and control of corrosion.

The corrosion of water treatment and supply systems is a very significant concern. Not only does it affect the aesthetic quality of the water but it also has an economic impact and poses adverse health implications. Corrosion by-products containing materials such as lead and cadmium have been associated with serious risks to the health of consumers of drinking water. In addition, corrosion-related contaminants commonly include compounds such as zinc, iron, and copper, which adversely affect the aesthetic aspects of the water.

The book is presented in two parts. Part I is basically a guidance manual for corrosion control with sections on how and why corrosion occurs and how best to handle it. Part II reviews the various materials used in the water works industry and their corrosion characteristics, as well as monitoring and detection techniques. Emphasis is placed on assessing the conditions and water quality characteristics due to the corrosion or deterioration of each of these materials.

The information in the book is from:

> *Corrosion Manual for Internal Corrosion of Water Distribution Systems* by J.E. Singley, B.A. Beaudet and P.H. Markey of Environmental Science and Engineering, Inc. under subcontract to Oak Ridge National Laboratory for the U.S. Department of Energy, April 1984.
>
> *Corrosion in Potable Water Systems* by David W. DeBerry, James R. Kidwell and David A. Malish of SumX Corporation for the U.S. Environmental Protection Agency, February 1982.

The table of contents is organized in such a way as to serve as a subject index and provides easy access to the information contained in the book.

Advanced composition and production methods developed by Noyes Publications are employed to bring this durably bound book to you in a minimum of time. Special techniques are used to close the gap between "manuscript" and "completed book." In order to keep the price of the book to a reasonable level, it has been partially reproduced by photo-offset directly from the original reports and the cost saving passed on to the reader. Due to this method of publishing, certain portions of the book may be less legible than desired.

NOTICE

The materials in this book were prepared as accounts of work sponsored by the U.S. Environmental Protection Agency and the U.S. Department of Energy. Publication does not signify that the contents necessarily reflect the views and policies of the contracting agencies or the publisher, nor does mention of trade names or commercial products constitute endorsement or recommendation for use.

Contents and Subject Index

PART I
GUIDANCE MANUAL FOR CORROSION CONTROL

PART II
REVIEW OF MONITORING, DETECTION, PREVENTION AND CONTROL TECHNIQUES

Part I

Guidance Manual for Corrosion Control

The information in Part I is from *Corrosion Manual for Internal Corrosion of Water Distribution Systems* by J.E. Singley, B.A. Beaudet and P.H. Markey of Environmental Science and Engineering, Inc. under subcontract to Oak Ridge National Laboratory for the U.S. Department of Energy, April 1984.

Acknowledgments

This manual was prepared by Environmental Science and Engineering, Inc. (ESE) of Gainesville, Florida. Dr. J. Edward Singley was Project Director and Senior Technical Advisor; Mr. Bevin A. Beaudet, P.E., was Project Manager; and Ms. Patricia H. Markey was Project Engineer. During the preparation of the manual, invaluable technical review and input were received from several individuals and agencies.

Appreciation is expressed to the Office of Drinking Water, U.S. Environmental Protection Agency (EPA), most particularly to Mr. Peter Lassovszky, Project Officer, for his direction and guidance through all stages of the writing.

Each draft of the manual was reviewed by a Blue Ribbon Panel of experts selected for their expertise and knowledge in the field of corrosion of potable water distribution systems. Special acknowledgment is due the following individuals, who served on this panel:

Mr. Russell W. Lane, P.E., Water Treatment Consultant; former head of the Illinois State Water Survey and professor, University of Illinois, Urbana-Champaign, Illinois.

Mr. Frank J. Baumann, P.E., Chief, Southern California Branch Laboratory, State of California Department of Health Services, Los Angeles, California.

Mr. Douglas Corey, South Dade Utilities, Miami, Florida; 1982 President of Florida Water and Pollution Control Operators Association, Inc.

Appreciation is expressed to Dr. Sidney Sussman, Technical Director of Olin Water Services for supplying several of the example photographs throughout the manual and for his contribution to the inhibitor treatment material in Section 7. Mr. Thomas F. Flynn, P.E., President of Shannon Chemical, also supplied valuable input to the section on inhibitor treatment. Dr. Jiterdra Saxena and Arthur Perler, Office of Drinking Water, provided a section on regulatory aspects associated with the use of inhibitors.

Acknowledgment is also due members of the American Water Works Association (AWWA) Research Foundation and individuals from EPA who reviewed the manual and provided technical assistance and input. Individuals deserving particular mention are Mr. James F. Manwaring, P.E., Executive Director, AWWA Research Foundation; Dr. Marvin Gardels, Mr. Michael R. Schock, and Dr. Gary S. Logsdon, from EPA Cincinnati; Mr. Peter Karalekas, P.E., EPA Region I; Dr. Mark A. McClanahan, EPA Region IV; Mr. Harry Von Huben, EPA Region V; Mr. Roy Jones, EPA Region X; and Mr. Hugh Hanson, Chief, Science and Technology Branch, Criteria and Standards Division, Office of Drinking Water, EPA.

Appreciation is also expressed to Dr. Joseph A. Cotruvo, Director, and Mr. Craig Vogt, Deputy Director, Criteria and Standards Division, Office of Drinking Water, EPA, for their support.

Acronyms

A-C	asbestos-cement
AI	Aggressive Index
ASTM	American Society for Testing and Materials
AWWA	American Water Works Association
CI	Riddick's Corrosion Index
CPW	Commissioners of Public Works
DFI	McCauley's Driving Force Index
DO	dissolved oxygen
DWRD	Drinking Water Research Division
EPA	U.S. Environmental Protection Agency
ESE	Environmental Science and Engineering, Inc.
ISWS	Illinois State Water Survey
LSI	Langelier Saturation Index
MCL	maximum contaminant level
MDC	Metropolitan District Commission
MWC	Middlesex Water Company
NACE	National Association of Corrosion Engineers
NAS	National Academy of Sciences
NIPDWR	National Interim Primary Drinking Water Regulations
ODW	Office of Drinking Water
ORNL	Oak Ridge National Laboratory
PCWS	Pinellas County Water System
PVC	polyvinyl chloride
RMICs	recommended maximum impurity concentrations
RSI	Ryznar Stability Index
SEM	scanning electron microscope
TDS	total dissolved solids

Frequently Used Units and Other Terms

MGD	million gallons per day
$CaCO_3$	calcium carbonate
H_2S	hydrogen sulfide
CO_2	carbon dioxide
NaOH	sodium hydroxide
SZP	sodium zinc phosphate
ZOP	zinc orthophosphate
gpm	gallons per minute
CaO	quicklime
mpy	mils per year
mg/cm^2	milligrams per centimeter square
mg/L	milligrams per liter

1. Purpose

This manual was written to give the operators of potable water treatment plants and distribution systems an understanding of the causes and control of corrosion. The many types of corrosion and the types of materials with which the water comes in contact make the problem more complicated. Because all operators have not had the opportunity to gain more than a basic understanding of chemistry and engineering, there is little of these disciplines included in the document.

The goal in writing the manual was to create a "how-to" guide that would contain additional information for those who want to study corrosion in more detail. Sections 3, 4, and 5 can be skipped in cases in which an immediate problem needs to be solved. Those sections, though, do help in understanding how and why corrosion occurs.

2. Introduction

Corrosion of distribution piping and of home plumbing and fixtures has been estimated to cost the public water supply industry more than $700 million per year. Two toxic metals that occur in tap water, almost entirely because of corrosion, are lead and cadmium. Three other metals, usually present because of corrosion, cause staining of fixtures, or metallic taste, or both. These are copper (blue stains and metallic taste), iron (red-brown stains and metallic taste), and zinc (metallic taste).

Since the Safe Drinking Water Act (P.L. 93-523) makes the supplying utility responsible for the water quality at the customer's tap, it is necessary to prevent these metals from getting into the water on the way to the tap.

The toxic metals lead and cadmium can cause serious health problems when present in quantities above the levels set by the National Interim Primary Drinkig Water Regulations (NIPDWR). The other metals—copper, iron, and zinc—are included in the Secondary Drinking Water Regulations because they cause the water to be less attractive to consumers and thus may cause them to use another, potentially less safe, source.

The corrosion products in the distribution system can also protect bacteria, yeasts, and other microorganisms. In a corroded environment, these organisms can reproduce and cause many problems such as bad tastes, odors, and slimes. Such organisms can also cause further corrosion themselves.

Corrosion-caused problems that add to the cost of water include

1. increased pumping costs due to corrosion products clogging the lines;
2. holes in the pipes, which cause loss of water and water pressure;
3. leaks and clogs, as well as water damage to the dwelling, which would require that pipes and fittings be replaced;
4. excessive corrosion, which would necessitate replacing hot water heaters; and
5. responding to customer complaints of "colored water," "stains," or "bad taste," which is expensive both in terms of money and public relations.

Corrosion is one of the most important problems in the water utility industry. It can affect public health, public acceptance of a water supply, and the cost of providing safe water. Many times the problem is not given the attention it needs until expensive changes or repairs are required.

Both the Primary and Secondary Regulations recognize that corrosion is a serious concern. However, the lack of a universal measurement or index for corrosivity has made it difficult to regulate. The United States Environmental Protection Agency (EPA) recognizes that corrosion problems are unique to each individual water supply system. As a result, the August 1980 amendments to the NIPDWR issued by EPA concentrate on identifying both potentially corrosive waters and finding out what materials are in distribution systems. The 1980 amendments to the regulations require that

1. All community water supply systems collect and analyze samples for the following corrosion characteristics: alkalinity, pH, hardness, temperature, total dissolved solids (TDS), and Langelier Saturation Index (LSI) [or Aggressive Index (AI) in certain cases]. "Corrosivity characteristics" need to be monitored and reported only once, unless individual states require additional sampling.
2. The samples be taken at a representative point in the distribution system. Two samples are to be taken within 1 year from each treatment plant, using a surface water source to account for extremes in seasonal variations. One sample per plant is required for plants using groundwater sources.

3. Community water supply systems identify whether the following construction materials are present in their distribution system, including service lines and home plumbing, and report their findings to the state: (a) lead from piping, solder, caulking, interior lining of distribution mains, alloys, and home plumbing; (b) copper from piping and alloys, service lines, and home plumbing; (c) galvanized piping, service lines, and home plumbing; (d) ferrous piping materials, such as cast iron and steel; and (e) asbestos-cement (A-C) pipe.

   In addition, states may require the identification and reporting of other construction materials present in distribution systems that may contribute contaminants to the drinking water, such as (f) vinyl-lined A-C pipe and (g) coal tar-lined pipes and tanks.

3. Definition of Corrosion and Basic Theory

3.1 DEFINITION

What is corrosion?

Corrosion is the deterioration of a substance or its properties due to a reaction with its environment. In the waterworks industry, the "substance" which deteriorates may be a metal pipe or fixture, the cement in a pipe lining, or an asbestos-cement (A-C) pipe. For internal corrosion, the "environment" of concern is water.

A common question is, "What type of water causes corrosion?" The correct answer is, "All waters are corrosive to some degree." A water's corrosive tendency will depend on its physical and chemical characteristics. Also, the nature of the material with which the water comes in contact is important. For example, water corrosive to galvanized iron pipe may be relatively noncorrosive to copper pipe in the same system.

3.2 BASIC THEORY

Physical and chemical actions between pipe material and water may cause corrosion. An example of a physical action is the erosion or wearing away of a pipe elbow because of excess flow velocity in the pipe. An example of a chemical action is the oxidation or rusting of an iron pipe. Biological growths in a distribution system can also cause corrosion by providing a suitable environment in which physical and chemical actions can occur. The actual mechanisms of corrosion in a water distribution system are usually a complex and interrelated combination of these physical, chemical, and biological actions.

Following is a discussion of the basic chemical reactions which cause corrosion in water distribution systems, for both metallic and nonmetallic pipes. Familiarity with these basic reactions will help users recognize and correct corrosion problems associated with water utilities.

A more detailed, yet relatively basic, discussion of the theory of corrosion can be found in an excellent book titled *NACE Basic Corrosion Course*, published by the National Association of Corrosion Engineers (NACE), which is now in its fifth printing.

Electrochemical Corrosion of Metal Pipes

Metals are generally most stable in their natural form. In most cases, this stable form is the same form in which they occur in native ores and from which they are extracted in processing. Iron ore, for instance, is essentially a form of iron oxide, as is rust from a corroded iron pipe. The primary cause of metallic corrosion is the tendency (also called activity) of a metal to return to its natural state. Some metals are more active than others and have a greater tendency to enter into solution as ions and to form various compounds. Table 3.1 lists the relative order of activity of several commonly used metals and alloys. Such a listing is also called a "galvanic series," for reasons which are discussed below.

When metals are chemically corroded in water, the mechanism involves some aspect of electrochemistry. When a metal goes into solution as an ion or reacts in water with another element to form a compound, electrons (electricity) will flow from certain areas of a metal surface to other areas through the metal.

The term "anode" is used to describe that part of the metal surface that is corroded and from which electric current, as electrons, flows through the metal to the other electrode. The term "cathode" is used to describe the metal surface from which current, as ions, leaves the metal and returns to the anode through the solution. Thus, the circuit is completed. All water solutions will conduct a current. "Conductivity" is a measure of that property.

Figure 3.1 is a simplified diagram of the anodic and cathodic reactions that occur when iron is in contact with water. The anode and cathode areas may be located in different areas of the pipe, as shown in Fig. 3.1, or they can be located right next to each other. The anode and cathode areas

Table 3.1. Galvanic series — Order of activity of common metals used in water distribution systems

Metal	Activity
Zinc	More active
Mild steel	↑
Cast iron	|
Lead	|
Brass	|
Copper	|
Stainless steel	Less active

Source: Environmental Science and Engineering, Inc., 1982.

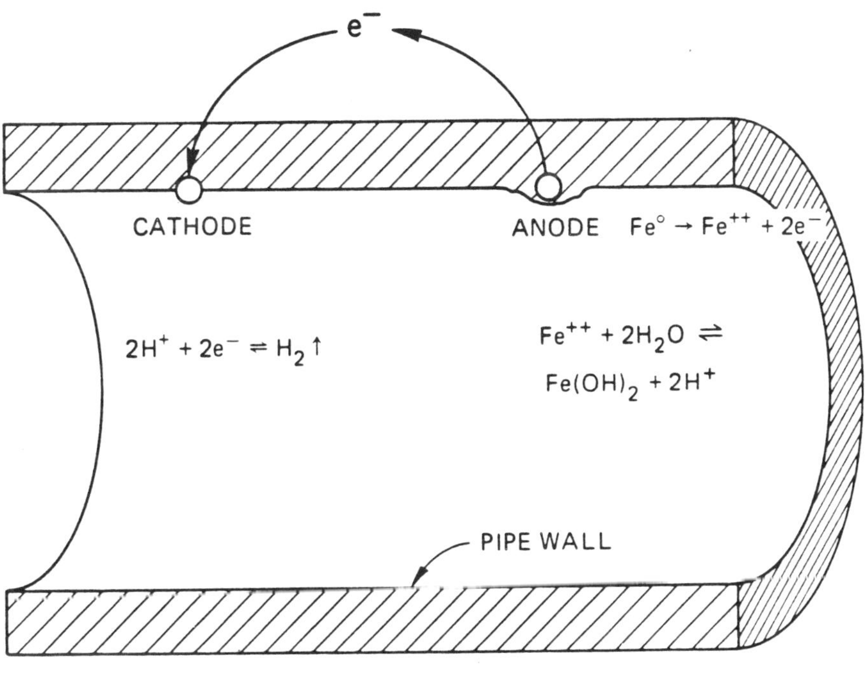

Fig. 3.1. Simplified anode and cathode reactions of iron in contact with water. Source of H^+ ions is the normal dissociation of water, $H_2O \rightleftarrows H^+ + OH^-$.

can set up a circuit in the same metal or between two different metals which are connected. In the case of iron corrosion, as the free iron metal goes into solution in the form Fe^{++} (ferrous) ion at the anode, two electrons are released. These electrons, having passed through the metal pipe, combine at the cathode with H^+ (hydrogen) ions that are always present due to the normal dissociation of water, according to ($H_2O \rightleftharpoons H^+ + OH^-$). This action forms hydrogen gas, which collects on the cathode and thus slows the reaction (polarization). The Fe^{++} ions released at the anode react further with the water to form ferrous hydroxide, $Fe(OH)_2$.

Oxygen plays a major role in the internal corrosion of water distribution systems. Oxygen dissolved in water reacts with the initial corrosion reaction products at both the anodic and cathodic regions. Ferrous (iron II) hydroxide formed at the anode reacts with oxygen to form ferric (iron III) hydroxide, $Fe(OH)_3$, or rust. Oxygen also reacts with the hydrogen gas evolved at the cathode to form water, thus allowing the initial anodic reaction to continue (depolarization).

The simplified equations that describe the role of oxygen in aiding iron corrosion are shown below. Similar equations could be shown for copper or other corroding metals. Equations (1) and (2) are for anodic reactions and Eq. (3) shows cathodic reactions.

$4Fe^{++}$	+	$10H_2O$	+	O_2	→	$4Fe(OH)_3$	+	$8H^+$	(1)
ferrous iron	+	water	+	free oxygen	→	ferric hydroxide	+	hydrogen	

or

$4Fe(OH)_2$	+	$2H_2O$	+	O_2	→	$4Fe(OH)_3$	(2)
ferrous hydroxide	+	water	+	free oxygen	→	ferric hydroxide	
$4H^+$	+	4e	+	O_2	→	$2H_2O$	(3)
hydrogen	+	electrons	+	oxygen	→	water	

The importance of dissolved oxygen (DO) in corrosion reactions of iron pipe is shown in Fig. 3.2.

A similar electrochemical reaction occurs when two dissimilar metals are in direct contact in a conducting solution. Such a connection is commonly called a "galvanic couple." An example of a galvanic couple would be a ductile iron nipple used to connect two pieces of copper pipe. In this case, the more active metal, iron, would corrode at the anode and give up electrons to the cathode. The net effect would be a slowing down or stopping of copper corrosion and an acceleration of iron corrosion where the metals are in contact. Figure 3.3 illustrates a typical galvanic cell. In addition, the farther apart the two dissimilar metals are in the galvanic series (see Table 3.1), the greater the corrosive tendencies. For example, a copper-to-zinc connection would be more likely to corrode than a copper-to-brass connection.

Corrosion of Metallic Lead

Metallic lead can be present in distribution systems either in the form of lead service pipes, found in many older systems, or in lead/tin solder used to join copper household plumbing. Lead is a stable metal of relatively low solubility and is structurally resistant to corrosion. However, the toxic effects of lead are pronounced [the NIPDWR maximum contaminant level (MCL) for lead is 0.05 milligram per liter (mg/L)]. Thus, even low levels of lead corrosion may be of major concern.

Metallic lead is frequently protected from corrosion by a thin layer of insoluble lead carbonates that forms on the surface of the metal. The solubility of metallic lead (plumbosolvency) is complicated and is related to the pH and the carbonate content (alkalinity) of the water. Consistent control of pH in the presence of sufficient alkalinity will generally minimize plumbosolvency in water distribution systems.

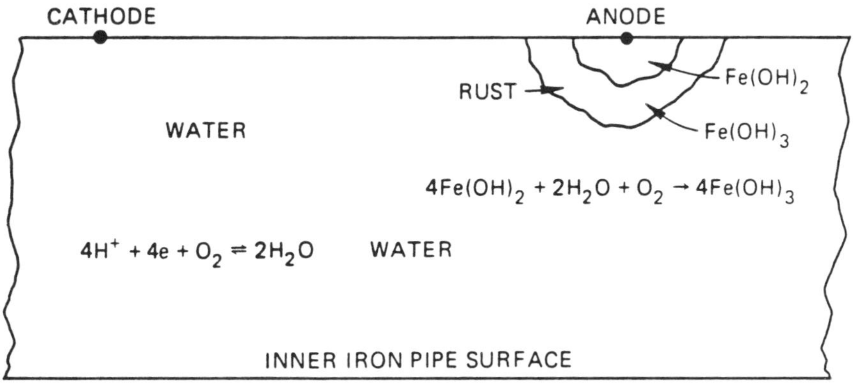

Fig. 3.2. Role of oxygen in iron corrosion. Source: ESE, 1982.

ORNL DWG 83-17053

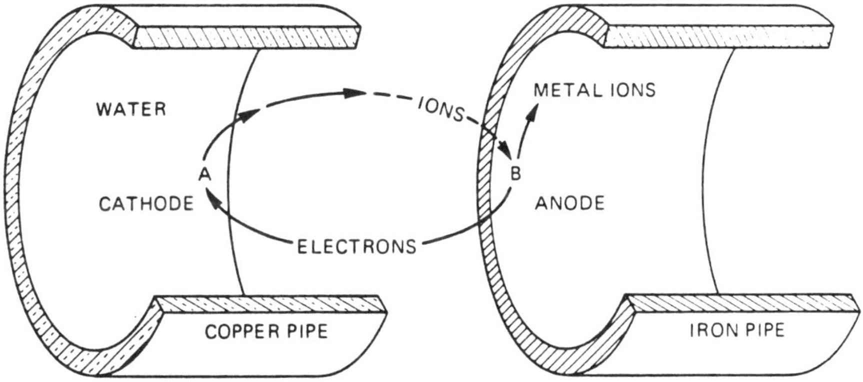

Fig. 3.3. Simplified galvanic cell. Note that areas A and B are located on the inner pipe surface.

Corrosion of Cement Materials

The corrosion of cement-lined pipe, concrete pipe, or A-C pipe is primarily a chemical reaction in which the cement is dissolved by water. Cement materials are made up of numerous, crystalline compounds which normally are hard, durable, and relatively insoluble in water.

Modern, autoclave-curved (Type II) A-C pipe is formed from a mixture of three main ingredients:

Ingredient	Percentage by weight
Asbestos fiber	15–20
Silica flour (ground sand or silicon dioxide)	34–37
Portland cement	51–48

The calcium-containing Portland cement serves as a binder, and the autoclaving process reduces free lime content to less than 1%. Silica flour acts as a reactive aggregate for the cement. The asbestos fibers give flexibility and structural strength to the finished product. When calcium is leached from the cement binder by the action of an aggressive (corrosive) water, the interior pipe surface is softened, and asbestos fibers may be released.

Type I A-C pipe was widely used before the 1950s and may be present in many older systems. Unlike Type II, Type I has no silica flour but contains 15 to 20% asbestos fibers, 80 to 85% Portland cement, and 12 to 20% free lime. Calcium leaching is more commonly observed in Type I A-C pipe.

The solubility of the calcium-containing cement compounds is pH dependent. At low pH (less than about 6.0), the leaching of these compounds from the pipe is much more pronounced than at a pH above 7.0. The solubility of a cement lining, concrete pipe, or an A-C pipe in a given water can be approximated by the tendency of that water to dissolve calcium carbonate ($CaCO_3$).

3.3 CHARACTERISTICS OF WATER THAT AFFECT CORROSIVITY

In Sect. 3.1, corrosion is defined as the deterioration of a material (or is properties) because of a reaction with its environment. In the waterworks industry, the materials of interest are the distribution and home water plumbing systems, and the environment that may cause internal pipe corrosion is drinking water.

For operators or managers of water utilities, the obvious question is, "What characteristics of this drinking water determine whether or not it is corrosive?" The answers to this question are important because waterworks personnel can control, to some extent, the characteristics of this drinking water environment.

Those characteristics of drinking water that affect the occurrence and rate of corrosion can be classified as (1) physical, (2) chemical, and (3) biological. In most cases, corrosion is caused or increased by a complex interaction among several factors. Some of the more common characteristics in each group are discussed in the following paragraphs to familiarize the reader with their potential effects. Controlling corrosion may require changing more than one of these because of their interrelationship.

Physical Characteristics

Flow velocity and temperature are the two main physical characteristics of water that affect corrosion.

Velocity. Flow velocity has seemingly contradictory effects. In waters with protective properties, such as those with scale-forming tendencies, high flow velocities can aid in the formation of protective coatings by transporting the protective material to the surfaces at a higher rate. However, high flow velocities are usually associated with erosion corrosion in copper pipes in which the protective wall coating or the pipe material itself is removed mechanically. High velocity waters combined with other corrosive characteristics can rapidly deteriorate pipe materials.

Another way in which high velocity flow can contribute to corrosion is by increasing the rate at which DO comes in contact with pipe surfaces. Oxygen often plays an important role in determining corrosion rates because it enters into many of the chemical reactions which occur during the corrosion process.

Extremely low velocity flows may also cause corrosion in water systems. Stagnant flows in water mains and household plumbing have occasionally been shown to promote tuberculation and pitting, especially in iron pipe, as well as biological growths. Therefore, one should avoid dead ends.

Proper hydraulic design of distribution and plumbing systems can prevent or minimize erosion corrosion of water lines. The NACE, the American Society for Testing and Materials (ASTM), and pipe manufacturers can provide guidance on design criteria for standard construction materials. A maximum value of 4 feet per second (ft/s), 9.8 gallons per minute (gal/min) in a 1-inch pipe for instance, is recommended for Type K copper tubing.

Temperature. Temperature effects are complex and depend on the water chemistry and type of construction material present in the system. Three basic effects of temperature change on corrosion rates are discussed here.

In general, the rate of all chemical reactions, including corrosion reactions, increases with increased temperature. All other aspects being equal, hot water should be more corrosive than cold. Water which shows no corrosive characteristics in the distribution system can cause severe damage to copper or galvanized iron hot water heaters at elevated temperatures. Figure 3.4 shows the inside of a water heater totally destroyed by pitting corrosion. The same water showed no corrosive characteristics in other parts of the distribution system.

Second, temperature significantly affects the dissolving of $CaCO_3$. Less $CaCO_3$ dissolves at higher temperatures, which means that $CaCO_3$ tends to come out of solution (precipitate) and form a protective scale more readily at higher temperatures. The protective coating resulting from this precipitation can reduce corrosion in a system. On the other hand, excessive deposition of $CaCO_3$ can clog hot water lines.

Finally, a temperature increase can change the entire nature of the corrosion. For example, a water which exhibits pitting at cold temperatures may cause uniform corrosion when hot. Although the total quantity of metal dissolved may increase, the attack is less acute, and the pipe will have a longer life. Another example in which the nature of the corrosion is changed as a result of changes in temperature involves a zinc-iron couple. Normally, the anodic zinc is sacrificed or corroded to prevent iron corrosion. In some waters, the normal potential of the zinc-iron couple may be reversed at temperatures above 140°F. In other words, the zinc becomes cathodic to the iron, and the corrosion rate of galvanized iron is much higher than is normally anticipated. Galvanized iron hot-water heaters can be especially susceptible to this change in potential at temperatures greater than 140°F.

Chemical characteristics

Most of the corrosion discussed in this manual involves the reaction of water with the piping. The substances dissolved in the water have an important effect on both corrosion and corrosion control. To understand these reactions thoroughly requires more knowledge of water chemistry than could be imparted here, but a brief overview will point out some of the most important factors. Table 3.2 lists some of the chemical factors that have been shown to have some effect on corrosion or corrosion control.

Several of these factors are closely related, and a change in one changes another. The most important example of this is the relationship between pH, carbon dioxide (CO_2), and alkalinity. Although it is frequently said that CO_2 is a factor in corrosion, no corrosion reactions include CO_2. The important corrosion effect results from pH, and pH is affected by a change in CO_2. It is not necessary to know all of the complex equations for these calculations, but it is useful to know that each of these factors plays some role in corrosion.

Following is a description of some of the corrosion-related effects of the factors listed in Table 3.2. A better understanding of their relationship to one another will aid in understanding corrosion and thus in choosing corrosion control methods.

pH. pH is a measure of the concentration of hydrogen ions, H^+, present in water. Since H^+ is one of the major substances that accepts the electrons given up by a metal when it corrodes, pH is an important factor to measure. At pH values below about 5, both iron and copper corrode rapidly and uniformly. At values higher than 9, both iron and copper are usually protected. However, under certain conditions corrosion may be greater at high pH values. Between pH 5 and 9, pitting is likely to occur if no protective film is present. The pH also affects the formation or solubility of protective films, as will be discussed later.

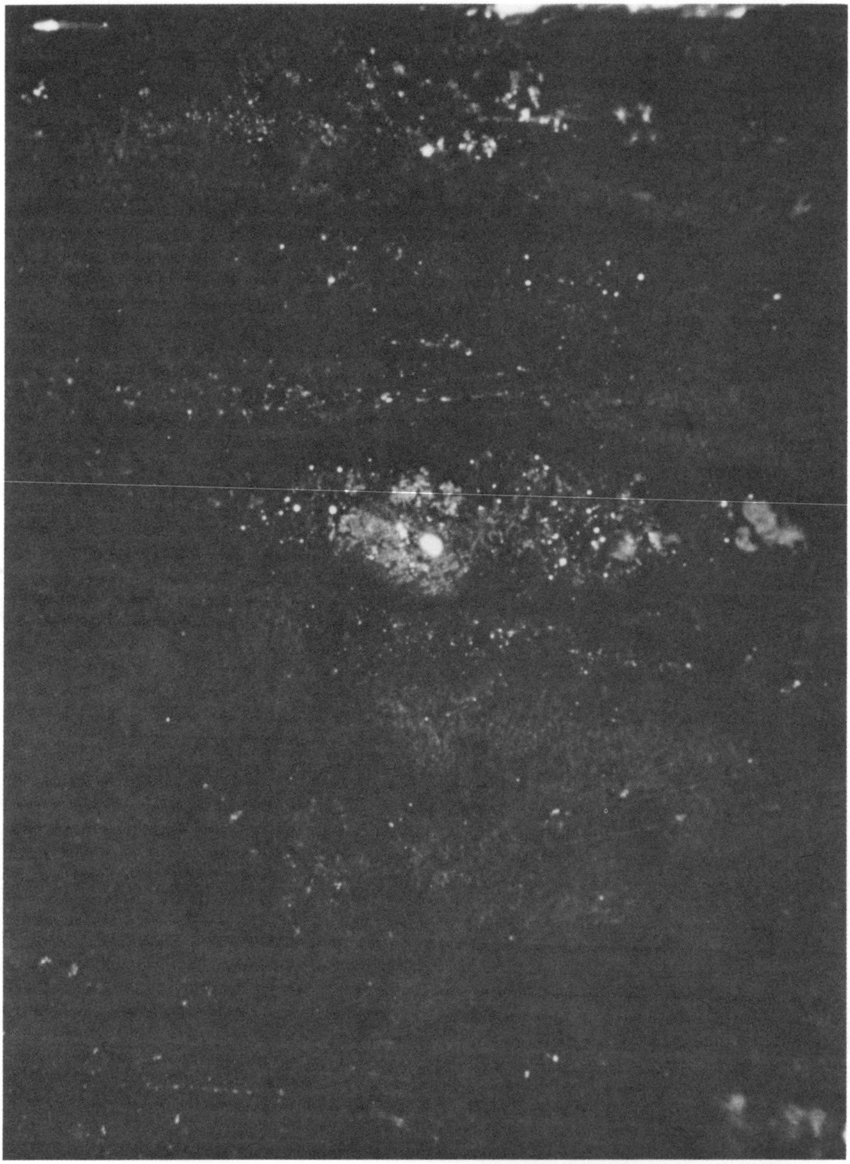

Fig. 3.4. Inside of hot-water heater destroyed by pitting.

Table 3.2. Chemical factors influencing corrosion and corrosion control

Factor	Effect
pH	Low pH may increase corrosion rate; high pH may protect pipes and decrease corrosion rates
Alkalinity	May help form protective $CaCO_3$ coating, helps control pH changes, reduces corrosion
DO	Increases rate of many corrosion reactions
Chlorine residual	Increases metallic corrosion
TDS	High TDS increases conductivity and corrosion rate
Hardness (Ca and Mg)	Ca may precipitate as $CaCO_3$ and thus provide protection and reduce corrosion rates
Chloride, sulfate	High levels increase corrosion of iron, copper, and galvanized steel
Hydrogen sulfide	Increases corrosion rates
Silicate, phosphates	May form protective films
Natural color, organic matter	May decrease corrosion
Iron, zinc, or manganese	May react with compounds on interior of A-C pipe to form protective coating

Source: Environmental Science and Engineering, Inc., 1982.

Alkalinity. Alkalinity is a measure of a water's ability to neutralize acids. In potable waters, alkalinity is mostly composed of carbonate, $CO_3^{=}$, and bicarbonates, HCO_3^-. The HCO_3^- portion of alkalinity can neutralize bases, also. Thus, the substances that normally contribute to alkalinity can neutralize acids, and any bicarbonate can neutralize bases. This property is called "buffering," and a measure of this property is called the "buffer capacity." Carbonate does not provide any buffer capacity for bases because it has no H^+ to react with the base. Buffer capacity can best be understood as resistance to change in pH.

The bicarbonate and carbonates present affect may important reactions in corrosion chemistry, including a water's ability to lay down a protective metallic carbonate coating. They also affect the concentration of calcium ions that can be present, which, in turn, affects the dissolving of calcium from cement-lined pipe or from A-C pipe. Alkalinity also reduces the dissolution of lead from lead pipes or lead-based solder by forming a protective coating of lead carbonate on the metallic surface.

DO. According to many corrosion experts, oxygen is the most common and the most important corrosive agent. In many cases, it is the substance that accepts the electrons given up by the corroding metal according to the following equation:

$$\underset{\text{free oxygen}}{O_2} + \underset{\text{water}}{2H_2O} + \underset{\text{electrons}}{4e^-} \rightarrow \underset{\text{hydroxide ions}}{4OH^-} \quad (4)$$

and so allows the corrosion reactions to continue.

Oxygen also reacts with hydrogen, H_2, released at the cathode. This reaction removes hydrogen gas from the cathode and allows the corrosion reactions to continue. The equation is

$$\underset{\text{hydrogen}}{2H_2} + \underset{\text{free oxygen}}{O_2} \rightarrow \underset{\text{water}}{2H_2O} \tag{5}$$

Hydrogen gas (H_2) usually covers the cathode and retards further reaction. This is called polarization of the cathode. The removal of the H_2 by the above reaction is called depolarization.

Oxygen also reacts with any ferrous iron ions and converts them to ferric iron. Ferrous iron ions, Fe^{+2}, are soluble in water, but ferric iron forms an insoluble hydroxide. Ferric iron accumulates at the point of corrosion, forming a tubercle, or settles out at some point in the pipe and interferes with flow. The reactions are

$$\underset{\text{metallic iron}}{Fe} \rightarrow \underset{\text{ferrous iron}}{Fe^{2+}} + \underset{\text{2 electrons}}{2e^-} \tag{6}$$

$$\underset{\text{ferrous iron}}{4Fe^{2+}} + \underset{\text{free oxygen}}{3O_2} + \underset{\text{water}}{6H_2O} \rightarrow \underset{\substack{\text{ferric hydroxide}\\ \text{(insoluble)}}}{4Fe(OH)_3} \tag{7}$$

When oxygen *is* present in water, tuberculation or pitting corrosion may take place. The pipes are affected both by the pits and by the tubercles and deposits. "Red water" may also occur, if velocities are sufficiently high to cause iron precipitates to be flushed out. In many cases when oxygen is not present, any corrosion of iron is usually noticed by the customer as "red water," because the soluble ferrous iron is carried along in the water, and the last reaction happens only after the water leaves the tap and is exposed to the oxygen in the air.

In some cases, oxygen may react with the metal surface to form a protective coating of the metal oxide.

Chlorine residual. Chlorine lowers the pH of the water by reacting with the water to form hydrochloric acid and hypochlorous acid:

$$\underset{\text{chlorine}}{Cl_2} + \underset{\text{water}}{H_2O} \rightarrow \underset{\text{hydrochloric acid}}{HCl} + \underset{\text{hypochlorous acid}}{HOCl} \tag{8}$$

This reaction makes the water potentially more corrosive. In waters with low alkalinity, the effect of chlorine on pH is greater because such waters have less capacity to resist pH changes. Tests show that the corrosion rate of steel is increased by free chlorine concentrations greater than 0.4 mg/L. Chlorine can act as a stronger oxidizing agent than oxygen in neutral (pH 7.0) waters.

Total dissolved solids (TDS). Higher TDS indicate a high ion concentration in the water, which increases conductivity. This increased conductivity in turn increases the water's ability to complete the electrochemical circuit and to conduct a corrosive current. The dissolved solids may affect the formation of protective films.

Hardness. Hardness is caused predominantly by the presence of calcium and magnesium ions and is expressed as the equivalent quantity of $CaCO_3$. Hard waters are generally less corrosive than soft waters if sufficient calcium ions and alkalinity are present to form a protective $CaCO_3$ lining on the pipe walls.

Chloride and sulfate. These two ions, Cl^- and $SO_4^=$, may cause pitting of metallic pipe by reacting with the metals in solution and causing them to stay soluble, thus preventing the formation of protective metallic oxide films. Chloride is about three times as active as sulfate in this effect. The ratio of the chloride plus the sulfate to the bicarbonate ($Cl^- + SO_4^=/HCO_3^-$) has been used by some corrosion experts to estimate the corrosivity of a water.

Hydrogen sulfide (H_2S). H_2S accelerates corrosion by reacting with the metallic ions to form insoluble sulfides. It attacks iron, steel, copper, and galvanized piping to form "black water," even in the absence of oxygen. An H_2S attack is often complex, and its effects may either begin immediately or may not become apparent for months and then will become suddenly severe.

Silicates and phosphates. Silicates and phosphates can form protective films which reduce or inhibit corrosion by providing a barrier between the water and the pipe wall. These chemicals are usually added to the water by the utility.

Natural color and organic matter. The presence of naturally occurring organic color and other organic substances may affect corrosion in several ways. Some natural organics can react with the metal surface and provide a protective film and reduce corrosion. Others have been shown to react with the corrosion products to increase corrosion. Organics may also tie up calcium ions and keep them from forming a protective $CaCO_3$ coating. In some cases, the organics have provided food for organisms growing in the distribution system. This can increase the corrosion rate in instances in which those organisms attack the surface as discussed in the section on biological characteristics. It has not been possible to tell which of these instances will occur for any specific water, so using color and organic matter as corrosion control methods is not recommended.

Iron, zinc, and manganese. Soluble iron, zinc and—to some extent—manganese, have been shown to play a role in reducing the corrosion rates of A-C pipe. Through a reaction which is not yet fully understood, these metallic compounds may combine with the pipe's cement matrix to form a protective coating on the surface of the pipe. Waters that contain natural amounts of iron have been shown to protect A-C pipe from corrosion. When zinc is added to water in the form of zinc chloride or zinc phosphate, a similar protection from corrosion has been demonstrated.

Biological Characteristics

Both aerobic and anaerobic bacteria can induce corrosion. Two common "corrosive" bacteria in water supply systems are iron-oxidizing and sulfate-reducing bacteria. Each can aid in the formation of tubercles in water pipes by releasing by-products which adhere to the pipe walls. In studies performed at the Columbia, Missouri, water distribution system, both sulfate-reducing and sulfur-oxidizing organisms were found where "red-water" problems were common.

Many organisms form precipitates with iron. Their activity can result in higher iron concentrations at certain points in the distribution system due to precipitation, as well as bioflocculation of the organisms.

Controlling these organisms can be difficult because many of the anaerobic bacteria exist under tubercles, where neither chlorine nor oxygen can get to them. In addition, they normally occur in dead ends or low-flow areas, in which a chlorine residual is not present or cannot be maintained.

4. Materials Used in Distribution Systems

This section discusses the types of materials commonly used by the waterworks industry for distribution and home service lines. Why should utility managers or operators be concerned with the materials used in their water distribution system? First, because the use of certain pipe materials in a system can affect both corrosion rates and the kind of contaminants or corrosion products added to the water. Second, because properly selected materials used to replace existing lines or to construct new ones can significantly reduce corrosion activity.

Another important reason to identify materials used in a distribution system is that certain types of construction materials in the system can affect the type of corrosion control program which should be used to reduce or prevent corrosion in the system. Control measures successful for A-C pipe may not be successful for copper pipe. When the system contains several different materials, care must be taken to prevent control measures used to reduce corrosion in one part of the system from causing corrosive action in another part of the system.

As is discussed in Sect. 3, internal pipe corrosion is initiated by a reaction between the pipe material and the water it conveys. The corrosion resistance of a pipe material depends on the particular water quality, as well as on the properties of the pipe. For a given water quality, some construction materials may be more corrosion resistant than others. Thus, a finished water may be noncorrosive to one part of a system and corrosive to another.

Table 4.1 lists the most common types of materials found in water supply systems and their uses. Service and home plumbing lines are usually constructed from different materials than transmission or distribution mains. The choice of materials depends on such factors as type of equipment, date equipment was put in service, and cost of materials. Often local building code requirements dictate the use of certain pipe materials.

Table 4.1. Common materials found in water supply systems and their uses

	In-plant systems		Other systems			
Material	Piping	Other	Storage	Transmission and distribution mains	Service lines	Residential and commercial buildings
Wrought iron	X	X		X	X	X
Cast/ductile	X	X		X	X	X
Steel	X	X	X	X	X	X
Galvanized iron	X			X	X	X
Stainless steel		X				X
Copper	X	(brass)			X	X
Lead		X (gaskets)		X	X	X
Asbestos-cement	X			X		
Concrete	X	X	X	X		
Plastic	X	X	X	X	X	X

Source: SUM X, 1981.

Older water systems are more likely to contain cast iron, lead, and vitrified clay pipe distribution lines. The introduction of newer pipe materials, however, has significantly changed pipe-usage trends. For example, ductile iron pipe, introduced in 1948, has completely replaced cast iron pipe, and, currently, all ductile iron pipe is lined with cement or another material, unless specified otherwise. The percentage of A-C pipe use increased from less than 6% to more than 13% between 1960 and 1975. The use of plastic pipe is also increasing, due partly to improvements in the manufacturing of larger-sized pipe and to greater acceptance of plastic pipe in building codes.

Many older systems still have lead service lines operating. Prior to 1960, copper and galvanized iron were the primary service line pipe materials. Although copper and galvanized iron service line pipes are still commonly used, recent trends show an increased use of plastic pipe.

Table 4.2 briefly relates various types of distribution line materials to corrosion resistance and the potential contaminants added to the water. In general, the more inert, nonmetallic pipe materials, such as concrete, A-C, and plastics, are more corrosion resistant.

Table 4.2. Corrosion properties of frequently used materials in water distribution systems

Distribution material	Corrosion resistance	Associated potential contaminants
Copper	Good overall corrosion resistance; subject to corrosive attack from high velocities, soft water, chlorine, dissolved oxygen, and low pH	Copper and possibly iron, zinc, tin, arsenic, cadmium, and lead from associated pipes and solder
Lead	Corrodes in soft water with low pH	Lead (can be well above MCL[a] for lead), arsenic, and cadmium
Mild steel	Subject to uniform corrosion; affected primarily by high dissolved oxygen levels	Iron, resulting in turbidity and red-water complaints
Cast or ductile iron (unlined)	Can be subject to surface erosion by aggressive waters	Iron, resulting in turbidity and red-water comp-plaints
Galvanized iron	Subject to galvanic corrosion of zinc by aggressive waters; corrosion is accelerated by contact with copper materials; corrosion is accelerated at higher temperatures as in hot water systems	Zinc and iron; cadmium and lead (impurities in galvanizing process may exceed primary MCLs)
Asbestos-cement	Good corrosion resistance; immune to electrolysis; aggressive waters can leach calcium from cement	Asbestos fibers
Plastic	Resistant to corrosion	

[a]MCL = Maximum contaminant levels.

Source: Environmental Science and Engineering, Inc., 1981.

How can the type of materials used throughout a distribution system be identified?

In older and larger systems, identifying the materials of construction may not be an easy task. Researching records, archives, and old blueprints is one approach. Other information sources may be surveys made by local, state, or national organizations, such as local or county health department surveys conducted to identify health-related contaminants in the water as a result of corrosion. The American Water Works Association (AWWA) has conducted several surveys regarding pipe usage. A good source of information about the older parts of the system can be former pipe and equipment installers for the system.

If practicable, utility personnel, such as meter readers or maintenance crews, can determine the type of material used for service and distribution lines, the former by checking the connections at the meter, the latter during routine maintenance checks of the main lines. When sections of pipe are being replaced or repaired, a utility should never pass up the opportunity to obtain samples of the old pipes. An examination of these samples can provide valuable information about the types of materials present in the system and can also aid in determining if the material has been subject to corrosive attack, and if so, to what kind. The sample pipe sections should be tagged and identified by type of material, location of pipe, age of pipe (if known), and date sample was obtained. The type of service (e.g., cold water, hot water, recirculating hot water, apartment, or home) should also be noted.

For small utilities with few connections, a house-to-house search to determine the types of materials in the distribution system may be feasible. In smaller communities, water, plumbing, and building contractors in the area could provide useful information about the use and service life of specific materials.

As information is obtained, the utility should keep accurate records which show the type and number of miles of each material used in the system, and its location and use.

A map of the distribution system indicating type, length, and size of pipe materials would be an excellent tool for cataloging this information and could be updated easily when necessary to show additions, alterations, and repairs to the system. As is discussed in Sect. 6.0, the map could also be used in conjunction with other utility records and surveys to identify particular areas and types of materials in the system that are more susceptible to corrosion than others.

5. Recognizing the Types of Corrosion

Previous sections have included discussions of the symptoms, basic characteristics, and chemical reactions of corrosion. The following questions will now be addressed.

How many types of corrosion are there? How can utility personnel recognize which type of corrosion is occurring in the system?

Literally dozens of types of corrosion exist. This section identifies the types of corrosion most commonly found in the waterworks industry and describes the basic characteristics of each. Illustrations are presented to help the reader identify each type by appearance. Recognizing the different types of corrosion often helps to identify their causes. Once the cause of the corrosion is diagnosed, it is easier to prescribe appropriate preventative or control measures to reduce the corrosive action.

Corrosion can be either uniform or nonuniform. Uniform corrosion results in an equal amount of material being lost over an entire pipe surface. Except in extreme cases, the loss is so minor that the service life of the pipe is not adversely affected. Nonuniform corrosion, on the other hand, attacks smaller, localized areas of the pipe causing holes, restricted flow, or structural failures. As a result, the piping will fail and will have to be replaced much sooner.

The most common types of corrosion in the waterworks industry are (1) galvanic corrosion, (2) pitting, (3) crevice corrosion, (4) erosion corrosion, and (5) biological corrosion.

Galvanic corrosion (as discussed in Sect. 3) is corrosion caused by two different metals or alloys coming in contact with each other. This usually occurs as joints and connections. Due to the differences in their activity, the more active metal corrodes. Galvanic corrosion is common in household plumbing systems where different types of metals are joined, such as a copper pipe to a galvanized iron pipe. Service line pipes are often of a different metal than household lines, so the point at which the two are joined is a prime target for galvanic corrosion. Galvanic corrosion is especially severe when pipes of different metals are joined at elbows, as is illustrated in Fig. 5.1.

This type of corrosion should be expected when different metals are used in the same system. It is common to use brass valves in galvanized lines or to use galvanized fittings in copper lines, especially at hot water heaters. An example is shown in Fig. 5.2, where a brass valve has been used in a galvanized line. Galvanic corrosion usually results in a localized attack and deep pitting. Often the threads of the pipe are the point of attack and show many holes all the way through the pipe wall. The outside of the pipe may show strong evidence of corrosion because some of the corrosion products will leak through and dry on the outside surface. Galvanic corrosion is particularly bad when a small part of the system is made up of the more active metal, such as a galvanized nipple in a copper line. In such cases, the galvanized nipple provides a small anode area which corrodes, and the copper lines provide a large cathode area to complete the reaction. Oxygen can also play a part in galvanic corrosion, as is discussed in Sect. 3.

Galvanic corrosion can be reduced by avoiding dissimilar metal connections or by using dielectric couplings to join the metals when this is not possible. Because galvanic corrosion is caused by the difference in activity or potential between two metals, the closer two metals are to each other in the galvanic series (Table 3.1), the less the chance for galvanic corrosion to occur. For this reason, a brass-to-copper connection is preferable to a zinc-to-copper connection.

Pitting is a damaging, localized, nonuniform corrosion that forms pits or holes in the pipe surface. It actually takes little metal loss to cause a hole in a pipe wall, and failure can be rapid. Pitting can begin or concentrate at a point of surface imperfections, scratches, or surface deposits. Frequently, pitting is caused by ions of a metal higher in the galvanic series plating out on the pipe surface. For example, steel and galvanized steel are subject to corrosion by small quantities (about 0.01 mg/L) of soluble metals, such as copper, which plate out and cause a galvanic type of corrosion. Chloride ions in the water commonly accelerate pitting. The presence of DO and/or high chlorine residuals in water may cause pitting corrosion of copper.

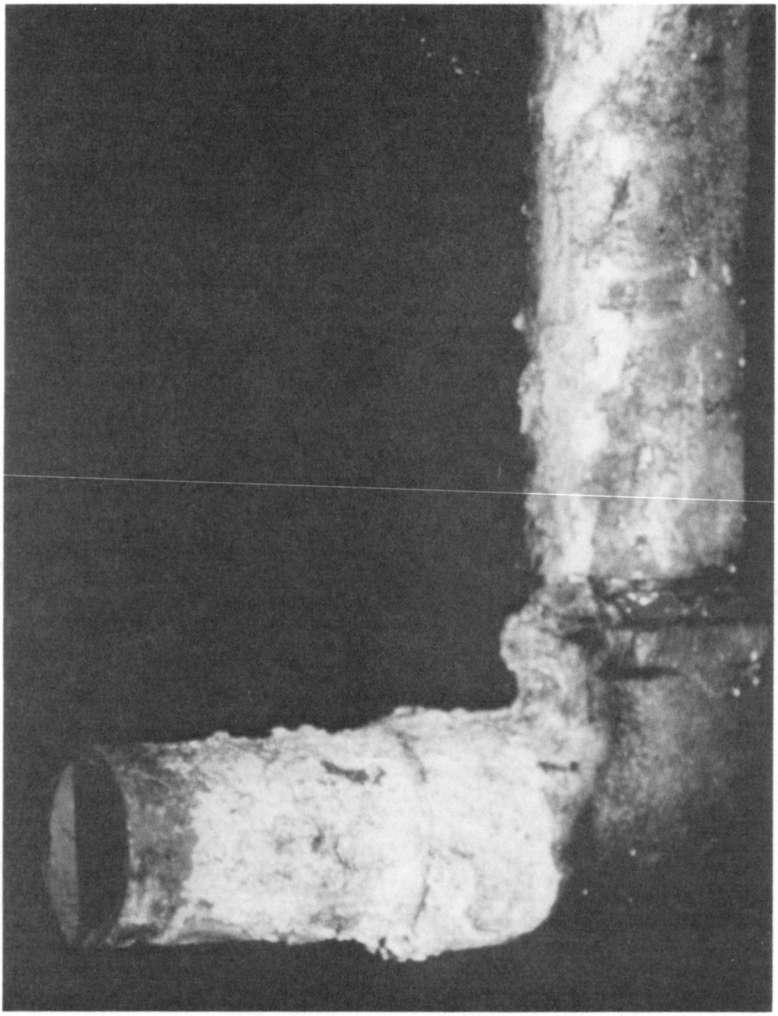

Fig. 5.1. Galvanic corrosion resulting from a galvanized pipe joined to a copper pipe by a brass elbow.

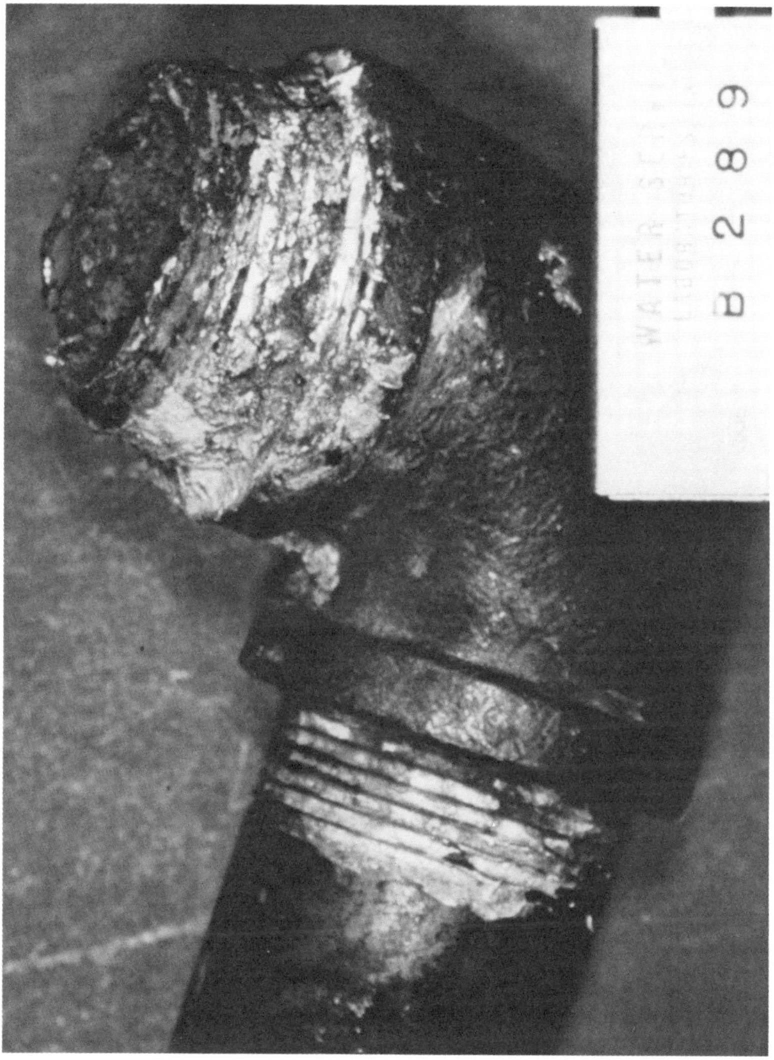

Fig. 5.2. Galvanic corrosion illustrated by severely corroded galvanized steel nipple in a brass elbow. *This was the only piece of steel pipe in an otherwise all brass domestic hot-water heater, illustrating the effects of a large cathodic area to a small anodic area*

Pitting is not usually noticed until the pipe wall gets a hole in it and the effect of the corrosion becomes obvious, as does the location of the pit. This type of corrosion also occurs in storage tanks at the water line, where the air and water come in contact and create corrosive conditions. Examples of pitting corrosion are shown in Figs. 5.3 and 5.4.

Tuberculation occurs when pitting corrosion products build up at the anode next to the pit, as illustrated in Fig. 5.5. In iron or steel pipes, the tubercles are made up of rust or iron oxide. These tubercles are usually rust colored and soft on the outside and are both harder and darker toward the inside. When copper pipe becomes pitted, the tubercle buildup is smaller and is a green to blue-green color. Examples of tuberculation are illustrated in Figs. 5.6 and 5.7.

Tuberculation is seen only when a piece of pipe is removed from the system because it rarely affects the water quality, although it is possible for some of the tubercles to break loose with changes in flow or when the pipes are hit hard enough to loosen them. This type of corrosion can be suspected, though, when the flow through a pipe is much less than should be expected, as tubercles add to the roughness of a main's interior and reduce the flow. In extreme cases, the flow can be completely stopped by tubercles.

Crevice corrosion is a form of localized corrosion usually caused by changes in acidity, oxygen depletion, dissolved ions, and the absence of an inhibitor. As the name implies, this corrosion occurs in crevices at gaskets, lap joints, rivets, and surface deposits.

Erosion corrosion mechanically removes protective films, such as metal oxides and $CaCO_3$, which serve as protective barriers against corrosive attack. It generally results from high flow velocities, turbulence, sudden changes in flow direction, and the abrasive action of suspended materials. Erosion is much worse at sharp bends, as is illustrated in Fig. 5.8. Erosion corrosion can be identified by grooves, waves, rounded holes, and valleys it causes on the pipe walls.

Cavitation corrosion is a type of erosion corrosion and is caused by a sudden drop in pressure to below vapor pressure at which time dissolved gases form vapor bubbles which collapse with an explosive effect as they move to a region of high pressure. These explosions create extremely high pressures which may blast off protective coatings and even the metal surface itself. Problems with cavitation occur at high flow velocities immediately following a constriction of the flow or a sudden change in direction. For these reasons cavitation is of greatest concern at pump impellers, partially closed valves, elbows and reducers. An example is shown in Fig. 5.9.

Biological corrosion results from a reaction between the pipe material and organisms such as bacteria, algae, and fungi. It is an important factor in the taste and odor problems that develop in a system, as well as in the degradation of the piping materials. Controlling such growths is complicated because they can take refuge in many protected areas, such as in mechanical crevices or in accumulations of corrosion products. The bacteria can exist under tubercles, where neither chlorine nor oxygen can destroy them. Mechanical cleaning may be necessary in some systems before control can be accomplished by residual disinfectants. Preventative methods include avoiding dead ends and stagnant water in the system.

Other types of corrosion in the waterworks industry that are not found as commonly as those discussed previously include (1) stray current corrosion and (2) dealloying or selective leaching.

Stray current corrosion is a type of localized corrosion usually caused by the grounding of home appliances or electrical circuits to the water pipes. Corrosion takes place at the anode, the point where the current leaves the metal to return to the power source or to ground. Stray current corrosion is difficult to diagnose since the point of corrosion does not necessarily occur near the current source. It occurs more often on the outside of pipes, but does show up in house faucets or other valves. Fig. 5.10 is an example of stray current corrosion.

Dealloying or selective leaching is the preferential removal of one or more metals from an alloy in a corrosive medium, such as the removal of zinc from brass (dezincification). This type of corrosion weakens the metals and can lead to pipe failure in severe cases. Dezincification is common in brasses containing 20% or more zinc and is rare in brasses containing less than 15% zinc. An example of this is shown in Fig. 5.11.

Fig. 5.3. Pitting of steel pipe.

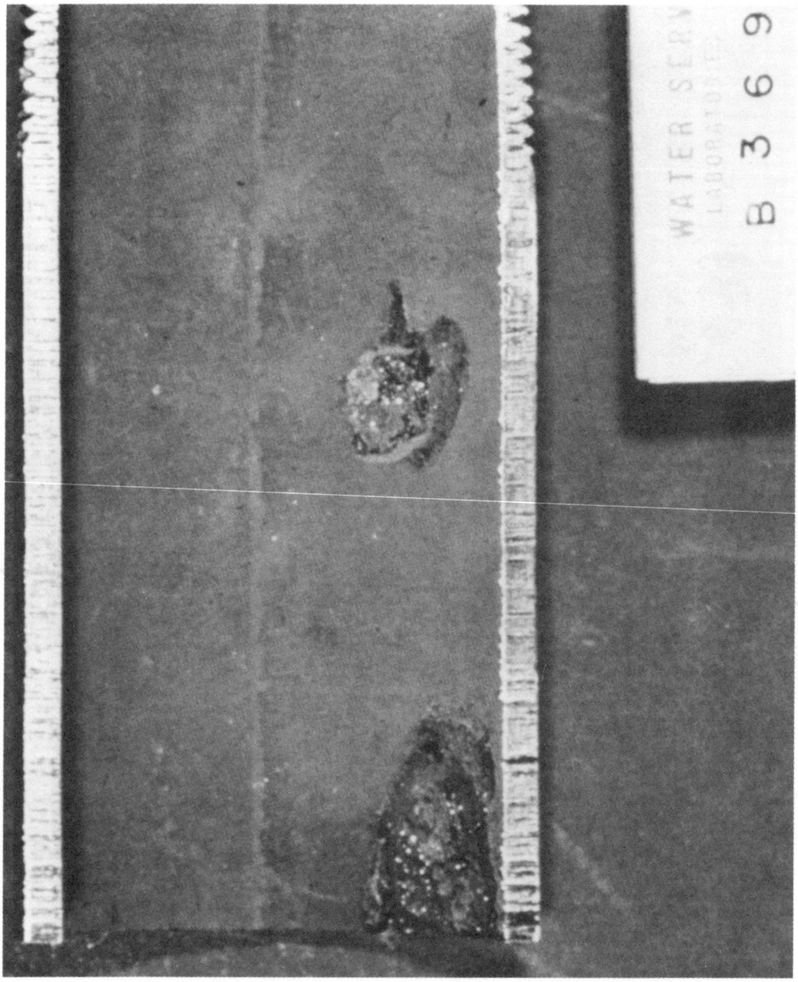

Fig. 5.4. Pitted red brass (85% copper) pipe from a domestic hot-water system.

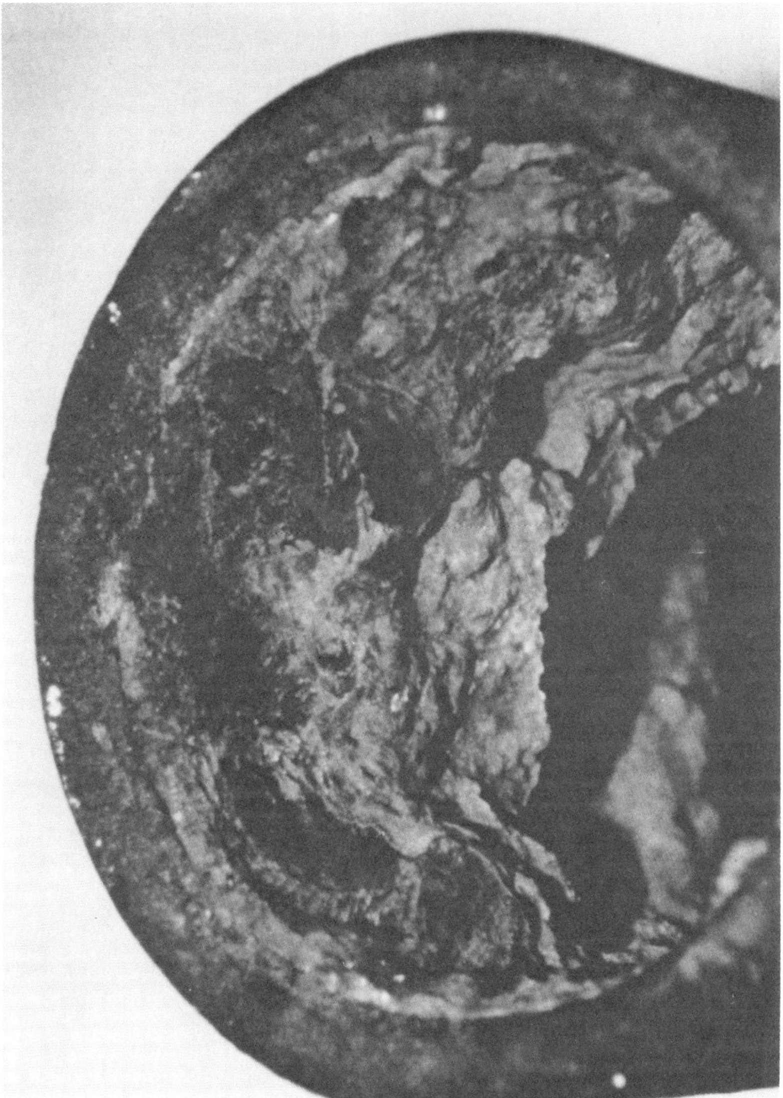

Fig. 5.5. Tuberculation in a cast iron pipe.

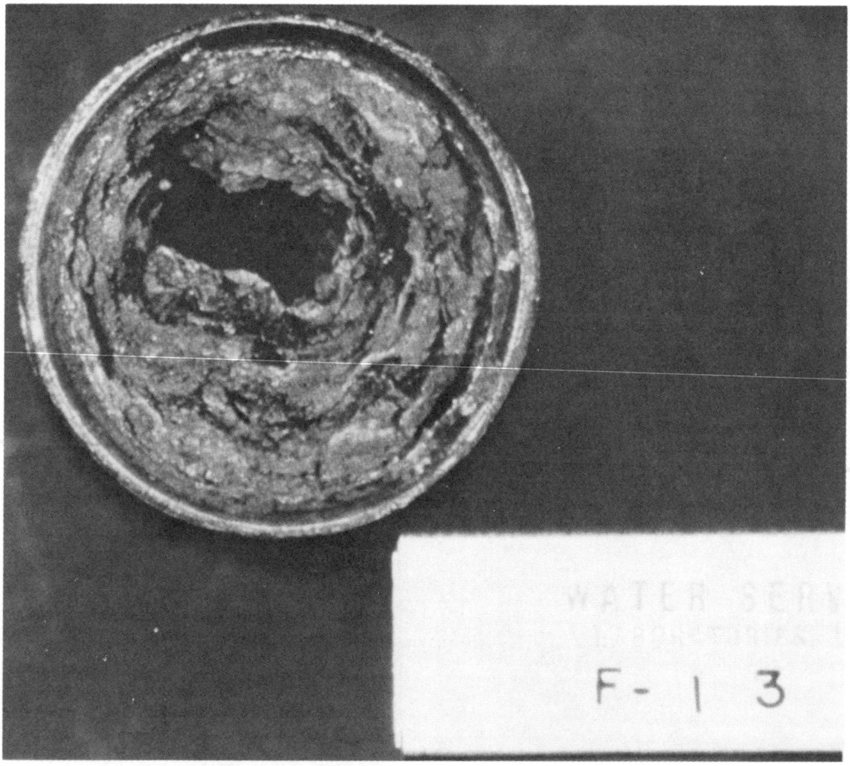

Fig. 5.6. Galvanized steel pipe from a domestic hot-water system showing almost complete clogging by corrosion products.

Fig. 5.7. Tuberculation in a cast iron pipe.

Fig. 5.8. Erosion corrosion of yellow brass impeller from domestic hot-water circulation pump.

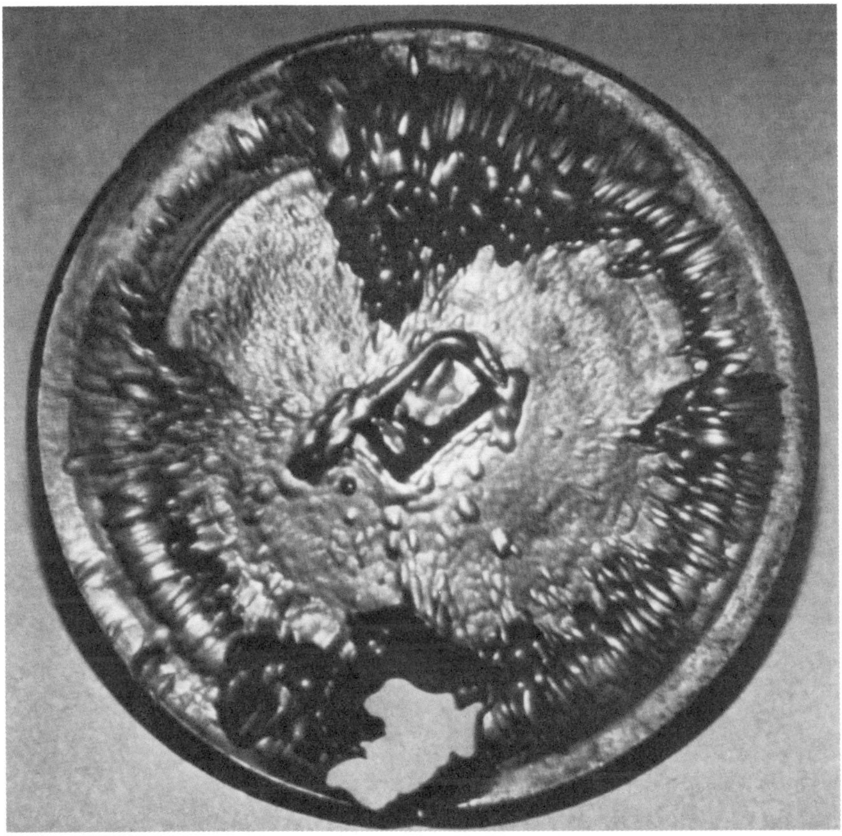

Fig. 5.9. Cavitation corrosion of brass impeller.

Fig. 5.10. *Extreme example of stray current corrosion in an outside water faucet caused by lightning leaving the pipe.*

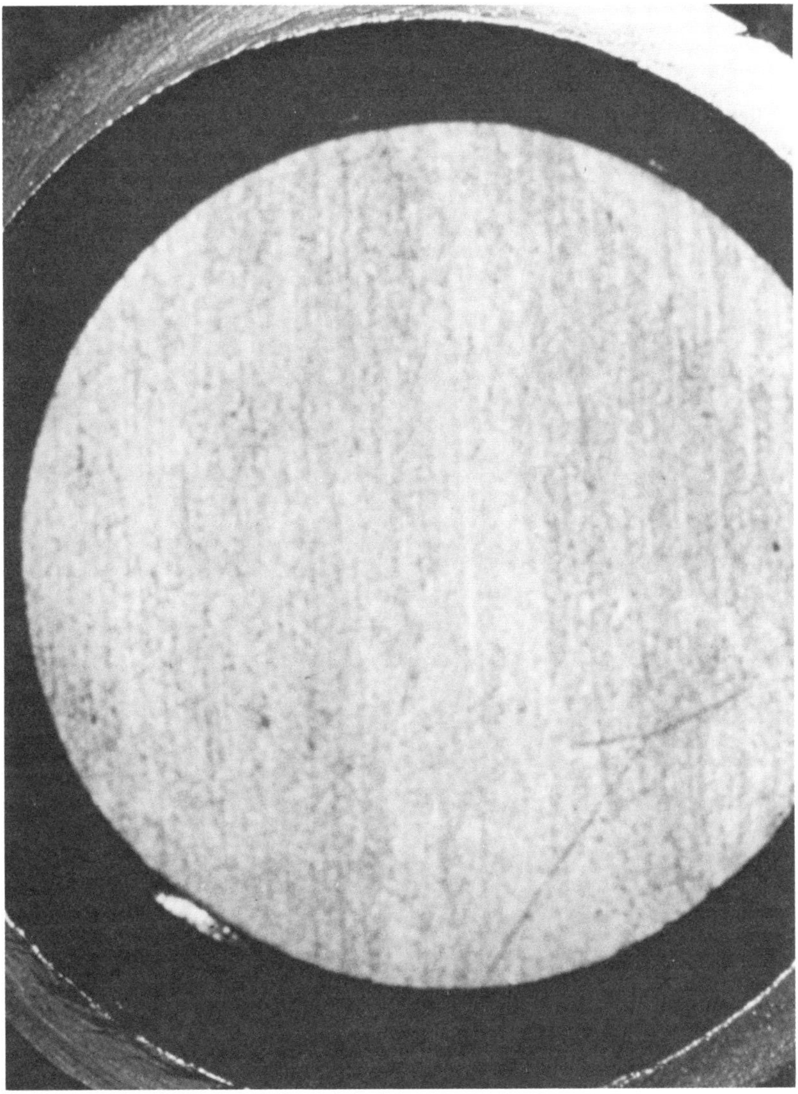

Fig. 5.11. Dezincification of yellow brass in domestic water pipe.

6. Corrosion Monitoring and Treatment

The previous sections of this manual have discussed what corrosion is and have briefly described how and why corrosion occurs in the waterworks industry. The purpose of this and the following sections is to point out some of the easiest, as well as the most effective, methods of identifying, monitoring, and correcting corrosion-related problems. In other words, these sections answer the questions how do you know if your utility has a corrosion problem, and what can you do to control or reduce the effects of the corrosion. The effects of corrosion, which may not be evident without monitoring, can be expensive and may even affect human health. Monitoring methods most useful to the small water utility are emphasized; that is, those methods which are the least expensive and the simplest to implement in terms of manpower and technical requirements. Methods for controlling or reducing corrosion are covered in the following section.

Just as there is no one cause of corrosion, there is no one way to measure or "cure" corrosion. Since corrosion in a system depends on a specific water and the reaction of that water with specific pipe materials, each utility is faced with a unique set of problems. There are, however, general methods of measuring and monitoring for corrosion that can provide a basis for a sound corrosion control program for any utility. Although no one method may provide an absolute or quantitative measure of corrosivity, several methods used together over a period of time will indicate if corrosion is occurring and will point out any undesirable effects on the system.

There are two different kinds of corrosion measurements—indirect and direct. The indirect methods do not measure corrosion rates. Rather, the data obtained from these methods must be compared and interpreted to determine trends or changes in the system. The indirect methods discussed here are (1) customer complaint logs, (2) corrosion indices, and (3) water sampling and chemical analyses. The direct corrosion measurements call for the actual examination of a corroded surface or the measurement of corrosion rates, particularly actual metal loss. The direct methods discussed here are (1) examination of pipe sections and (2) rate measurements.

6.1 INDIRECT METHODS

Customer Complaint Logs

Usually, customer complaints will be the first evidence of a corrosion problem in a water system. The most common symptoms are listed in Table 6.1, along with their possible causes. The

Table 6.1. Typical customer complaints due to corrosion

Customer complaint	Possible cause
Red water or reddish-brown staining of fixtures and laundry	Corrosion of iron pipes or presence of iron in raw water
Bluish stains on fixtures	Corrosion of copper lines
Black water	Sulfide corrosion of copper or iron lines
Foul taste and/or odors	By-products from microbial activity
Loss of pressure	Excessive scaling, tubercle build-up from pitting corrosion, leak in system from pitting or other type of corrosion
Lack of hot water	Build-up of mineral deposits in hot water system (can be reduced by setting thermostat to under 140°F)
Short service life of household plumbing	Rapid deterioration of pipes from pitting or other types of corrosion

Source: Environmental Science and Engineering, Inc., 1982.

complaints may not always be due to corrosion. For example, red water may also be caused by iron in the raw water that is not removed in treatment. Therefore, in some cases, further investigation is necessary before attributing the complaint to corrosion in the system.

Complaints can be a valuable corrosion monitoring tool if records of the complaints are organized. The complaint record should include the customer's name and address, date the complaint was made, and nature of the complaint. The following information should also be recorded:

1. Type of material (copper, galvanized iron, plastic, etc.) used in the customer's system;
2. Whether the customer uses home treatment devices prior to consumption (softening, carbon filters, etc.);
3. Whether the complaint is related to the hot water system and, if so, what type of material is used in the hot water tank and its associated appurtenances; and
4. Any follow-up action taken by the utility or customer.

These records can be used to monitor changes in water quality due to system or treatment changes.

The development of a complaint map is useful in pinpointing problem areas. The complaint map would be most useful when combined with the materials map discussed in Sect. 4.0, which indicates the location, type, age, and use of a particular type of construction material. If complaints are recorded on the same map, the utility can determine if there is a relationship between complaints and the materials used. To supplement the customer complaint records, it might be useful to send questionnaires to a random sampling of customers. These questionnaires should be short but thorough. A sample questionnaire used by the city of Seattle is shown in Fig. 6.1.

Customer complaint records and questionnaires are useful monitoring tools that can be used as part of any corrosion monitoring and control program. The low costs associated with keeping a good record of complaints can be well worth the time. The resulting information would indicate the real effect of water quality at the customer's tap and would show the effect of any process changes made as part of a corrosion control program.

Corrosion Indices

Many attempts have been made to develop an index that would predict whether or not a water is corrosive; unfortunately, none of these attempts has been entirely successful. However, several of

Do you ever have rusty water? Yes_____ No_____

If so, how often? Every Morning_____ Once/week_____ Seldom_____

Do you have blue-green stains on your sink or bathtub? Yes_____ No_____

What type of plumbing do you have in your house? All Copper_____,
Some Copper_____, Iron_____, All Galvanized_____,
Some Galvanized_____, Not Certain_____

Do you have low pressure problems? Yes_____ No_____

Where? Front Hose Bib_____, Bathtub_____, Kitchen Sink_____,
Everywhere_____

Fig. 6.1. Sample questionnaire. Source: City of Seattle, 1981.

the indices can be useful for predicting corrosion. These indices can be calculated by all small utilities and can be used in an overall corrosion control program. In addition, the 1980 amendments to the NIPDWR require all community water supply systems to determine either the Langelier Saturation Index (LSI) or the Aggressive Index (AI) and report these values to the state regulatory agencies.

Since the LSI and AI are the two most commonly used corrosion indices in the waterworks industry, they are the only indices discussed in detail in the following paragraphs. However, several of the less frequently used indices are briefly described to acquaint the reader with their usefulness and method of calculation.

The LSI and the AI indices estimate the tendency of a water to "lay down" or precipitate a protective coating of $CaCO_3$ on the pipe wall. A thin layer of $CaCO_3$ is desirable, as it keeps the water from contacting the pipe and reduces the chance of corrosion.

"Scaling" occurs when thick layers of $CaCO_3$ are deposited. Although the pipe is protected from corrosion, excessive scaling can result in loss of carrying capacity in the system, as is shown in Fig. 6.2.

The equation for the deposition of $CaCO_3$ scale is

$$\underset{\text{Calcium}}{Ca^{++}} + \underset{\text{Bicarbonate}}{HCO_3^-} \rightleftharpoons \underset{\text{Calcium carbonate}}{CaCO_3} + \underset{\text{Hydrogen Ion}}{H^+} \tag{9}$$

If the reaction proceeds to the right, a protective scale of $CaCO_3$ is deposited. If the reaction proceeds to the left, the scale is dissolved, leaving the surfaces that had been protected exposed to corrosion. When the water is exactly saturated with $CaCO_3$, it will neither dissolve nor deposit $CaCO_3$. The saturation value of the water with respect to $CaCO_3$ depends on the calcium ion concentration, alkalinity, temperature, pH, and the presence of other dissolved materials, such as chlorides and sulfates.

Langelier Saturation Index. The LSI is the most widely used and misused index in the water treatment and distribution field. The index is based on the effect of pH on the solubility of $CaCO_3$. The pH at which a water is saturated with $CaCO_3$ is known as the pH of saturation or pH_s. At pH_s, a protective scale will neither be deposited nor dissolved. The LSI is defined by the following equation:

$$LSI = pH - pH_s \tag{10}$$

The results of the equation are interpreted as follows:

$LSI > 0$ Water is supersaturated and tends to precipitate a scale layer of $CaCO_3$.

$LSI = 0$ Water is saturated (in equilibrium) with $CaCO_3$; a scale layer of $CaCO_3$ is neither precipitated nor dissolved.

$LSI < 0$ Water is undersaturated, tends to dissolve solid $CaCO_3$.

To calculate the LSI, the following information is needed:

1. Total alkalinity (milligrams per liter) as $CaCO_3$;
2. Calcium, mg/L, as $CaCO_3$;
3. Total dissolved solids, mg/L;
4. pH;
5. Temperature; and
6. pH_s.

The value of pH_s can be calculated by the following equation:

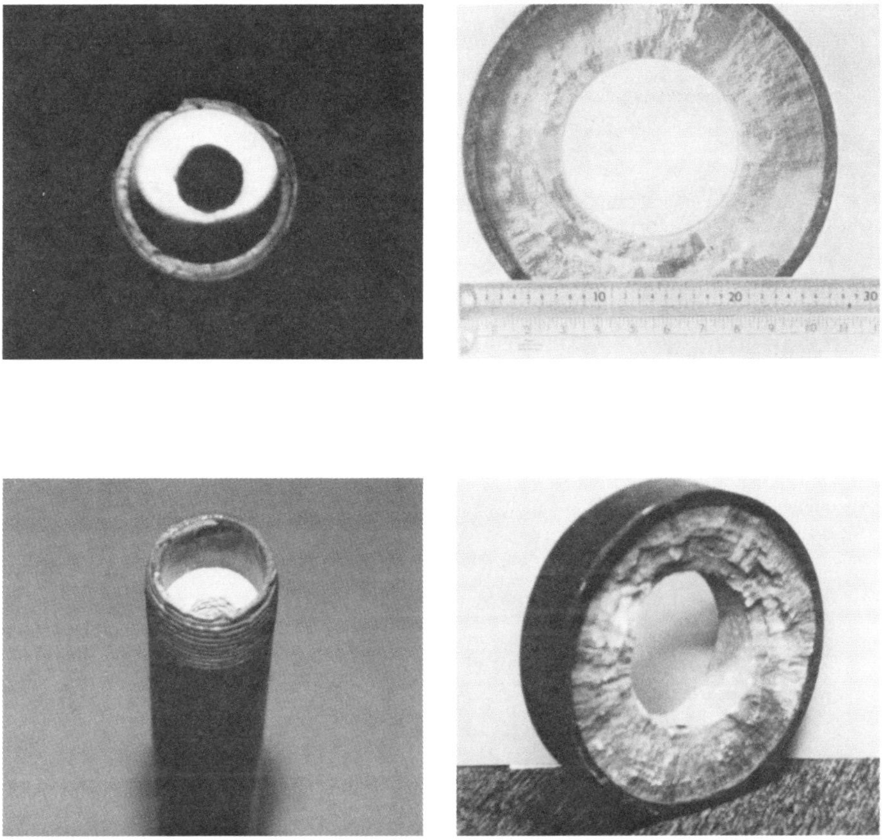

Fig. 6.2. Excessive $CaCO_3$ scaling resulting in loss of carrying capacity.

$$pH_s = A+B - \log [Ca^{++}] - \log \textit{total alkalinity}, \qquad (11)$$

where both A and B are constants related to the temperature and dissolved solids of the water. Values for A and B are tabulated in Tables 6.2 and 6.3.

The log of the calcium and alkalinity is obtained from Table 6.4.

Now, let's take as an example Chicago's tap water, which has the following characteristics:

Calcium (as $CaCO_3$), 88.0 mg/L
Total Alkalinity (as $CaCO_3$), 110.0 mg/L
Total dissolved solids, 170.0 mg/L
Case I: pH = 8.20; Temperature (T) = 25°C (77°F)
Case II: pH = 8.05; Temperature (T) = 57°C (135°F).

The step-by-step calculation of the LSI, using Tables 6.2 through 6.4, is as follows:

Case 1: pH = 8.2, T - 25°C (77°F)

$pH_s = A + B - \log[Ca^{++}] - \log$ alkalinity
$= 2.00 + 9.81 - 1.94 - 2.04$
$pH_s = 7.83$
$LSI = pH - pH_s$
$= 8.20 - 7.83$
$= 0.37$

Case 2: pH = 8.05, T = 57°C (135°F)

If the same water used in Case 1 were heated to 57°C (135°F), as is typical in hot water tanks, the calculation of the LSI would be as follows:

$pH_s = 1.45 + 9.81 - 1.94 - 2.04 = 7.28$
$LSI = 8.05 - 7.28 = 0.77$

Table 6.2 Constant "A" as function of water temperature

Water temperature		
°F	°C	Constant[a]
32	0	2.60
39.2	4	2.50
46.4	8	2.40
53.6	12	2.30
60.8	16	2.20
68	20	2.10
77	25	2.00
86	30	1.90
104	40	1.70
122	50	1.55
140	60	1.40
158	70	1.25
176	80	1.15

[a]Calculated from K_2 as reported by Harned and Scholes and K_2 as reported by Larson and Buswell. Values above 40°C have been extrapolated.

Source: *Federal Register*, 1980.

Table 6.3. Constant "B" as function of total filterable residue

Total dissolved solids (mg/L)	Constant
0	9.70
100	9.77
200	9.83
400	9.86
800	9.89
1000	9.90

Source: *Federal Register*, 1980.

Table 6.4. Logarithms of calcium and alkalinity concentrations

Ca^{+2} or Alkalinity ($mg/L\ CaCO_3$)	Log
10	1.00
20	1.30
30	1.48
40	1.60
50	1.70
60	1.78
70	1.84
80	1.90
100	2.00
200	2.30
300	2.48
400	2.60
500	2.70
600	2.78
700	2.84
800	2.90
900	2.95
1000	3.00

Source: *Federal Register*, 1980.

The results of the above calculations may be interpreted as follows:

Case I: LSI = +0.37, water tends to form a scale
Case II: LSI = +0.77, water definitely tends to form a scale.

The above examples show two important factors. First, they show the effect of the change in temperature and pH on the calculated LSI value. This demonstrates the need for accurate, onsite pH and temperature measurements. Second, a water which may deposit a thin protective scale in the distribution system at T = 25°C may form excessive scaling in the hot water system; therefore, the customer's hot water heaters may have to be descaled or replaced sooner than expected.

There are several limitations to the LSI. First, it is generally agreed that the LSI may only be used to estimate corrosive tendencies of waters within a pH range of 6.5 to 9.5. More importantly, the LSI only indicates the tendency for corrosion to occur. It is not a measurement of corrosivity.

Table 6.5 shows examples of water sources with different pH, pH_s, and Langelier index results. Pipe sections were physically examined to establish whether or not the water was corrosive. The results confirm that the LSI, by itself, does not indicate corrosiveness. It is, however, a valuable monitoring tool where a protective $CaCO_3$ film is being used or when used in conjunction with other indirect or direct corrosion monitoring methods.

A useful procedure for estimating the pH_s is an experimental method commonly called the Marble Test. In this test, duplicate samples of the water are collected. $CaCO_3$ (about 1 g/L) is added to one of the samples and shaken. After a time interval (usually 1 h or longer), aliquots from both samples are filtered and analyzed for alkalinity or pH. If the alkalinity or pH in the untreated sample is greater than that of the sample with $CaCO_3$, the water is supersaturated with $CaCO_3$ and may be scale forming. If the alkalinity or pH of the untreated sample is less than that of the $CaCO_3$-treated sample, the water is undersaturated with $CaCO_3$. If the alkalinities or pHs of the two samples are equal, the water is just saturated with $CaCO_3$.

Aggressive Index (AI). The AI was developed at the request of consulting engineers to govern the selection of the proper type (I or II) A-C pipe and to ensure long-term structural integrity. The AI is defined by the AWWA Standard C-400 as follows:

$$AI = pH + \log [(A)(H)] \tag{12}$$

where

pH = Hydrogen ion concentration, pH units
A = Total alkalinity, milligrams per liter as $CaCO_3$
H = Calcium hardness, mg/L as $CaCO_3$

The values obtained are interpreted as follows:

$AI < 10 \rightarrow$ very aggressive (corrosive)
$AI = 10\text{–}12 \rightarrow$ moderately aggressive
$AI > 12 \rightarrow$ nonaggressive

The AI is based on pH and the solubility of $CaCO_3$. It is a simplified form of the LSI and only approximates the solubility of $CaCO_3$, not the corrosivity. However, it can be a useful tool in selecting materials or treatment options for corrosion control.

A sample calculation for the AI follows.

Given: pH = 7.4
A = 199 mg/L, as $CaCO_3$
H = 153 mg/L calcium hardness, as $CaCO_3$

Table 6.5. Corrosivity of waters versus the Langelier Saturation Index (LSI)

Source	pH	pH_s	LSI	Corrosive
Well water	7.30	7.20	+0.10	No
Well water	7.40	7.25	+0.15	Yes
Well water	7.10	7.14	−0.04	No
Well water	7.50	7.10	+0.40	Yes
Spring	7.30	8.08	−0.78	No
Deep well	6.30	8.27	−1.97	No
Deep well	6.80	7.88	−1.08	No
Spring	7.80	8.90	−1.10	No

Source: Singley, 1981.

Sample calculation:

$$\begin{aligned} AI &= pH + \log[(A)(H)] \\ &= 7.4 + \log(199 \times 153) \\ &= 7.4 + \log(199) + \log(153) \\ &= 7.4 + 2.3 + 2.1 \\ &= 11.8 \end{aligned}$$

In this example, the water should be classified as "moderately aggressive."

Other Corrosion Indices. Other corrosion indices commonly seen in the literature are

1. **Ryznar Stability Index (RSI)**—For this index, Ryznar used the same parameters as the LSI, but reversed the signs and doubled the pH_s, such that

$$RSI = 2\ pH_s - pH \tag{13}$$

   Ryznar also developed a curve based on these field observations, showing the scaling or corrosion of steel mains as a function of the index. This curve is shown in Fig. 6.3.

2. **Riddick's Corrosion Index (CI)**—Riddick's Index is based on actual field observations. The values obtained apply to the soft waters of the eastern seaboard of the United States, but not to the harder waters of the middle part of the country. The major contribution of this index is that it introduces factors other than $CaCO_3$ solubility, such as dissolved oxygen, chloride ion, and noncarbonate hardness, as well as the useful effect of silica.

3. **McCauley's Driving Force Index (DFI)**—This index is also based on $CaCO_3$ solubility and attempts to predict the amount of $CaCO_3$ that will precipitate. It can be useful in estimating the amount of precipitate that may be formed.

Table 6.6 lists the equation used to calculate each index, the analytical parameters required to perform the calculation, and the meaning of each index.

There have been attempts to use other water quality parameters to predict the tendency of a water to attack metal pipes. The classic studies of the Illinois State Water Survey by Larson, Sollo, and their co-workers have shown that other factors, such as the ratios of various anions, velocity, pH, and calcium ion concentration, affect the rates of corrosion of mild steel and cast iron. It was shown that increasing the Cl^- to HCO_3^- ratio, particularly above 0.3, increased the corrosion rate.

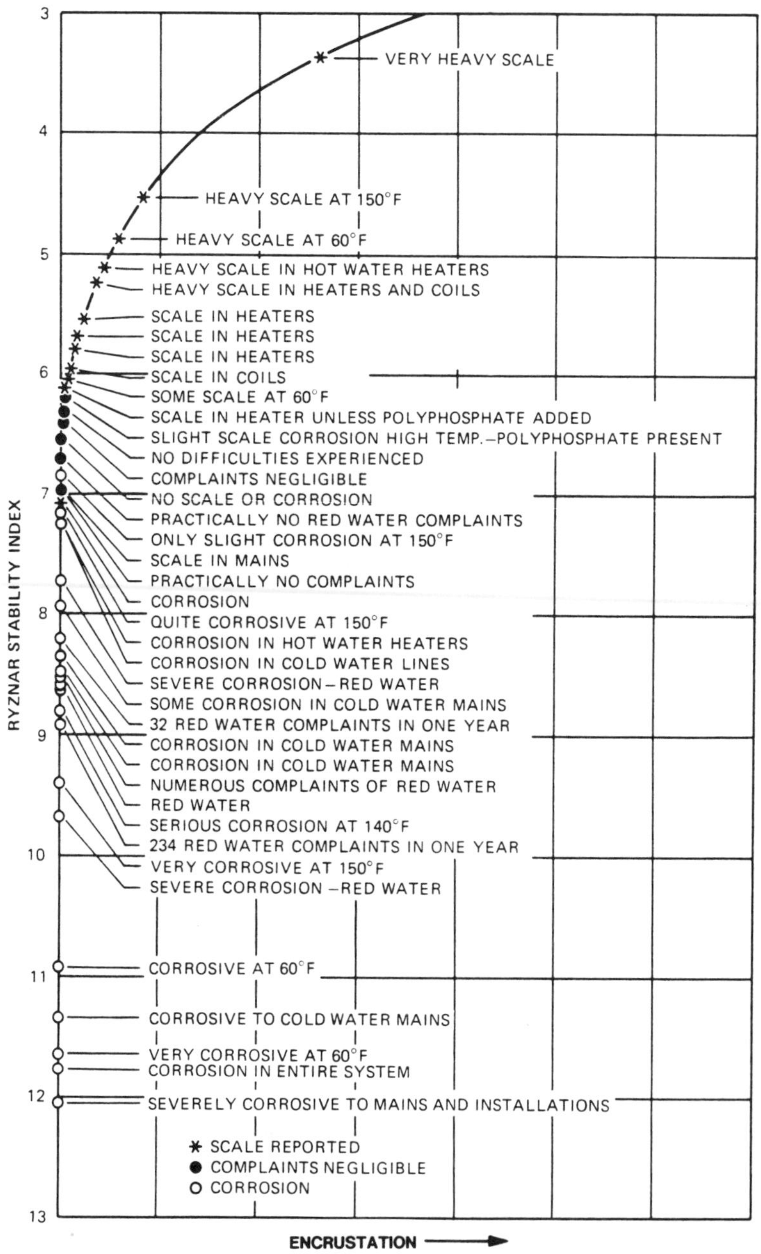

Fig. 6.3. Graphic representation of the various degrees of corrosion and encrustation.

Table 6.6. Summary of corrosion indices

Index	Equation	Parameters	Meaning
Langelier Saturation Index (LSI)	$LSI = pH - pH_s$	Total alkalinity, mg/L as $CaCO_3$ Calcium, mg/L as $CaCO_3$ Hardness, mg/L as $CaCO_3$ Total dissolved solids, mg/L Onsite pH Onsite temperature	LSI > 0 = Water is supersaturated; tends to precipitate $CaCO_3$ LSI = 0 = Water is saturated (in equilibrium); $CaCO_3$ scale is neither dissolved nor deposited LSI < 0 = Water is undersaturated; tends to dissolve solid $CaCO_3$
Aggressive Index (AI) (for use with asbestos cement)	$AI = pH + log[(A)(H)]$	Total alkalinity, mg/L as $CaCO_3$ Hardness, mg/L as $CaCO_3$ Onsite pH	AI < 10 = Very aggressive AI = 10–12 = Moderately aggressive AL > 12 = Nonaggressive
Ryznar Stability Index (RSI)	$RSI = 2pH_s - pH$	Total alkalinity, mg/L as $CaCO_3$ Calcium, mg/L as $CaCO_3$ Hardness, mg/L as $CaCO_3$ Total dissolved solids, mg/L Onsite pH Onsite temperature	RSI < 6.5 = Water is supersaturated; tends to precipitate $CaCO_3$ 6.5 < RSI < 7.0 = Water is saturated (in equilibrium); $CaCO_3$ scale is neither dissolved nor deposited RSI > 7.0 = Water is undersaturated; tends to dissolve solid $CaCO_3$
Riddick's Corrosion Index (CI)	$\frac{75}{Alk}\left[CO_2 + \frac{1}{2}\left(Hardness - Alk\right) + Cl^- + 2N\right] \times \left(\frac{10}{SiO_2}\right)\left(\frac{DO+2}{Sa+DO}\right)$	CO_2, mg/L Hardness, mg/L as $CaCO_3$ Alkalinity, mg/L as $CaCO_3$ Cl^-, mg/L N, mg/L DO, mg/L Saturation DO[a] (value for oxygen saturation), mg/L	CI = 0–5 Scale forming 6–25 Noncorrosive 26–50 Moderately corrosive 51–75 Corrosive 76–100 Very corrosive 101 + Extremely corrosive
Driving Force Index (DFI) $CaCO_3$	$\frac{Ca^{++}(ppm) \times CO_3{=}(ppm)}{K_{SO} \times 10^{10}}$	Calcium, mg/L as $CaCO_3$ $CO_3^=$ = mg/L as $CaCO_3$ K_{SO} = solubility product of $CaCO_3$	DFI < 1 = Water supersaturated; tends to precipitate DFI = 1 = Water saturated (in equilibrium); $CaCO_3$ scale is neither dissolved nor deposited DFI < 1 = Water undersaturated; tends to dissolve $CaCO_3$

[a]DO = dissolved oxygen

The presence of both calcium ion and alkalinity was shown to reduce the corrosion rate. These studies have led to a much better understanding of corrosion but have not resulted in a corrosion index.

Sampling and Chemical Analysis

Since corrosion is affected by the chemical composition of a water, sampling and chemical analysis of the water can provide valuable corrosion-related information. Some waters tend to be more aggressive or corrosive than others because of the quality of the water. For example, waters having a low pH (<6.0), low alkalinity (<40 mg/L), and high carbon dioxide (CO_2) tend to be more corrosive than waters with a pH greater than 7.0, high alkalinity, and low CO_2. Whether corrosion is occurring in the system, however, depends on the action of the water on the pipe material.

Most utilities routinely analyze their water (1) to ensure that they are providing a safe water to their customers and (2) to meet regulatory requirements. The 1980 Amendments to the NIPDWR require all community water supply systems to sample for certain "corrosive characteristics." Table 6.7 summarizes the sampling and analytical requirements of the 1980 amendments. The purpose of this sampling and analysis is to identify potentially corrosive waters throughout the country.

The amendments also require the water utility to identify the type of construction material used throughout the system, including service lines and home plumbing, and report the findings to the state. A water with "corrosive characteristics" may or may not be corrosive to a specific pipe material. Either way, sampling and analyzing for these "corrosive characteristics" can tell a utility if the water is potentially corrosive and alert the utility to potential problems.

Although the minimal sampling and analysis required by the 1980 amendments to the NIPDWR will provide an initial indication of the corrosive tendency of a finished water, additional sampling and chemical analysis performed over a period of time are necessary to indicate if corrosion is taking place and what materials are being corroded.

Table 6.7. 1980 Amendments to the NIPDWR: Sampling and analytical requirements
Individual states may add requirements

Parameters required	Sampling location	Number of samples	
		Water supply source	Number of samples per year
Alkalinity (mg/L as $CaCO_3$) pH (pH units) Hardness (mg/L as $CaCO_3$) Temperature (°C) Total dissolved solids (mg/L) Langelier or Aggressive Index[a]	Sample(s) are to be taken at one representative point as the water enters the distribution system	Groundwater only	1
		Surface water only or groundwater and surface water	2 samples, taken at different times of the year to account for seasonal variations in surface water supplies, such as mid-summer high temperatures and mid-winter low temperatures, or high flow and low flow conditions.

[a]The Langelier Saturation and Aggressive indices are calculated from the results of the chemical parameters. These indices are discussed on pages 36-41.

Source: *Federal Register*, August 1980.

Recommended Sampling Locations for Additional Corrosion Monitoring. It is generally desirable to collect water samples at the following locations within the system:

1. Water entering the distribution system (i.e., high-service pumping),
2. Water at various locations in the distribution system prior to household service lines,
3. Water in several household service lines throughout the system, and
4. Water at the customer's taps.

Water entering the distribution system at the plant can be conveniently sampled from the clearwell, the storage tank, or a sample tap on a pipe before or after the high-service pump.

To represent conditions at the customer's tap, "standing" samples should be taken from an interior faucet in which the water has remained for several hours (i.e., overnight). The sample should be collected as soon as the tap is opened.

A representative sample from the household service line (between the distribution system and the house itself) can be obtained by collecting a "running" sample from the customer's faucet after letting the tap run for a few minutes to flush the household lines. Frequently, the water temperature noticeably decreases when water in the service line reaches the tap. By letting the same faucet run for several minutes following the initial temperature change, the running water sample at the tap is representative of the water recently in the distribution main itself.

If a comparison of the sampling results shows a change in the water quality, corrosion may be occurring between the sampling locations.

Analysis of Corrosion By-product Material. Valuable information about probable corrosion causes can be found by chemically analyzing the corrosion by-product material. Scraping off a portion of the corrosion by-products, dissolving the material in acid, and qualitatively analyzing the solution for the presence of suspected metals or compounds can indicate the type or cause of corrosion. These analyses are relatively quick and inexpensive. If a utility does not have its own laboratory, samples of the pipe sections can be sent to an outside laboratory for analysis. The numerical results of these analyses cannot be quantitatively related to the amount of corrosion occurring since only a portion of the pipe is being analyzed. However, such analyses can give the utility a good overview of the type of corrosion that is taking place. The compounds for which the samples should be analyzed depend on the type of pipe material in the system and the appearance of the corrosion products. For example, brown or reddish-brown scales should be analyzed for iron and for trace amounts of copper. Greenish mineral deposits should be analyzed for copper. Black scales should be analyzed for iron and copper.

Sampling Technique. Since many important decisions are likely to be made based on the sampling and chemical analyses performed by a utility, it is important that care be taken during the sampling and analysis to obtain the best data. Samples should be collected without adding air, as air tends to remove CO_2 and also affects the oxygen content in the sample. To collect a sample without additional air, fill the same container to the top so that a meniscus is formed at the opening and no bubbles are present. The sample bottle should be filled below the surface of the water. To do this, slowly run water down the side of a larger container and immerse the sample bottle in the larger container. Cap the sample bottle as soon as possible.

Recommended Analyses for Additional Corrosion Monitoring. The parameters which should be analyzed for in a thorough corrosion monitoring program depend to a large extent on the materials present in the system's distribution, service, and household plumbing lines. In all cases, temperature and pH should be measured in situ (in the field). Dissolved gases, such as hydrogen sulfide (H_2S), oxygen, CO_2, and chlorine residual, also should be measured as part of a corrosion monitoring program. These parameters can be measured in situ or fixed for laboratory measurement. Total hardness, calcium, alkalinity, and TDS (or conductivity) must be measured if a protective coating of $CaCO_3$ is used for corrosion control or if cement-lined or A-C pipe is present in the system. These analyses are also necessary to calculate the $CaCO_3$-based corrosion indices. Heavy metals analyses should be conducted for the specific metals used in the distribution, service, and household plumbing lines. Measurement of anions, such as chloride and sulfate, may also indicate corrosion potential. Table 6.8 summarizes parameters recommended to be analyzed in a thorough corrosion monitoring program.

Frequency of analysis depends on the extent of the corrosion problems experienced in the system, the degree of variability in raw and finished water quality, the type of treatment and corrosion control practiced by the water utility and cost considerations.

Interpretation of Sampling and Analysis Data. Comparing sampling data from various locations within the distribution system can isolate sections of pipe that may be corroding. Increases in levels of metals such as iron or zinc, for instance, indicate potential corrosion occurring in sections of iron and galvanized iron pipe, respectively. The presence of cadmium, a minute contaminant in the zinc alloy used for galvanized pipe, also indicates the probable corrosion of a galvanized iron pipe.

Corrosion of cement-lined or A-C pipe is generally accompanied by an increase in both pH and calcium throughout the system, sometimes in conjunction with an elevated asbestos fiber count.

The following example illustrates the changes that can take place between a distribution system and a customer's tap. The analytical results in Table 6.9 were obtained from a small water supply system in Florida and the customer's hot water taps. In this case, A-C pipe is used throughout the distribution system. The home plumbing systems are mostly copper.

The water in the distribution system had no traces of copper or lead, and the LSI, calculated from the data as the water entered the distribution system, was slightly positive or potentially noncorrosive. Data in Table 6.9 show that high levels of copper from the household pipes and lead from the solder joints were being added to the customer's water through corrosion of the household plumbing. Further investigation of the household plumbing showed that the customer's hot water system was corroding.

Another example of the importance of data interpretation to an overall corrosion monitoring program is discussed below for A-C pipe. According to EPA's Drinking Water Research Division (DWRD), calculating the AI alone is not sufficient to predict the corrosive behavior of water to A-C pipe. For A-C pipe, additional sampling and data interpretations are recommended by DWRD for determining the corrosivity of a water to A-C pipe.

Table 6.8. Recommended analyses for a thorough corrosion monitoring program

In situ measurements	pH, temperature
Dissolved gases	Oxygen, hydrogen sulfide, carbon dioxide, free chlorine
Parameters required to calculate $CaCO_3$-based indices, or required for cement-lined or A-C pipe	Calcium, total hardness, alkalinity, total dissolved solids, fiber count (A-C pipe only)
Heavy Metals	
Iron or steel pipe	Iron
Lead pipe or lead-based solder	Lead
Copper pipe	Copper, lead
Galvanized iron pipe	Zinc, iron, cadmium, lead
Anions	Chloride, sulfate

Source: Environmental Science and Engineering, Inc., 1982.

Table 6.9. Water quality data from a Florida water utility

Sample location	Cu (mg/L)	Pb (mg/L)
Water entering distribution system	0	0
Water in distribution system	0	0
Water at customer's tap		
Sample set 1	5.0	0
Sample set 2	1.66	3.26

Source: Environmental Science and Engineering, Inc., 1982.

The following conditions indicate situations in which the water *may not attack* A-C pipe:

1. An initial AI above about 11;
2. No significant change in the pH or the concentration of calcium at different locations in the system;
3. No asbestos fibers *consistently* found in *representative* water samples after passage through A-C pipe;
   a. Significant asbestos fiber counts being found in *representative* water samples *at one time but not another* at a location where water flow is sufficient to clean the pipe of tapping debris (recent tapping can cause high fiber counts not related to pipe attack) and
   b. Significant asbestos fiber counts being found only in water samples collected from low-flow dead ends or from fire hydrants *(nonrepresentative samples)* and nowhere else in the system.

The following conditions indicate situations in which the water *may be attacking* A-C pipe:

1. An initial AI below about 11,
2. A significant increase in pH and the concentration of calcium at different locations in the system,
3. Significant asbestos fiber counts being found consistently in representative water samples collected from locations where (a) the flow is sufficient to clean the pipe of debris and (b) the pipe has been neither drilled nor tapped near or during the sampling period, and
4. Inlet water screens at coin-operated laundries become plugged with fibers.

The data obtained by sampling for corrosive characteristics can be used as a guide to water quality changes that might be required to reduce or control corrosion, such as pH adjustment or the addition of silicates or phosphates. Results of additional sampling, conducted after starting a corrosion control program, can indicate the success of any water quality changes.

6.2 DIRECT METHODS

Scale or Pipe Surface Examination

Examining the scale found inside a pipe is a direct monitoring and measuring corrosion control method that can tell a great deal about water quality and system conditions. It can be used as a tool to determine why a pipe is deteriorating or why it is protected and can be used to monitor the results of any corrosion control program. For example, a high concentration of calcium in a scale may shield the pipe wall from DO diffusion and thereby reduce the corrosion rate.

Methods used to examine scale on pipe walls include physical inspection [both macroscopic (human eye) and microscopic], X-ray diffraction, and Raman spectroscopy. Physical inspection is the only method of practical use to utility personnel, as X-ray diffraction and Raman spectroscopy require expensive, complicated instruments and experienced personnel to interpret the results.

Physical Inspection. Physical inspection is usually the most useful inspection tool to a utility because of the low cost. Both macroscopic (human eye) and microscopic observations of scale on the inside of the pipe are valuable tools in diagnosing the type and extent of corrosion. Macroscopic studies can be used to determine the amount of tuberculation and pitting and the number of crevices. The sample should be examined also for the presence of foreign materials and for corrosion at joints.

Utility personnel should try to obtain pipe sections from the distribution or customer plumbing systems whenever possible, such as when old lines and equipment are replaced. If a scale is not found in the pipe, an examination of the pipe wall can yield valuable information about the type and extent of corrosion and corrosion-product formation, (such as tubercles), though it may not indicate the most probable cause.

Examination under a microscope can yield even more information, such as hairline cracks and local corrosion too small to be seen by the unaided eye. Such an examination may provide additional clues to the underlying cause of corrosion by relating the type of corrosion to the metallurgical structure of the pipe.

Photographs of specimens should be taken for comparison with future visual examinations. High magnification photographs should be taken, if possible.

X-ray Diffraction. The diffraction patterns of X-rays of scale material can be used to identify scale constituents. The diffraction of the X-rays will produce a pattern on a film strip which can be compared with X-ray diffraction patterns of known materials. It is possible to identify complex chemical structures by their X-ray "fingerprint."

Raman Spectroscopy. Raman spectroscopy is a technique for identifying compounds present in corrosion scale and films without removing a metal sample. In Raman spectroscopy, an infrared beam is reflected off the surface to be analyzed, and the change in frequency of the beam is recorded as the Raman spectrum. This spectrum, which is different for all compounds, is compared with Raman spectra of known materials to identify the constituents of the corrosion film.

Raman spectroscopy and X-ray diffraction are useful in corrosion research and in corrosion studies where the nature of the scale is unknown. However, the cost of the analyses makes them too expensive to be used in solving most corrosion problems. Nearly all corrosion problems can be solved without the detailed information provided by these techniques.

Rate Measurements

Rate measurements are another method frequently used to identify and monitor corrosion. The corrosion rate of a material is commonly expressed in mils (0.001/inch) penetration per year (mpy). Common methods used to measure corrosion rates include (1) weight-loss methods (coupon testing and loop studies) and (2) electrochemical methods. Weight-loss methods measure corrosion over a period of time. Electrochemical methods measure either instantaneous corrosion rates or rates over a period of time, depending on the method used.

Coupon Weight-Loss Method. This method uses "coupons" or pipe sections as test specimens. It is used for field, pilot-, and bench-scale studies, provided the samples are cleaned and installed in the corrosive environment in such a way that the attack is not influenced by the pipe or container. The coupons usually are placed in the middle of the pipe section.

The weight of the specimen or coupon is measured on an analytical balance before and after immersion in the test water. The weight loss due to corrosion is converted to a uniform corrosion rate by the following formula (as per ASTM Method D2688 Method B):

$$Corrosion\ rate\ in\ mils/year = \frac{534\ W}{DAT} \tag{14}$$

where
W = weight loss [milligrams (mg)],
D = density of specimen [grams per cubic centimeter (g/cm^3)],
A = surface area of specimen [square inches (in^3)], and
T = exposure time [hour (h)].

Coupon weight-loss test results do not measure localized corrosion but are an excellent method for measuring general or uniform corrosion. Coupons are most useful when corrosion rates are high so that weight loss data can be obtained in a reasonable time. The ASTM method above should be followed.

Following are lists of the advantages and disadvantages of the coupon method:

Advantages

1. provides information on the amount of material attacked by corrosion over a specified period of time and under specified operating conditions,
2. coupons can be placed in actual distribution systems for monitoring purposes, and
3. the method is relatively inexpensive.

Disadvantages

1. rate determinations may take a long time (i.e., months, if corrosion rates are moderate or low);
2. the method will not indicate any variations in the corrosion rate that occurred during the test;
3. the specimen or coupon may not be representative of the actual material for which the test is being performed;
4. the reaction between the metal coupon and the water may not be the same as the reaction at the pipe wall due to friction or flow velocity, since the coupon is placed in the middle of the pipe section; and
5. there may be difficulty in removing the corrosion products without removing some of the metal.

Loop System Weight-Loss Method. Another method for determining water quality effects on materials in the distribution system is the use of a pipe loop or sections of pipe. Either the loop or sections can be used to measure the extent of corrosion and the effect of corrosion control methods. Pipe loop sections can be used also to determine the effects of different water qualities on a specific pipe material. The advantage is that actual pipe is used as the corrosion specimen. The loop may be made from long or short sections of pipe.

Water flow through the loop may be either continuous or shut off with a timer part of the time to duplicate the flow pattern of a household. Pipe sections can be removed for weight-loss measurements and then opened for visual examination. This method is called the Illinois State Water Survey (ISWS) method and is an ASTM standard method (D2688, Method C) and should be followed closely.

Following are lists of the advantages and disadvantages of a loop system:

Advantages

1. actual pipe is used as the corrosion specimen;
2. loops can be placed at several points in the distribution system;
3. loops can be set up in the laboratory to test the corrosive effects of different water qualities on pipe materials;

4. the method provides information on the amount of material attacked by corrosion over a specified period of time and under specified operating conditions; and

5. the method is relatively inexpensive, as many corrosive effects can be examined visually.

Disadvantages

1. determination of corrosive rates can take a long time (i.e., months, if corrosion rates are moderate or low), and

2. the method does not indicate variations in the corrosion rate that occur during the test.

Electrochemical Rate Measurements. These methods are based on the electrochemical nature of corrosion of metals in water. An increasing number of these instruments are now on the market. However, they are relatively expensive and probably not widely used by smaller utilities. They are discussed here for completeness.

One type of electrochemical rate instrument has probes with two or three metal electrodes that are connected to an instrument meter to read corrosion in mpy. The electrode materials can be made of the material to be studied and inserted into the pipe or corrosive environment. For the other type, the loss of material over time is detected by an increase in the resistance of an electrode made of the metal of interest. Measurements made over a period of time can be used to estimate corrosion rates.

Following are lists of the advantages and disadvantages of electrical resistance measurements:

Advantages

1. data may provide a graphic history of corrosion rate as it occurs,

2. measurements are rapid, and

3. short-term changes can be measured using linear polarization.

Disadvantages

1. probes may not represent actual material;

2. it is difficult to measure low corrosion rates by the resistance method;

3. they are useful only for metals;

4. the corrosion of a metal often depends on the amount of time it is exposed; therefore, the "instantaneous" corrosion rates given by these methods may not be the same as true long-term corrosion rates

5. as with all monitoring methods, many factors can affect the results; therefore, it is important not to jump to conclusions; and

6. trained, experienced personnel are needed to obtain and interpret data.

7. Corrosion Control

What can a water utility do to control corrosion in its water distribution system?

A schematic representation of a general approach to solving corrosion problems is shown in Fig. 7.1. To completely eliminate corrosion is difficult if not impossible. There are, however, several ways to reduce or inhibit corrosion that are within the capability of most water utilities. This section describes several methods most commonly used to control corrosion. The utility operator should use common sense in selecting the best and most economical method for successful corrosion control in a particular system. Because corrosion depends on both the specific water quality and pipe material in a system, a particular method may be successful in one system and not in another.

Corrosion is caused by a reaction between the pipe material and the water in direct contact with each other. Consequently, there are three basic approaches to corrosion control:

1. modify the water quality so that it is less corrosive to the pipe material,
2. place a protective barrier or lining between the water and the pipe, and
3. use pipe materials and design the system so that it is not corroded by a given water.

The most common ways of achieving corrosion control are to

1. properly select system materials and adequate system design;
2. modify water quality;
3. use inhibitors;
4. provide cathodic protection; and
5. use corrosion-resistant linings, coatings, and paints.

7.1 PROPER SELECTION OF SYSTEM MATERIALS AND ADEQUATE SYSTEM DESIGN

In many cases, corrosion can be reduced by properly selecting system materials and having a good engineering design. As discussed in Sect. 4, some pipe materials are more corrosion resistant than others in a specific environment. In general, the less reactive the material is with its environment, the more resistant the material is to corrosion. When selecting materials for replacing old lines or putting new lines in service, the utility should select a material that will not corrode in the water it contacts. Admittedly, this provides a limited solution since few utilities can select materials based on corrosion resistance alone. Usually several alternative materials must be compared and evaluated based on cost, availability, use, ease of installation, and maintenance, as well as resistance to corrosion. In addition, the utility owner may not have control over the selection and installation of the materials for household plumbing. There are, however, several guidelines that can be used in selecting materials.

First, some materials are known to be more corrosion resistant than others in a given environment. For, example, a low pH water that contains high DO levels will cause more corrosion damage in a copper pipe than in a concrete or cement-lined cast iron pipe. Other guidelines relating water quality to material selection are given in Table 4.3.

A good description of the proper selection of materials can be found in *The Prevention and Control of Water-caused Problems in Building Potable Water Systems*, published by the NACE.

Second, compatible materials should be used throughout the system. Two metal pipes having different activities, such as copper and galvanized iron, that come in direct contact with others can set up a galvanic cell and cause corrosion. The causes and mechanisms of galvanic corrosion are discussed in Sect. 3.0. As much as possible, systems should be designed to use the same metal throughout or to use metals having a similar position in the galvanic series (Table 3.1). Galvanic corrosion can be avoided by placing dielectric (insulating) couplings between dissimilar metals.

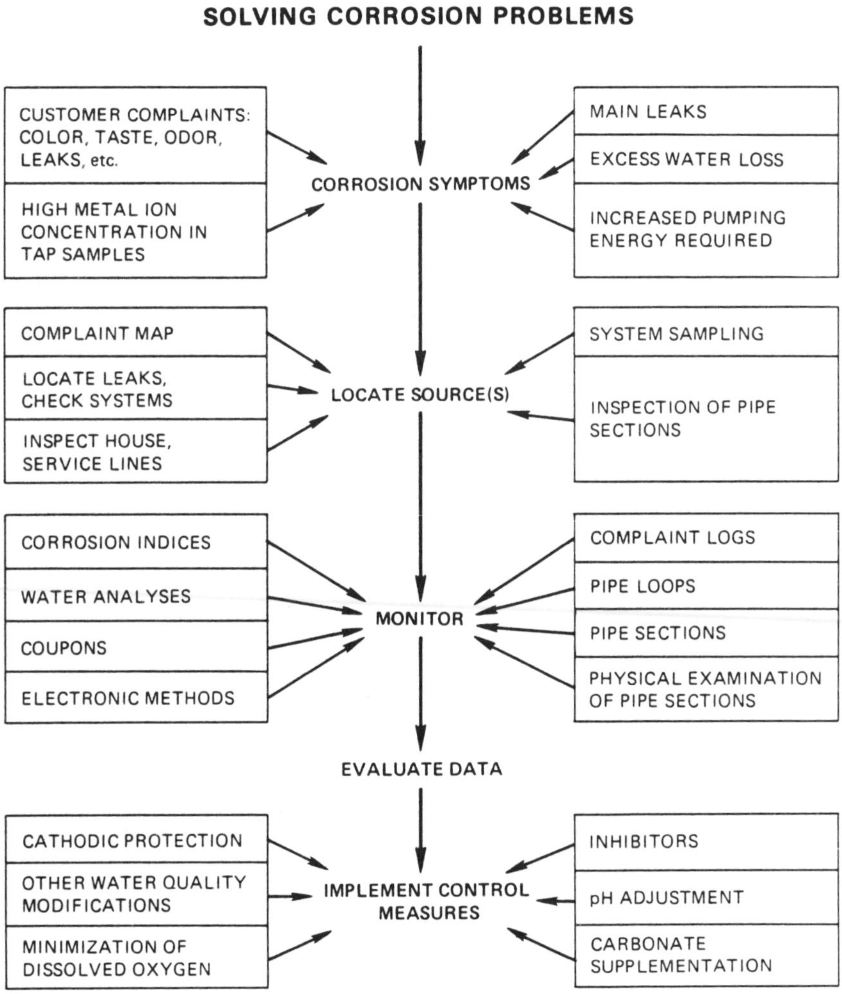

Fig. 7.1. Steps toward solving corrosion problems.

The design of the pipes and structures is as important as the choice of construction materials. A faulty design may cause severe corrosion, even in materials that may be highly corrosion resistant. Some of the important design considerations include

1. avoiding dead ends and stagnant areas;
2. using welds instead of rivets;
3. providing adequate drainage where it is needed;
4. selecting an appropriate flow velocity;
5. selecting an appropriate metal thickness;
6. eliminating shielded areas;
7. reducing mechanical stresses;
8. avoiding uneven heat distribution;
9. avoiding sharp turns and elbows;
10. providing adequate insulation;
11. choosing a proper shape and geometry for the system;
12. providing easy access to the structure for periodic inspection, maintenance, and replacement of damaged parts; and
13. eliminating grounding of electrical circuits to the system.

Many plumbing codes are outdated and allow undesirable situations to exist. Such codes may even create problems, for example, by requiring lead joints in some piping. Where such problems exist, it may be helpful for the utility to work with the responsible government agency to modify outdated codes.

7.2 MODIFICATION OF WATER QUALITY

In many cases, the easiest and most practical way to make a water noncorrosive is to modify the water quality at the treatment plant. Because of the differences among raw water sources, the effectiveness of any water quality modification technique will vary widely from one water source to another. However, where applicable, water quality modification can often result in an economical method of corrosion control.

pH Adjustment

pH adjustment is the most common method of reducing corrosion in water distribution systems. pH plays a critical role in corrosion control for several reasons:

1. Hydrogen ions (H^+) act as electron acceptors and enter readily into electrochemical corrosion reactions. Acid waters are generally corrosive because of their high concentration of hydrogen ions. When corrosion takes place below pH 6.5, it is generally uniform corrosion. In the range between pH 6.5 and 8.0, the type of attack is more likely to be pitting.
2. pH is the major factor that determines the solubility of most pipe materials. Most materials used in water distribution systems (copper, zinc, iron, lead, and cement) dissolve more readily at a lower pH. Increasing the pH increases the hydroxide ion (OH^-) concentration, which, in turn, decreases the solubility of metals that have insoluble hydroxides, including copper, zinc, iron, and lead. When carbonate alkalinity is present, increasing the pH, up to a point, increases the amount of carbonate ion in solution. This may control the solubility of metals that have

insoluble carbonates, such as lead and copper. The cement matrix of A-C pipe or cement-lined pipe is also more soluble at a low pH. Increasing the pH is a major factor in limiting the dissolution of the cement binder and thus controlling corrosion in these types of pipes.

3. The relationship between pH and other water quality parameters, such as alkalinity, carbon dioxide (CO_2), and TDS, governs the solubility of calcium carbonate ($CaCO_3$), which is commonly used to provide a protective scale on interior pipe surfaces. To deposit this protective scale, the pH of the water must be slightly above the pH of saturation for $CaCO_3$, provided sufficient alkalinity and calcium are present.

pH adjustment alone is often insufficient to control corrosion in waters that are low in carbonate or bicarbonate alkalinity. A protective coating of $CaCO_3$, for instance, will not form unless a sufficient number of carbonate and calcium ions are in the water.

Some metals, notably lead and copper, form a layer of insoluble carbonate, which minimizes corrosion rates and the dissolution of these metals. In low alkalinity waters, carbonate ion must be added to form these insoluble carbonates. For such waters, soda ash (Na_2CO_3) or sodium bicarbonate ($NaHCO_3$) are the preferred chemicals generally used to adjust pH because they also contribute carbonate ($CO_3^=$) or bicarbonate ions (HCO_3^-). The number of carbonate ions available is a complex function of pH, temperature, and other water quality parameters. Bicarbonate alkalinity can be converted to carbonate alkalinity by increasing the pH. If carbonate supplementing is necessary to control corrosion in a water system, pH also must be carefully adjusted to ensure that the desired result is obtained.

The proper pH for any given water distribution system is so specific to its water quality and system materials that a manual of this type can provide only general guidance. If the water contains a moderate amount of carbonate alkalinity and hardness (approximately 40 mg/L as $CaCO_3$ or more of carbonate or bicarbonate alkalinity and calcium hardness), the utility should first calculate the LSI and/or AI to determine at what pH the water is stable with regard to $CaCO_3$. Other indices can be used to check this value. To start, the pH of the water should be adjusted such that the LSI is slightly positive, no more than 0.5 unit above the pH_s. If the AI is used as a guide, an initial AI value equal to or greater than 12 is desirable. If no other evidence is available, such as a good history of the effect of pH on the laying down of a protective coating of $CaCO_3$ or laboratory or field test results, then the LSI and/or AI provide a good starting point. Keeping the pH above the pH_s should cause a protective coating to develop. If no coating forms, then the pH should be increased another 0.1 to 0.2 unit until a coating begins to form. It is important to watch the pressure in the system carefully as too much scale build-up near the plant could seriously clog the transmission lines.

There is a strong tendency to overestimate the accuracy of the calculated values of the LSI or AI. Soft, low alkalinity waters cannot become supersaturated with $CaCO_3$ regardless of how high the pH is raised. In fact, raising the pH to values greater than about 10.3 is useless because no more carbonate ions can be made available. Excess hydroxide alkalinity is of no value since it does not aid in $CaCO_3$ precipitation.

For systems that do not rely on $CaCO_3$ deposition for corrosion control, it is more difficult to estimate the optimum pH. If lead and/or copper corrosion is a problem, adjusting the pH to values of from 7.5 to 8.0 or higher may be required. Practical minimum lead solubility occurs at a pH of about 8.5 in the presence of 30 to 40 mg/L of alkalinity. pH adjustment coupled with carbonate supplementing may be required to minimize lead corrosion problems.

Phosphates and other corrosion inhibitors often require a narrow pH range for maximum effectiveness. If such an inhibitor is used, consideration must be given to adjusting the pH to within the recommended range.

Chemicals commonly used for pH adjustment and/or carbonate supplementing, recommended dosages, and equipment requirements are summarized in Table 7.1. Schematics of typical chemical feed systems are shown in Fig. 7.2. The pH should be adjusted after filtration since waters having higher pHs need larger doses of alum for optimum coagulation.

Table 7.1. Chemicals for pH adjustment and/or carbonate supplementation

pH adjustment chemical	Typical feed rate	1 mg/L adds ___mg/L alkalinity[a]	Equipment required
Lime, as $Ca(OH)_2$	1–20 mg/L (8–170 lb/MG)	1.35	Quicklime-slaker, hydrated lime-solution tank, and feed pump with erosion-resistant lining as eductor
Caustic soda, NaOH (50% solution)	1–29 mg/L (8–170 lb/MG)	1.25	Proportioning pump or rotameter
Soda ash, Na_2CO_3	1–40 mg/L (8–350 lb/MG)	0.94	Solution tank, proportioning pump, or rotameter
Sodium bicarbonate, $NaHCO_3$	5–30 mg/L (40–250 lb/MG)	0.59	Solution tank, proportioning pump, or rotameter

[a]Caustic soda and lime add only hydroxide alkalinity. Soda ash and sodium bicarbonate add carbonate or bicarbonate alkalinity, depending on pH.

It is recommended that a corrosion monitoring program, such as that described in Sect. 6.0, be initiated to monitor the effects of this pH change over time. Evaluating the performance of chemical feed systems for pH adjustment is the key to an effective corrosion control program. Addition of lime, soda ash, or other chemicals for pH control can be evaluated by continuous readout pH recorders. The recorders monitor the pH of the water as it leaves the utility and can be wired to send a signal to the feed mechanism to add more or fewer chemicals as necessary. The pH levels at the outer reaches of the distribution system should be checked periodically for indications of any changes occurring within the system that might be due to corrosion.

Keep in mind that although pH adjustment can aid in reducing corrosion, it cannot eliminate corrosion in every case. However, pH adjustment is the least costly and most easily implemented method of achieving some corrosion control, and utilities should use it if at all possible.

Reduction of Oxygen

As explained in Sect. 3.0, oxygen is an important corrosive agent for the following reasons:

1. oxygen can act as an electron acceptor, allowing corrosion to continue;
2. oxygen reacts with hydrogen to depolarize the cathode and thus speeds up corrosive reaction rates; and
3. oxygen reacts with iron ions to form tubercles and leads to pitting in copper.

If oxygen could be removed from water economically, the chances of corrosion starting, and also the corrosion rate once it had started, would be reduced. Unfortunately, oxygen removal is too expensive for municipal water systems and is not a practical control method. However, there are ways to minimize the addition of oxygen to the raw water, particularly to groundwaters.

Often, aeration is the first step in treating groundwaters having high iron, hydrogen sulfide (H_2S) or CO_2 content. Though aeration helps remove these substances from raw water, it can also cause more serious corrosion problems by saturating the water with oxygen. In lime-soda softening plants for treating groundwater, the water is often aerated first to save on the cost of lime by eliminating free CO_2. Iron is oxidized and precipitated in this step, but this is incidental, because the

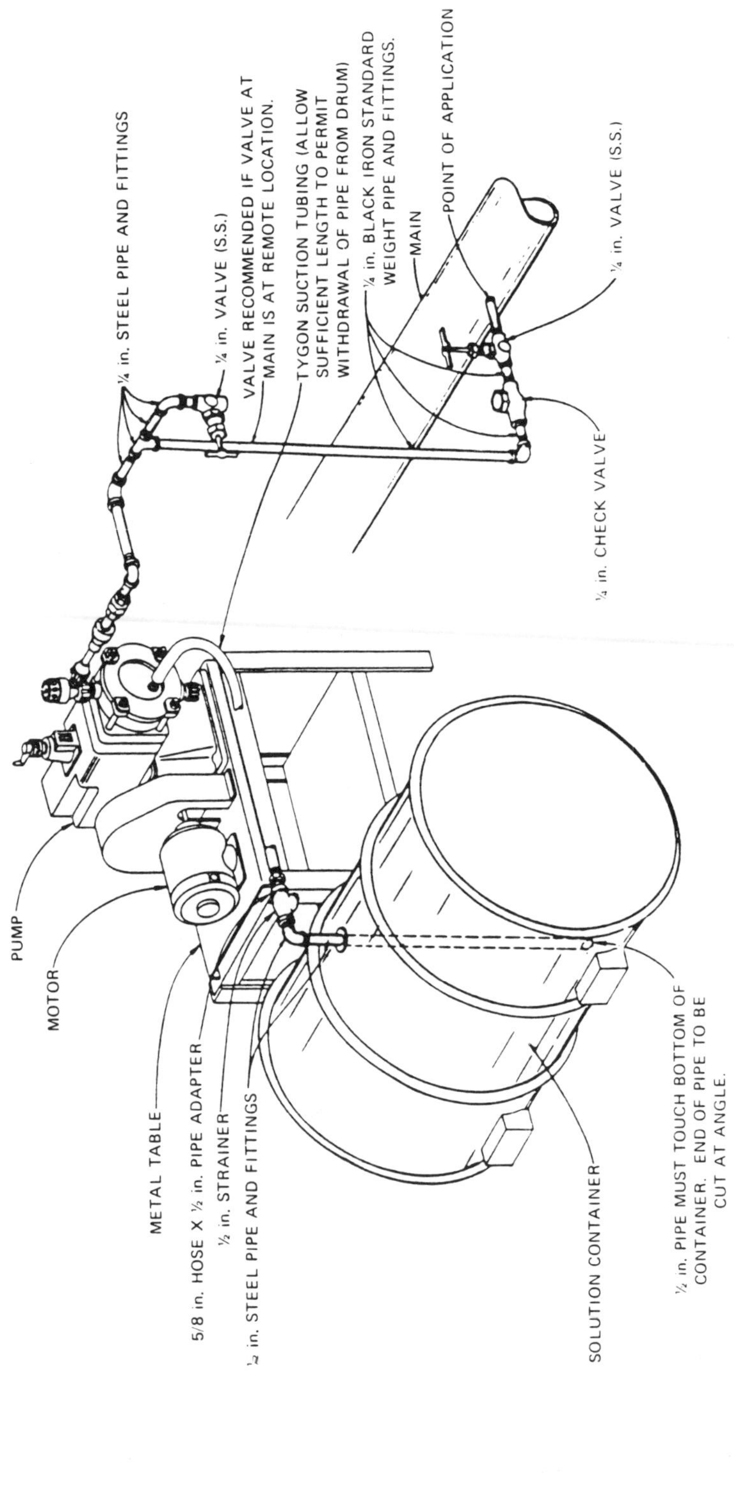

Fig. 7.2. Schematic of a chemical feed system.

iron would be removed in the subsequent softening process even if the water were not aerated. The actual result is that DO increases to near saturation, and corrosion problems are increased. Thus, the attempt to save on lime addition may actually end up costing a great deal more in corrosion damage.

Measures that help keep the DO levels as low as possible include (1) sizing well pumps and distribution pumps so as to avoid air entrainment and (2) using as little aeration as possible when aerating for H_2S or CO_2 removal. This can be achieved by by-passing the aerators with part of the raw water. It has even been possible to completely eliminate the use of aerators if enough detention time is available in the reservoir so that enough oxygen can be absorbed at the surface to oxidize the H_2S or to let the CO_2 escape. DO levels can be kept as low as 0.5 to 2.0 mg/L by this method. This is low enough, in many cases, to reduce corrosion rates considerably.

7.3 USE OF INHIBITORS

Corrosion can be controlled by adding to the water chemicals which form a protective film on the surface of a pipe and provide a barrier between the water and the pipe. These chemicals, called inhibitors, reduce corrosion but do not totally prevent it.

The three types of chemical inhibitors commonly approved for use in potable water systems are chemicals which cause $CaCO_3$ scale formation, inorganic phosphates, and sodium silicate. There are currently several hundred commercially available products listed with various state and federal agencies for this use (see Sect. 7.6).

The success of any inhibitor in controlling corrosion depends upon three basic requirements. First, it is best to start the treatment at two or three times the normal inhibitor concentration to build up the protective film as fast as possible. This minimizes the opportunity for pitting to start before the entire metal surface has been covered by a protective film. Usually it takes several weeks for the coating to develop.

Second, the inhibitor may be fed continuously and at a sufficiently high concentration. Interruptions in the feed can cause loss of the protective film by re-dissolving it, and too low concentrations may prevent the formation of a protective film on all parts of the surface. Both interrupted feeding and low dosages can lead to pitting. On the other hand, excessive use of some alkaline inhibitors over a period of time can cause an undesirable build-up of scale, particularly in harder waters. The key to good corrosion inhibitor treatment is feed control.

Third, flow rates must be sufficient to continuously transport the inhibitor to all parts of the metal surface, otherwise an effective protective film will not be formed and maintained. Corrosion will then be free to take place. For example, corrosion inhibitors often can not reduce corrosion in storage tanks because the water is not flowing, and the inhibitor is not fed continuously. To avoid corrosion of the tanks, it is necessary to use a protective coating, cathodic protection, or both. Similarly, corrosion inhibitors are not as effective in protecting dead ends as they are in those sections of mains which have a reasonably continuous flow.

$CaCO_3$ Deposition

Under certain conditions, a layer of $CaCO_3$ will deposit on the surface of the pipe and serve as a protective barrier between the pipe wall and the water. This process is discussed in Sect. 6.0. It is mentioned again here because the addition of lime or alkalinity is a kind of inhibitor treatment.

Inorganic Phosphates

Phosphates are used to control corrosion in two ways: to prevent scale or excess $CaCO_3$ build-up and to prevent corrosive attack of a metal by forming a protective film on the surface of the pipe wall. Phosphates inhibit the deposition of a $CaC0_3$ scale on the pipe walls, which is an advantage only in the waters in which excessive scaling occurs. The mechanism by which phosphates form a protective film and inhibit corrosive attack, though not completely understood, is known to depend on flow velocity, phosphate concentration, temperature, pH, calcium, and carbonate levels.

There are several different types of phosphates used for corrosion control, including polyphosphates, orthophosphates, glassy polyphosphates, and bimetallic polyphosphates. Recent developments in corrosion control include the use of zinc along with a polyphosphate or orthophosphate.

Low dosages (about 2 to 4 mg/L) of glassy phosphates, such as sodium hexametaphosphate, have long been used to solve red water problems. In such cases, the addition of glassy phosphates masks the color, and the water appears clear because the iron is tied up as a complex ion. The corrosive symptoms are removed, but the corrosion rates are not reduced. Controlling actual metal loss requires dosages up to 10 times higher (20 to 40 mg/L) of the glassy phosphates. Other glassy phosphates which contain calcium as well as sodium are more effective as corrosion inhibitors. Adding zinc along with a phosphate has been successfully used to both inhibit corrosion and control red water at dosages of about 2 mg/L. The zinc phosphate treatment has also been used to eliminate rusty water, blue-green staining, lead pickup, and to reduce measured corrosion rates of metals.

The choice of a particular type of phosphate to use in a corrosion control program depends on the specific water quality. Some phosphates work better than others in a given environment. It is usually advisable to conduct laboratory or field tests of one or more phosphate inhibitors before long-term use is initiated. The case histories in Sect. 8.0 contain several examples of how such tests are performed and evaluated.

For smaller water utility plants [up to 1 million gallons per day (MGD)], phosphate feed solutions can be made up easily by batch as needed. A maximum phosphate solution concentration of 10 wt.% or 0.834 pound per gallon (lb/gal) is normally recommended. For a phosphate dose of 3 mg/L and a flow of 1 MGD, the volume of phosphate solution fed can be calculated as follows:

$$1\,MGD \times 3\,mg/L \times \frac{8.34\,lb/MG}{mg/L} \times \frac{1\,gal}{0.834\,lb} = \frac{30\,gal}{day} \tag{15}$$

The equipment needed to feed phosphates to the water includes a 55-gal solution feed tank; a drum mixer; a chemical metering feed pump; and associated piping, feed lines, valves, and drains. The capital expenditure required is usually less than $2000 and is, therefore, within the means of most small water utilities.

Sodium Silicate

Sodium silicate (water glass) has been used for over 50 years to reduce corrosivity. The way in which sodium silicate acts to form a protective film is still not completely understood. However, it can effectively reduce corrosion and red water complaints in galvanized iron, yellow brass, and copper plumbing systems in both hot and cold water.

The effectiveness of sodium silicate as a corrosion inhibitor depends on water quality properties such as pH and bicarbonate concentration.

As a general rule, feed rates of 2 to 8 mg/L and possibly up to 12 mg/L of sodium silicate are sufficient to control corrosion in a system once a protective film is formed. Silicate has been found to be particularly useful in waters having very low hardness and alkalinity and a pH of less than 8.4. It is also more effective under higher velocity flow conditions. The equipment needed to feed sodium silicate is the same as that needed to add phosphate. The application of sodium silicate requires the use of solution feeders and small positive displacement pumps that deliver a specific volume of chemical solution for each piston stroke or impeller rotation. Figure 7.3 shows an example of a commercially available phosphate and/or silicate feed system for small water utilities.

Monitoring Inhibitor Systems

When phosphates or silicates are added to the water, samples should be collected at the far reaches of the system and analyzed for polyphosphates, orthophosphates, and sodium silicate, as appropriate. If no residual phosphate or silicate is found, the feed rate should be increased. Usually,

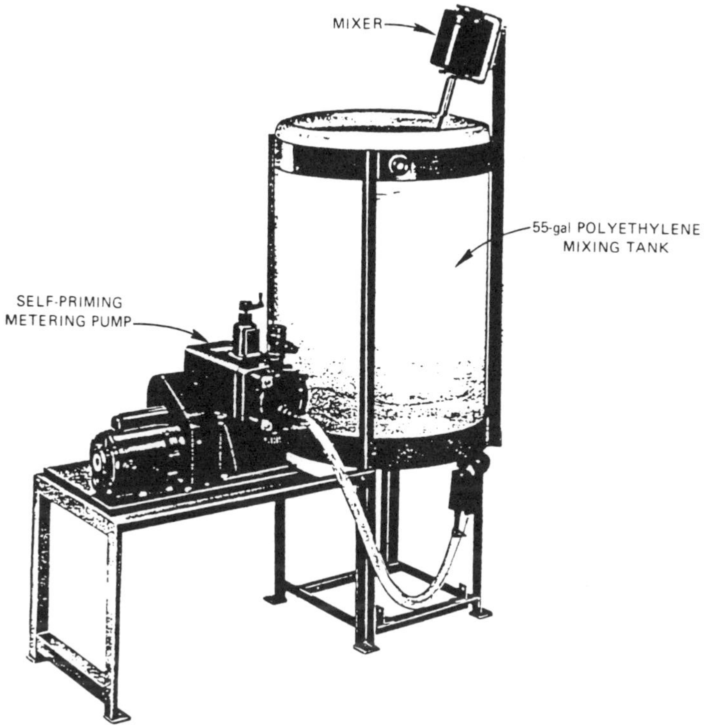

Fig. 7.3. Commercially available phosphate or silicate feed system.

only a residual is necessary to inhibit corrosion. If the concentration at the far reaches of the system is the same as that applied at the utility (e.g., 2 ppm), the utility may wish to decrease the chemical feed rate to save on costs for chemicals.

As previously discussed, initial inhibitor feed rates (for the first 2 weeks) should be 5 to 10 times higher than normal. During this time, water from the far reaches of the system should be sampled about twice a week to determine if corrosion products are leaching from the pipe wall. If the pipes are heavily tubercled, the tubercles are frequently broken loose by the inhibiting chemical. Where pitting has occurred, the system may be suddenly plagued with leaks as a result, and other corrective action must be initiated.

After the system has stabilized, sampling frequency can be reduced to about once a month or quarterly, depending on the resources available to the utility.

Feed Pumps for Inhibitor Systems

Chemical feed pumps. Most metering pumps used to add phosphate or silicate are positive displacement pumps. Pumping action for this type of pump is achieved by means of a piston, plunger, or diaphragm in which movement in one direction draws in a liquid through a valve, and movement in the opposite direction forces the liquid out through a second valve, causing a positive displacement of the liquid during each stroke of the unit. These types of pumps are generally used for chemical feeding when liquids heavier than water are being added. Chemical feed rates can be adjusted by changing the length and speed of the piston or diaphragm stroke. Usually, the water is pumped from a well or storage tank by centrifugal pumps throughout the distribution system. A signal can be wired from the centrifugal pump to the feed pump so that the feed pump is activated only when water is being pumped to the distribution system. Chemical feed pumps can be single or dual headed so that one or two chemicals can be added at the same time. The advantage of these pumps is that they are both accurate and reliable in feeding a specified amount of chemical to the system. The feed pumps should be calibrated about once a week to ensure that the desired amount of chemical is added.

7.4 CATHODIC PROTECTION

Cathodic protection is an electrical method for preventing corrosion of metallic structures. As discussed in Sect. 3.0, metallic corrosion involves contact between a metal and an electrically conductive solution which produces a flow of electrons or current from the metal to the solution. Cathodic protection stops the current by overpowering it with a stronger current from some outside source. This forces the metal that is being protected to become a cathode; that is, it has a large excess of electrons and cannot release any of its own. There are two basic methods of applying cathodic protection. One method uses inert electrodes, such as high-silicon cast iron or graphite, that are powered by an external source of direct current. The current impressed on the inert electrodes forces them to act as anodes, thus minimizing the possibility that the metal surface being protected will become an anode and corrode. The second method uses a sacrificial galvanic anode. Magnesium or zinc anodes produce a galvanic action with iron such that they are sacrificed (or corrode) while the iron structure they are connected to is protected from corrosion. This type of system is common to small hot water heaters. Another form of sacrificial anode is galvanizing where zinc is used to coat iron or steel. The zinc becomes the anode and corrodes, protecting the steel, which is forced to be the cathode.

The primary reason for applying cathodic protection in water utilities is to prevent internal corrosion in water storage tanks. Because of the high cost, cathodic protection is not a practical corrosion control method for use throughout a distribution piping system. Another limitation of cathodic protection is that it is almost impossible for cathodic protection to reach down into holes, crevices, or internal corners.

7.5 LININGS, COATINGS, AND PAINTS

Another way to keep corrosive water away from the pipe wall is to line the wall with a protective coating. These linings are usually mechanically applied, either when the pipe is manufactured or in the field before it is installed. Some linings can be applied even after the pipe is in service, though this method is much more expensive. The most common pipe linings are coal-tar enamels, epoxy paint, cement mortar, and polyethylene.

Water storage tanks are most commonly lined to protect the inner tank walls from corrosion. Common water tank linings include coal-tar enamels and paints, vinyls, and epoxy.

Although coal-tar-based products have been widely used in the past for contact with drinking water, currently there is concern at EPA about their use because of the presence of polynuclear aromatic hydrocarbons and other hazardous compounds in coal tar and the potential for their migration in water. Table 7.2 summarizes the most commonly used pipe linings and coatings and lists the advantages and disadvantages of each. Common water tank linings are summarized in Table 7.3.

Table 7.2. Pipe wall linings

Material	Use	Advantages	Disadvantages
Hot applied coal tar enamel	Lining for steel pipes (used in 50 to 80% of steel pipes in distribution systems)	Long service life (>50 years) Good erosion resistance to silt or sand Resistant to biological attachment	Need to reapply to welded areas Extreme heat may cause cracking Extreme cold may cause brittleness May cause an increase in trace organics in water
Epoxy	Lining for steel and ductile iron pipes (can be applied in the field or in a foundry)	Smooth surface results in reduced pumping costs Formulated from components approved by the Food and Drug Administration	Relatively expensive Less resistant to abrasion than coal tar enamel Service life <15 years
Cement mortar	Standard lining for ductile iron pipes, sometimes used in steel or cast-iron pipes	Relatively inexpensive Easy to apply (can be applied in place or in pipe manufacturing process) Calcium hydroxide release may protect uncoated metal at pipe joints	Rigidity of lining may lead to cracking or sloughing Thickness of coating reduces cross-sectional area of pipe and reduces carrying capacity
Polyethylene	Lining used in ductile iron and steel pipe (applied at foundry)	Long service life (50 years) Good erosion resistance to abrasives (silt and sand) Good resistance to bacterial corrosion Smooth surface results in reduced pumping costs	Relatively expensive

Source: Environmental Science and Engineering, Inc., 1981.

Table 7.3. Water storage tank linings and coatings

Material	Comments
Hot applied coal tar enamel	Most common coal-tar based coating used in water tanks; tends to sag or ripple when applied above the waterline when tank walls are heated
Coal tar paints	Most commonly used to reline existing water tanks; those paints containing xylene and naphtha solvents give the water an unpleasant taste and odor and should be used only above the waterline
	Other coal tar paints containing no solvent bases can be used below the waterline but should not be exposed to sunlight or ice; service life of 5 to 10 years
Coal tar epoxy paints	Less resistant to abrasion than coal tar enamel; can cause taste and odor problems in the water; and service life of about 20 years
Coal tar emulsion paint	Good adhesive characteristics, odorless, and resists sunlight degradation but not as watertight as other coal tar paints, which limits use below waterline
Vinyl	Nonreactive; hard, smooth surface; service life (about 20 years) is reduced by soft water conditions
Epoxy	Forms hard, smooth surface; low water permeability; good adhesive characteristics if properly formulated and applied
Hot and cold wax coatings	Applied directly over rust or old paint, short service life (about 5 years)
Metallic-sprayed zinc coating	Relatively expensive process that requires special skills and equipment, good rust inhibition, and service life of up to 50 years
Zinc-rich paints	Hard surface; resistant to rust and abrasion; relatively expensive
Chlorinated rubber paints	Used when controlling fumes from application of other linings is difficult or where their use is specified
Asphalt-based linings	Use is generally limited to relining existing asphalt-lined tanks

7.6 REGULATORY CONCERNS IN THE SELECTION OF PRODUCTS USED FOR CORROSION CONTROL

The need for government involvement in the use of corrosion control products stems from the possibility that potable water may become contaminated with potentially harmful substances when these products are used. Concerns about the public health risks focus on the residual amounts of water treatment chemicals in drinking water and the impurities found in them and on the potentially hazardous chemicals which could leach from materials and substances in contact with the water.

The EPA, operating in cooperation with the States and under the authority of the Safe Drinking Water Act, is charged with assuring that the public is provided with safe drinking water. Under the auspices of that charge, EPA assists the States and the public by providing scientific advice on the health safety of chemicals and other substances in and in contact with drinking water.

In rendering advisory opinions on corrosion control products, EPA does not "authorize," "approve," or otherwise control the use of such additives. However, in practice, many state health departments have relied heavily on EPA's opinions in their approval of products and equipment for use in treatment and distribution systems of public utilities. These opinions on product safety are handled through a voluntary product safety evaluation program at EPA.

Additionally, the National Academy of Sciences (NAS), under contract to ODW, recently published the first edition of the "Water Chemicals Codex," which sets recommended maximum impurity concentrations (RMICs) for harmful substances found in many common direct additives (bulk treatment chemicals). EPA has adopted the specifications in the "Codex" as informal guidelines for evaluating treatment chemicals, including corrosion inhibitors.

8. Case Histories

This section presents several case histories of corrosion problems experienced by water utilities or commercial complexes responsible for providing potable water. Methods used to monitor and control corrosion in the distribution systems are presented. The case histories are as follows:

Case 1. Pinellas County Water System (PCWS), Pinellas County, Florida;
Case 2. Mandarin Utilities, Jacksonville, Florida;
Case 3. Middlesex Water Company (MWC), Woodbridge, New Jersey;
Case 4. A Small Hospital, Sierra Nevada, California;
Case 5. Boston Metropolitan Area Water System, Boston, Massachusetts;
Case 6. Galvanized Pipe and the Effects of Copper—A Composite of Incidents Experienced in California; and
Case 7. Greenwood Commissioners of Public Works (CPW), Greenwood, South Carolina.

Each case presents a corrosion problem unique to that utility or complex because of a specific water quality in a given environment. In each case, the source and the effects of the corrosion are different, and the control methods implemented also are unique to each system. However, the approaches to the problems are similar and relevant to most utilities, regardless of size or the nature of the corrosion problem. Each case is presented in some detail to emphasize the different steps used in corrosion control, such as investigating the extent and cause of the problem, sampling and analyzing to further evaluate the problem, testing different control alternatives, and implementing the corrective actions.

In addition to the case histories discused here, another excellent case history is the corrosion monitoring and control program implemented by Seattle, Washington. The Seattle experience has been described in several journals but is not included here because of the complexity and length of the study. Interested readers are referred to the report written by J.E. Courthene and G.J. Kirmeyer, "Seattle Internal Corrosion Control Plan—Summary Report," published in the AWWA Seminar Proceedings, June 25, 1978. The reader also will benefit by referring to the recent summary report released by EPA titled "Seattle Distribution System Corrosion Control Study, Vol. I, Cedar River Water Pilot Plant Study" (Hogt, Herrera, and Kirmeyer 1982).

Many corrosion problems can be solved by the water utility itself. Sometimes, however, in-house diagnosis may lead to wrong conclusions and ineffective treatment. There is often no substitute for consulting with experienced corrosion engineers, the local health department, or state water treatment personnel for assistance in solving corrosion problems.

8.1 PINELLAS COUNTY WATER SYSTEM

This study, excerpted from a paper presented by J.A. Nelson and F.J. Kingery at the AWWA Conference in June 1978, illustrates

1. the problems associated with copper pitting;
2. the effects of pH, CO_2, DO, and phosphate inhibitors on corrosion rates; and
3. the use of coupon tests to evaluate several control strategies.

Background

The PCWS, located on the west coast of Florida, includes two plants, serving about 350,000 consumers. Water production averages about 40 MGD. The water source is wells averaging 350 ft in depth from a typical lime rock formation known as the Floridan Aquifer. Water treatment originally involved aeration to remove H_2S, chlorination to give a free chlorine residual to 2.0 mg/L, and stabilization with sodium hydroxide to adjust the pH. Table 8.1 shows the results of a typical effluent water analysis from the plant.

Table 8.1. PCWS typical effluent water analysis

Parameter	mg/L
Total hardness as $CaCO_3$	214
Calcium as $CaCO_3$	198
Magnesium as $CaCO_3$	16
Total alkalinity as $CaCO_3$	200
Carbonate hardness as $CaCO_3$	200
Noncarbonate hardness as $CaCO_3$	14
Specific conductance	400
TDS	284
Iron as Fe	0.04
Carbon dioxide as CO_2	9
Chloride as Cl^-	22
Sulfate as SO_4	2
Turbidity (NTU)	0.12
pH	7.65
pH_s	7.45
Saturation index	+0.20

Source: AWWA Journal, June 1978, *AWWA Proceedings*.

Reports of leaking copper pipes in numerous homes and apartment complexes alerted PCWS personnel to its copper corrosion problem. To determine the cause and extent of the corrosion and correct deficiencies, the PCWS initiated an investigative monitoring program.

Initial Investigation and Monitoring Program

Procedure. To determine the extent of copper corrosion and acquire background information for evaluating future treatment modifications, the following investigation and monitoring program was instituted before any changes in plant operation were made:

1. Approximately 25 random samples were collected from customers' residences.
2. Twenty residents' homes were monitored weekly beginning in September 1974 for copper, pH, DO, and chlorine residual. Weekly sampling continued through May of 1980.
3. Drinking fountains throughout Pinellas County were monitored for copper content and found to average 1.35 mg/L.

Results. The results of the investigation indicated that not only was there a pitting problem, but also that copper levels averaged 1.5 mg/L. In some isolated points, 5.0 mg/L of copper was found in water left standing overnight in customers' copper service lines. It became evident that it was necessary to reduce the pitting action and to reduce the copper level to under 1.0 mg/L.

Testing of Alternative Control Methods

Alternative 1: Adjustment of pH and CO_2

Procedure. To determine the degree of copper corrosion caused by low pH and thus high CO_2, the pH was increased to 7.9 by increasing the sodium hydroxide feed to 18 mg/L. Raising the pH reduced the CO_2 level from about 8.0 mg/L to 3.0 mg/L.

Results. The average copper content was reduced by 0.33 mg/L, but after 1 month, excessive scaling of pipes and pumps occurred throughout the plant near the point of chemical addition, and pH had to be reduced to 7.65.

This demonstrates that in an effort to control an existing problem, one frequently creates another, possibly worse, problem. Especially when using pH adjustment as a means of controlling corrosion, $CaCO_3$ solubility must be kept in mind. A typical water shows a Langelier shift of +0.8 unit when heated from 60°F (15.5°C) to 140°F (60°C). By adjusting to slightly positive in the distribution system, the utility frequently runs the risk of scaling consumer water heaters or other equipment in the system.

Alternative 2: Reduction of DO

Procedure. To determine the degree of copper corrosion caused by DO, the Plant 1 aerators were by-passed. Plant 1 supplies one area of distribution exclusively before blending with water from Plant 2 about 10 miles away at a 20-million gallon storage and booster station.

The service area fed by Plant 1 consisted of 5 of the original 20 distribution sample points and provided an excellent opportunity to compare results of further treatment changes. Also, a 50-ft coil of ½-in. copper tubing was placed in the effluent water of each plant for additional monitoring.

Results. After by-passing the Plant 1 aerators, the DO of the finished water was reduced from 7.5 to 0.5 mg/L. Sodium hydroxide was increased to 24 mg/L in oder to maintain a pH of 7.65.

Daily samples were taken of both plant effluents and within the distribution system. The copper level in the Plant 1 effluent at the 50-ft copper tubing dropped from 2.5 mg/L to an average of 0.15 mg/L. Oxygen levels averaged 1.0 mg/L within the distribution system as a result of an open clearwell and tank storage.

Alternative 3: Sodium Zinc Phosphate (SZP) Pilot Test

SZP was considered as a possible inhibitor of copper corrosion. Figure 8.1 illustrates methods used for a 3-month pilot test.

Procedure. A micropump was used to feed a stock solution of SZP at the rate of 1.0 mg/L into the water flowing through a 50-ft coil of ½-in. copper tubing. Water was controlled at 1½ ft/s by use of a constant-head device. An untreated section of copper pipe was used as a control.

Water dosed with SZP was allowed to flow through one section of copper tubing for 8 h. Both the untreated and dosed water were then turned off and allowed to stand in the copper pipe for up to 24 h before testing. The CO_2 content was 9.0 mg/L, and oxygen averaged 7.5 mg/L throughout the test period.

Samples were taken from each tap and analyzed for their copper content. Sequestering with 2.5 mg/L of SZP for 2 d preceded the test run.

Results. Over a period of 90 d, the average copper reduction was 0.5 mg/L, approximately 30%.

Alternative 4: SZP Started on Plant 1

Procedure. Based on the results of the pilot test using SZP to control copper corrosion, it was decided to use this inhibitor in water from Plant 1 for a 3-month trial period.

The SZP was fed at the rate of 1.0 mg/L using a diaphragm proportioning pump. Because a lower pH was recommended, the pH of the finished water was reduced to 7.4, which increased the CO_2 to 14.0 mg/L.

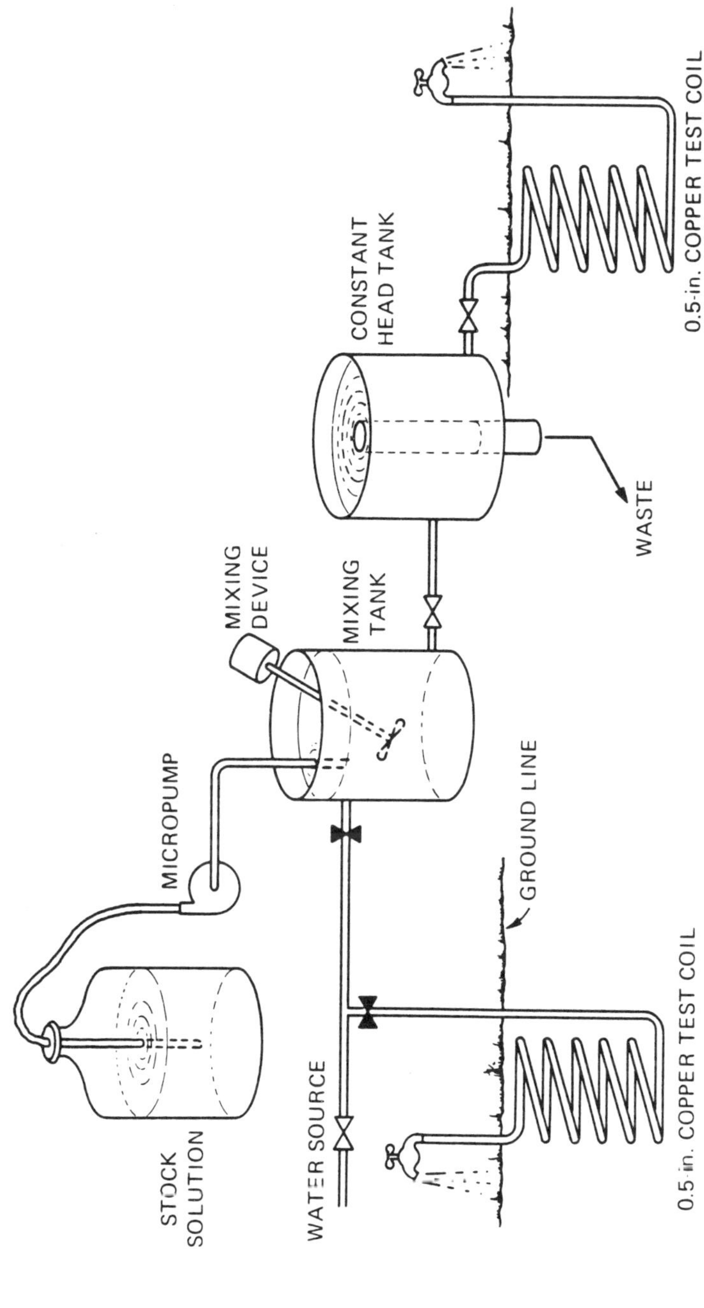

Fig. 8.1. Inhibitor pilot test.

The distribution sampling points that had been selected previously were monitored weekly for copper content. Over a period of 3 months, the average copper content of these 65 samples was 1.51 mg/L.

Results. Results of the test were questionable because copper levels did not compare with those of the pilot test program. The pilot test resulted in 1.10 mg/L of copper, while the actual results of monitoring points averaged 1.51 mg/L of copper, about a 30% difference.

Excluding minor variations in plant operation, it is probable that the lower pH and higher CO_2 content were the principal reasons for the higher copper levels recorded at the distribution monitoring stations. Higher feed rates of SZP may be necessary to achieve favorable results.

Alternative 5: Zinc Orthophosphate (ZOP)

To find the best inhibitor of copper corrosion, ZOP was investigated in a bench study using both mild steel and copper coupons. A 30-d test using both SZP and ZOP was compared with conventional stabilization using sodium hydroxide (NaOH).

Procedure. The test was based on one reported by E.D. Mullen in the ***AWWA Journal*** (August 1974), except that in this test the copper and mild steel coupons were used simultaneously.

The two test units were plastic assemblies of three cylindrical cells each, connected in series. The inlet cell held the copper and steel control coupons; the water then flowed to the center cell for chemical addition and mixing and then through the last cell holding the test copper and steel coupons.

Both units were connected to the same plant effluent line that was fitted with ½-gallon-per-minute (gpm) ball valves for flow control.

Inhibitors were fed to the center mixing cell using a controlled siphon. The chemical feed rates and flow rates were checked daily for 30 d.

Results. Figure 8.2 compares the corrosion rates of each inhibitor. ZOP reduced the corrosion rate of copper by 51%, and SZP reduced the corrosion rate of copper by 5%.

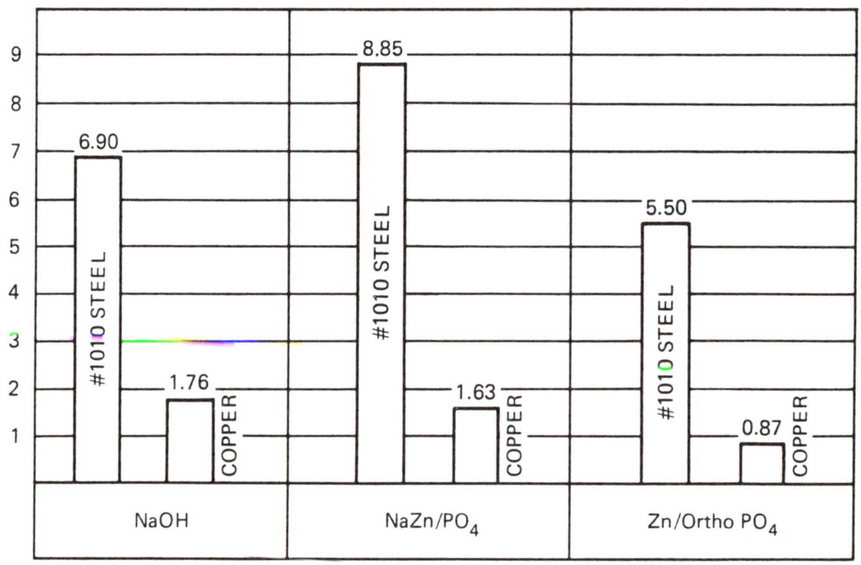

Fig. 8.2. Coupon corrosion rates of NaOH and inhibitors.

Additional Studies

To evaluate the effects of lime-softened water on copper pipe, a study was conducted to compare samples of pipe from a neighboring city, which has used lime softening for over 40 years, to the PCWS pipe.

A number of miscellaneous samples of copper tubing and water meter screens were sent to The University of Florida in Gainesville, Florida, for X-ray examination to determine corrosion products. It was possible to separate the deposits on the meter screens into several layers, varying in color and texture.

The screens generally had a yellowish-white outer deposit and bluish or greenish underlying deposits. Although it was not possible to identify all the compounds present in the reaction products on the various samples of pipe and screen, several observations could be made.

1. The most significant difference in the composition of deposits on the screens from a lime-softened water compared to that of the PCWS is the amount of calcium present. Calcium was present in far greater amounts on the softened-water screens. Presumably, the calcium is largely in the form of carbonate. There is no certain way of determining exactly when or at what rate calcium was deposited. However, calcium was a major constituent in all layers of the deposits of the softened-water screens examined.
2. The relative lack of calcium on the PCWS screens suggests the absence of protective $CaCO_3$ films over some extended period. This would explain the relatively higher corrosion observed on the PCWS screens.
3. The use of ZOP appears to favor deposition of calcium as well as zinc and phosphorus.

Current corrosion control methods. After full-scale implementation of phosphate inhibitor treatment, the Pinellas County utility found that its copper corrosion problem could be controlled just by adjusting the pH and reducing DO in the system. Currently, the utility carefully controls the pH at 7.65. The water by-passes the aerators completely and flows directly into the clearwell under the aerators. This reduces H_2S and maintains the DO level at less than 1 mg/L, which does not appear to be corroding the copper in the system.

Conclusions

Several suggestions are offered by PCWS utility personnel for monitoring the extent of corrosion within the distribution system.

1. Collect weekly samples from several remote sections of the distribution system; run tests for pH, alkalinity, specific conductance, iron, and copper and compare with plant effluent analyses for deterioration of water quality.
2. Check copper meter screens; observe any discoloration or corrosion products. Submit samples for X-ray analysis if needed.
3. Check with local plumbing shops for frequency and types of plumbing repairs.
4. Examine pipe coupons where large taps are made; inspect and gage for a protective calcium layer.
5. Purchase and install corrosivity meters, now available, which can accurately measure corrosion rate.
6. Use both copper and mild steel coupons at the plant and within the distribution system. The AWWA's Water Quality Goals suggest a weight loss of 5 milligrams per square centimeter (mg/cm^2) for a 90-d period, using galvanized wrought-iron coupons. The rate, when calculated as mils per year and compared to mild steel, corresponds to a corrosion rate of 1.0 mpy. (Generally accepted guidelines consider that 5 to 10 mpy will provide an acceptable water quality and corrosion protection.)

8.2 MANDARIN UTILITIES

This case history, which summarizes a study performed by consultants to the utility, illustrates (1) how a small utility company solved a copper corrosion ("black water") problem and (2) the benefits of actively logging and investigating consumer complaints about corrosion.

Background

Mandarin Utilities is a private utility in Jacksonville, Florida, that provides drinking water to several residential and commercial subdivisions.

The Mandarin Utilities system consists of six plants located throughout the utility's service area, with a total production of about 1.5 MGD. The water source for the plants is groundwater from wells averaging 175 ft in depth from the Floridan Aquifer. Corrosion problems were occurring only in the area served by the Pickwick Park plant, which produces about 0.9 MGD.

Currently, treatment consists of aeration to remove about 1 mg/L of dissolved H_2S and chlorination before storage and distribution. Prior to November 1980, no aeration facilities for removing H_2S existed at the Pickwick Park plant. All other plants serving the Mandarin system had aerators installed for H_2S removal. During this time, customers served by the Pickwick Park plant experienced severe "black-water" corrosion of their copper household plumbing as a result of the reaction of sulfides with the copper plumbing. Elemental sulfur, which forms when sulfides are oxidized by chlorine or oxygen, can also react with copper plumbing to cause corrosion and black water.

Typical finished water quality at Pickwick Park prior to installation of the aerator is shown in Table 8.2. When the aerators were installed, Mandarin Utilities instituted a comprehensive program for logging and investigating each consumer complaint. Before November 1980 (when Pickwick Park had no aerator), complaints of black-water corrosion numbered about 25 per month and were primarily confined to the Pickwick Park service area. The black-water problem at several residences served by Mandarin Utilities exhibited the classic symptom of black-water copper corrosion: a gritty, dark precipitate of copper sulfide, occurring predominantly on the hot-water side at the farthest point from the water heater.

Mandarin Utilities' managers determined that aeration to remove H_2S at Pickwick Park was necessary to solve the black-water problem. A cone-type aerator was installed between the wells and the ground-level storage tank at Pickwick Park. This additional treatment step effectively removed nearly all the dissolved sulfide from the finished water. Black-water complaints decreased from more than 25 to fewer than 5 per month in the 6 months following installation of the aerator.

However, a few customers continued to complain about persistent black-water problems. At this point, Mandarin Utilities hired an outside consultant to investigate the causes of the continuing problems and recommend corrective action.

Corrosion investigation and monitoring of the water supply procedure. Historical information such as complaint logs, plant operating data, and water quality data was evaluated to determine the cause and extent of the continuing corrosion problem.

Measured DO concentrations of between 3 and 6 mg/L throughout the Pickwick Park service area confirmed that the aerator was successfully eliminating sulfides from the treated water. An in situ test conducted to determine the extent of elemental sulfur present in the treated water indicated that less than 0.25 mg/L of colloidal sulfur was present. Particulate sulfur can accumulate in low-flow areas of a distribution system and cause localized corrosion problems, thus requiring continual vigilance. The amount of sulfur present in the Pickwick Park system during the test was too low to be a direct cause of black-water corrosion problems in the system.

Along with the elemental sulfur deposited on the filter, a small amount (0.04 mg/L) of oxidized iron was also present. This amount of iron oxide also would not be expected to cause problems in the system.

A finished water analysis was performed on 3 consecutive days. An LSI of -0.1 was calculated for these analyses, indicating that the water had a slightly corrosive tendency.

Table 8.2. Mandarin Utilities' finished water quality at Pickwick Park prior to aeration installation

TDS (mg/L)	452
Total hardness (mg/L as $CaCO_3$)	282
Alkalinity (mg/L as $CaCO_3$)	105
Calcium (mg/L)	61.8
Magnesium (mg/L)	30.8
pH, in situ	7.4
Iron (mg/L)	0.3
CO_2 (mg/L)	8.0
Temperature, in situ (°C)	25
H_2S (mg/L)	1.0
DO (mg/L)	None
LSI	−0.35

Source: Mandarin Utilities, 1981.

In addition, several residential connections that had been the source of numerous recurring complaints were visited by consulting engineers and utility personnel. One of the residences was found to have several galvanized-steel nipples coupled with copper elbows in the hot-water system. The galvanized nipples were removed and were found to be heavily tuberculated. Black copper-sulfide precipitate was found in the hot-water plumbing. The precipitate appeared to have accumulated over a long time in the crevices and tubercles caused by the iron corrosion. Other residences had similar galvanized connections on the hot-water side or on home water softeners preceding the water heaters. Most of the complaintants had not flushed their hot-water systems since the aerators were installed at Pickwick Park.

Results. Upon completion of the corrosion investigation conducted at residences with black-water problems, it was apparent that current complaints were due to a combination of improper plumbing practices (galvanic connections in household plumbing) and residual problems from the high sulfide water at Pickwick Park prior to installation of the aerator. Once copper corrosion is well established, corrosion products, which fill cracks, crevices, and tuberculated areas in pipes and water heaters, often set up "concentration cells." These cells continue to cause copper corrosion problems and can persist even in the original water quality problems are remedied. Accumulated copper sulfide tubercles can harbor bacteria which continue to corrode the copper plumbing. Even in the absence of continuing corrosion, residual corrosion products can take months or years to be completely eliminated because of the concentration cells or bacteria.

Recommended Control Methods

The following recommendations were proposed by the consulting engineers and are currently being implemented:

1. Add caustic soda (NaOH) to raise the pH by 0.3 to 0.5 unit to attain a positive LSI.
2. Baffle the storage tank outlet to obtain maximum use of the storage tank for settling of sulfur and iron particles and to reduce residual copper corrosion problems. (Although particulate elemental sulfur is not currently a major problem, sulfur accumulation in slow-moving sections of the system could compound residual copper corrosion problems.)
3. By-pass the aerator with part of the water to reduce the DO level to less than 1.0 mg/L. This water should be pumped directly into the clearwell so that it discharges below the water level into the storage tank to help reduce the DO of the finished water to acceptable levels.
4. Assist customers with residual copper corrosion problems through an aggressive program of repeated cleaning and flushing of household plumbing fixtures. Consider assisting residence owners with such corrective action by disconnecting water meters during major flushing efforts. Flushing with high chlorine residual water may be effective if bacterial action is adding to the residual copper corrosion. One or more test cases of flushing with high chlorine residuals should be attempted and the results monitored to determine the effectiveness of this remedy.
5. Continue the complaint response program, which involves inspection of galvanic connections at water heaters, water softener problems, and hot-water heaters and plumbing fixtures. Actions which the homeowner can take to reduce or eliminate residual copper corrosion (e.g., flushing, cleaning hot-water heaters and fixtures, or removing galvanic connections) should be identified at the time of inspection.

The engineers further advised the utility that residual corrosion problems at houses which have galvanic (copper to iron or steel) connections at water heaters are likely to resist correction until the galvanic connections are removed. Due to corrosion and tuberculation of the iron pipe or nipple, the rough surfaces provide, locations for residual copper corrosion to continue in spite of water quality improvement.

Mandarin Utilities currently is implementing the modifications suggested in recommendations 1 through 3 and has aggressively pursued recommendations 4 and 5, as well as the additional advice of the engineers, through an effective public information program, which has significantly reduced the number of corrosion-related complaints.

8.3 MIDDLESEX WATER COMPANY

This case history is excerpted from a publication by E.D. Mullen and J.A. Ritter, published in the May 1980 *AWWA Journal*, and it illustrates the following:

1. corrosion control by phosphate inhibitors;
2. a relationship between pH, temperature, inhibitor dose, and corrosion rate for a specific system; and
3. the use of coupon testing to evaluate several control strategies.

Background

Prior to 1969, MWC, located in Woodbridge, New Jersey, relied mainly on groundwater sources. To meet the growth in water demand, a new 20-MGD plant was built in 1969 to treat surface water from the Delaware and Raritan canals. Average water analyses for groundwater and surface water supplies are given in Table 8.3. More than half the MWC water distribution system consists of unlined cast iron mains. After the change from hard well water to soft surface water, MWC began receiving consumer complaints of discolored water in the areas where the iron mains were located. The discoloration was due to corrosion of the cast iron. Treating the surface water with caustic soda (NaOH) to obtain an LSI of +0.5 to +0.8 did not significantly reduce the red water complaints.

Table 8.3. Average water analyses

Chemical parameter	Surface water	Groundwater
Alkalinity (mg/L)	45	170
Hardness (mg/L)	68	260
DO (mg/L)	10	
Sulfate (mg/L)	49	83
Total solids (mg/L)	158	370
pH	7.7	7.7
Temperature (°C)	0.5–28	13

Source: Mullen and Ritter, 1980.

Initial Investigation and Monitoring Program

To monitor for corrosion and find a control method that would reduce customer complaints as well as protect the utility's mains and consumer's pipes, MWC initiated coupon (weight-loss) bench-scale laboratory studies. After contacting the ASTM and the NACE for information on coupon testing, MWC established a corrosion monitoring program according to NACE Standard TM-01-69.

MWC built two acrylic bench-test units, each consisting of three cylindrical cells connected in series, as shown in Fig. 8.3. The water entered through the inlet cell, which contained a control coupon. The center cell was the chemical-dosing cell, and the outlet cell contained the test coupon that measured the effects of chemical addition.

As discussed in Sect. 6.0, coupon tests measure the metal loss from corrosion over a specific time interval. The coupons are carefully weighed before and after they are placed in the water. Coupons are thoroughly cleaned of corrosion by-products and other foreign matter prior to being weighed. The difference in the coupon weights is the loss from corrosion, which can be converted into a corrosion rate by using the following formula:

$$\textit{Corrosion rate (mpy)} = \frac{\textit{weight loss (mg)} \times 534}{\textit{area of coupon (sq. in)} \times \textit{time (h)} \times \textit{metal density } (g/cm^3)} \quad (16)$$

Testing of Alternative Control Methods

Alternative 1: Inhibitor treatment

Procedure. Initial bench-scale studies tested the effectiveness of adding two phosphate inhibitors to the pH-adjusted water.

Two milligrams per liter of a sodium-zinc-glass phosphate was added to one test unit, and 2.5 mg/L of ZOP was added to the second test unit. The test was conducted for 2 months.

Results. At the end of each month, the control and test coupons in each unit were weighed, and the corrosion rates were calculated from the measured weight losses. The bimetallic sodium-zinc-glass phosphate averaged a 13% reduction in corrosion rate. The ZOP averaged a 55% reduction in corrosion rate.

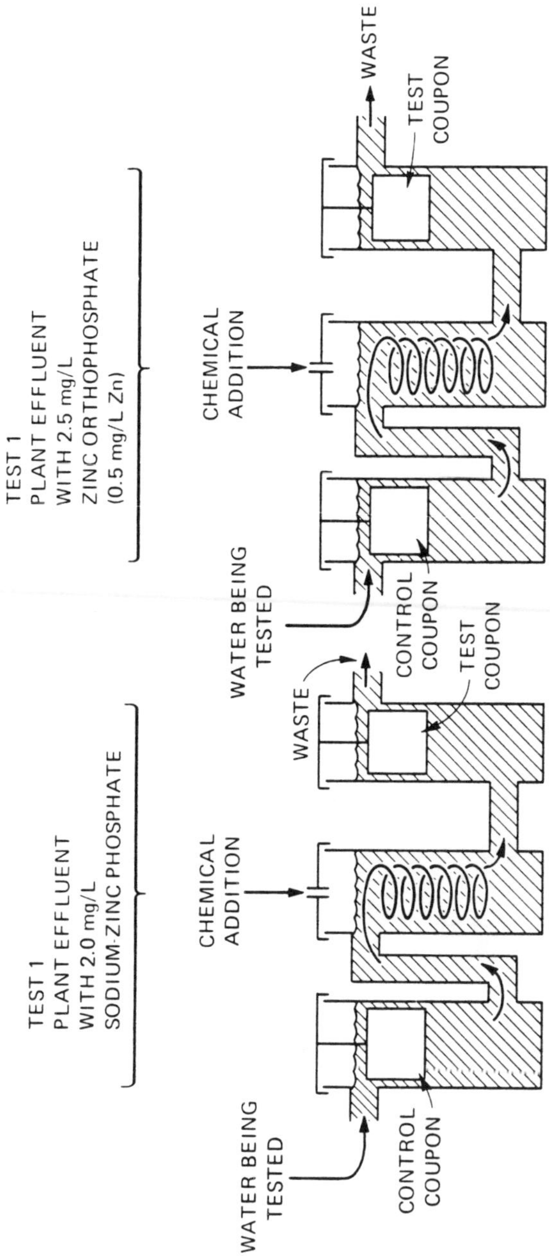

Fig. 8.3. Coupon testing cell assembly.

Alternative 2: Addition of zinc orthophosphate with and without pH adjustment

Procedure. In these tests, 2.5 mg/L of ZOP (0.5 mg/L of zinc) was added to each of the two test units. Water in one unit was supplied by a line from the plant filter effluent (pH 6.8). Water in the other unit was supplied by plant effluent (pH 7.8). When the water temperature was higher than 18°C (65°F), the plant effluent was maintained at the pH of saturation, pH_s.

Results. Inhibitor treatment without pH adjustment reduced corrosion by 54%. Inhibitor treatment with pH adjustment reduced corrosion by 79%. During these tests, the following relationship between pH adjustment, inhibitor treatment, and temperature changes was discovered:

1. At temperatures below 13°C (55°F), inhibitor treatment without pH adjustment was more effective than inhibitor treatment with pH adjustment.
2. At higher temperatures, inhibitor treatment without pH adjustment increased corrosion.

Alternative 3: Testing of zinc orthophosphate addition and pH adjustment in the distribution system

Procedure. Coupons were placed at six locations in the distribution system. Monitoring started 5 months before the plant began inhibitor treatment. The liquid ZOP was stored in a 23-kL (6,000-gal) underground fiberglass tank. Chemical metering pumps inside the plant discharged to the clearwell reaction chamber. Capital investment totaled $11,500. A schematic of the inhibitor installation is shown in Fig. 8.4.

Results. Two areas were identified in which treatment could be improved to produce better water and reduce costs. It was found that during the winter, lower zinc dosages could be used, and the caustic soda pH adjustment could be reduced. Annual posttreatment caustic soda requirements have been reduced 60% from 15.2 mg/L in 1970 to 1971 to 6.1 mg/L in 1978. Peak corrosion rates (July and August) could be suppressed by increasing the zinc dosages, based on water temperature. The maximum summer zinc dosage needed in July was about 0.54 mg/L as zinc. In the cooler months, when the corrosion rate drops naturally as the water temperature drops, inhibitor treatment is continued at a lower dosage. The minimum wintertime zinc dosage is about 0.2 mg/L. MWC considered discontinuing the inhibitor treatment in the winter, but since the zinc phosphate film is constantly dissolving and being laid down, the film inhibitor treatment must be maintained.

In 1974, the six monthly distribution coupons were reduced to one monthly coupon. In 1975, MWC began the current program of measuring one coupon every 3 months. Inhibitor dosages and pH adjustments are increased or decreased with water temperature changes, which results in cost savings from lower corrosion rates and lower chemical costs. Between 1973 and 1978, corrosion rates were reduced by about 70 to 80%.

8.4 SMALL HOSPITAL SYSTEM

This study, conducted by a private consultant, illustrates an economical, low maintenance solution to copper corrosion in a small system.

Background

Prior to the opening of a small 15-bed hospital in the eastern Sierra Nevada Mountains of California, blue staining from copper was apparent in every water fixture. Chemical analyses showed up to 10 mg/L of copper in the water. The corrosion appeared to be general or uniform, without evidence of pitting. The water supply to the hospital is surface lake water, containing 20 to 40 mg/L total dissolved solids (TDS) at about pH 6. The LSI of the water averages -2.0.

Initial Investigation and Monitoring Program

Procedure. The task was to make the water less aggressive by adjusting the pH. Mechanical feeders could not be used to adjust the pH because they are not accurate or reliable at low-flow rates.

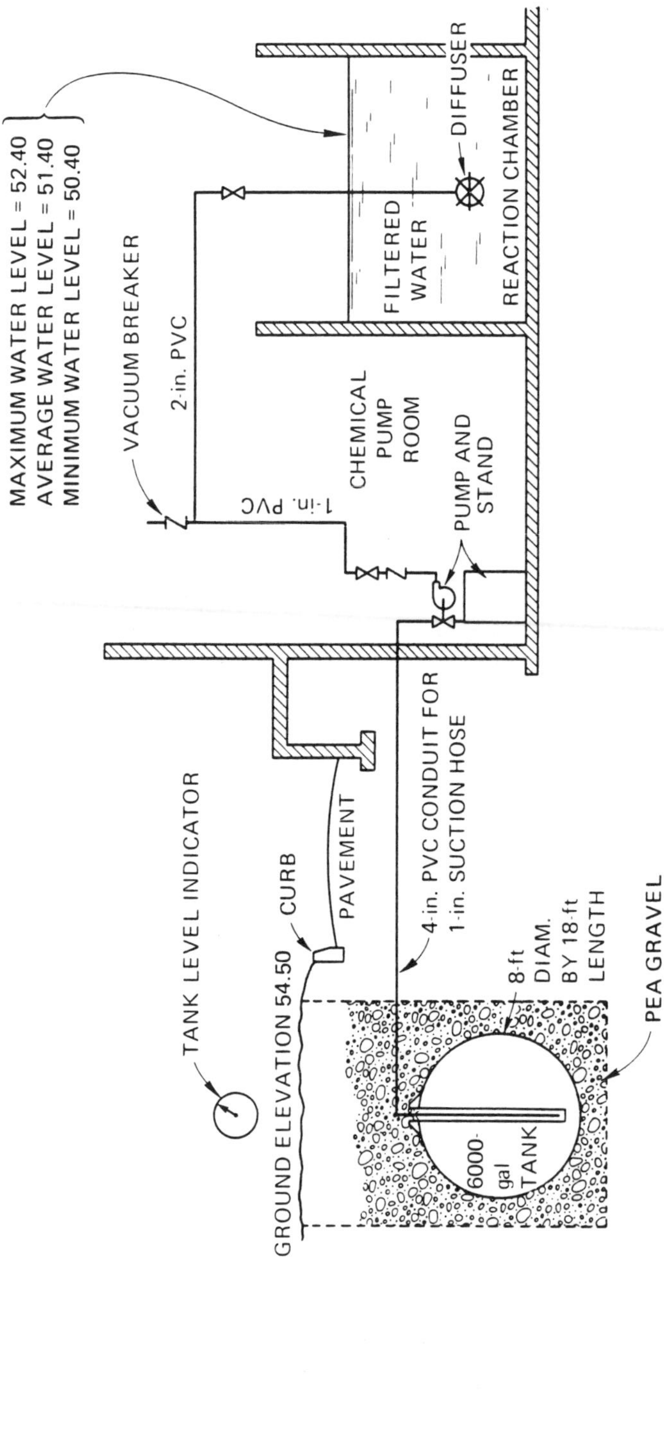

Fig. 8.4. Schematic of inhibitor installation.

To solve the copper corrosion problem, a 5-ft × 24-in. tank was installed on the incoming-water line. The tank was filled with crushed calcite ($CaCO_3$), approximately ⅜ in. in diameter. Empty bed contact time at maximum flow was about 5 min.

Results. The water picked up about 4 to 6 mg/L of calcium while in contact with the limestone. Alkalinity increased by 10 to 15 mg/L, and the pH rose to about 7.2. The water became less aggressive, and the staining stopped. The system contains no moving parts and requires no maintenance other than the addition of calcite about once a year.

8.5 BOSTON METROPOLITAN AREA WATER SYSTEM

This case history, excerpted from a paper presented by P.C. Karalekas, C.R. Ryan, and F.B. Taylor at the 1982 AWWA Miami Conference illustrates the following:

1. the problems associated with lead corrosion in an old distribution system containing lead piping,
2. the effects of phosphate inhibitor and pH control programs on lead corrosion rates, and
3. the benefits of a good monitoring program for evaluating corrosion control methods.

Background

Studies prior to that by Karalekas et al. had shown that lead concentrations at customer's taps in the Boston metropolitan area were consistently above the NIPDWR acceptable level (0.5 mg/L).

Boston and the surrounding communities purchase water wholesale from the Metropolitan District Commission (MDC), a state agency. The MDC pipes water from Quabbin Reservoir to the Wachusett Reservoir and then to the metropolitan area. The watersheds of these two large reservoirs are well protected from pollution sources.

The MDC serves about 1.8 million people in the entire Boston metropolitan area, having an average daily demand of about 300 MGD.

Prior to the start of corrosion control, treatment consisted of only chlorination and ammoniation. Table 8.4 lists various raw and finished water quality parameters. Raw water is low in hardness, alkalinity, TDS, and pH, all of which indicate soft corrosive water. Copper, iron, zinc, and lead are consistently below detection limits in both raw and finished water. Finished water represents water after treatment with chlorine, ammonia, hydorfluosilicic acid, and NaOH. The major difference between raw and finished water is the increase in pH from 6.7 to 8.5. Alkalinity and sodium also increase.

Lead in Boston water results from a combination of a soft corrosive water, which is quite acidic and low in hardness and alkalinity, and the extensive use in the past of lead pipe for service lines and plumbing.

In a 1975 study conducted in the Boston metropolitan area, Karalekas et al. found 15.4% of the water samples collected at consumer's taps exceeded the lead standard. Furthermore, more than 70% of the 383 homes surveyed had detectable levels of lead in their drinking water, which indicated the widespread nature and seriousness of the problem.

Finding high lead concentrations from the corrosion of lead pipe and the association between lead in water and blood prompted the MDC to embark on a treatment program to protect public health by reducing corrosion.

Initial Investigations and Monitoring

Procedure. Before the MDC began treating their water to reduce corrosion, EPA developed a monitoring program which involved sampling for trace metals at consumer's taps known to be supplied through lead service lines. The purpose of this sampling program was to evaluate water quality both prior to and after implementing corrosion control. This sampling has been done regularly since 1976. At the outset, 23 homes with lead service lines were included in the sampling. During

Table 8.4. Metropolitan District Commission water quality data

Parameters	Shaft 4 (Southborough, MA) Raw water	Norumbega Reservoir (Weston, MA) Finished water
Hardness (as $CaCO_3$)	12	12
Alkalinity (as $CaCO_3$)	8	12
TDS	37	46
Calcium	3.2	3.4
Sodium	5.5	9.7
Sulfate	<15	<15
Chloride	<10	<10
Specific conductance (micromhos)	59	78
pH (units)	6.7	8.5
Copper	<0.02	<0.02
Iron	<0.10	<0.10
Zinc	<0.02	<0.02
Lead	<0.005	<0.005

All values in mg/L unless otherwise specified.

the intervening period, a number have been dropped or have missed months because the occupants moved. Currently, 14 of the original 23 homes are being monitored.

To assess the variation in lead concentration in drinking water that had been standing for varying lengths of time in piping, three samples were collected at each home, by the homeowner, using the instructions in Table 8.5. Water was collected at the kitchen sink the first thing in the morning, before any water was used in the house.

Sample 1, the interior plumbing sample, was collected immediately upon opening the faucet. This sample represented water that had been standing overnight in the fixture and the interior plumbing serving the faucet. Sample 2, the service line sample, was collected after the sample collector felt the water temperature change from warm to cold. Since water would be expected to warm slightly after standing in interior plumbing, this cold water would represent water that had been standing overnight just outside the foundation of the house and in contact with the interior of the lead service line underground. Sample 3, the water main sample, was collected after allowing the water to run for several minutes. This sample represented water that would have a minimum contact with the service line and the interior plumbing.

Results. Monitoring results showed that lead concentrations at the customer's taps were consistently well above the NIPDWR level of 0.05 mg/L.

Testing of Alternative Control Methods

Because lead pipe was used extensively throughout the system, its removal would have been prohibitively expensive and consequently was not a feasible option. Therefore, two alternative methods of controlling lead corrosion were implemented and evaluated between 1976 and 1981.

Table 8.5. Sampling instructions

After 11:00 p.m., do not use the kitchen cold water faucet until collecting the water samples the next morning. Using the following directions, in the morning, collect the water samples at the faucet before using any faucet or flushing any toilets in the house. Fill the provided containers to 1 inch below the top and put the caps on tightly.

Sample 1: Open the cold water faucet, immediately fill bottle #1, and turn off the water. Recap this bottle.

Sample 2: Turn the faucet on and place your hand under the running water, and immediately upon noticing that the water turns colder, fill bottle #2. Recap this bottle.

Sample 3: Allow the water to run for 3 additional minutes and then fill bottle #3. Recap this bottle.

The representative from EPA will stop at your house on the morning of ____________, to pick up the samples. If you do not expect to be home, please leave the samples outside the front door.

Alternative 1: Treatment with ZOP

Procedure. In June 1976, MDC began treating the water with ZOP, a commercial corrosion inhibitor. ZOP was tried because of its reported effectiveness in controlling iron corrosion and its potential for controlling lead corrosion. After an initial ZOP dose of about 13 mg/L for several weeks, the dosage was reduced to between 3.2 and 4.5 mg/L for the remainder of the trial, which ended in December 1976.

Results. Figure 8.5 illustrates the variation in lead concentrations over time for all lead samples collected from February 1976 to mid 1981. Each point represents the average of 39 to 69 separate samples collected in the distribution system.

As can be seen from the graph, the average lead concentration was consistently above 0.05 mg/L between February 1976 and May 1977. During the period in which ZOP was used, there appeared to be an initial increase in lead followed by a subsequent decrease. This may have been due to adding the ZOP or to some other factor, such as water temperature. In Fig. 8.6, which plots water temperature versus time, a close parallel between temperature and lead can be noted. Water temperature increases in the summer months appear to be followed by increases in lead, and water temperature decreases appear to be closely followed by decreases in lead. It should be noted that there is more than a 30°F seasonal change in water temperature, which can certainly account for part of the change because of the resulting increase in chemical reaction rates and, thus, the increased corrosion potential. Whatever the reason for the fluctuation in lead, it is clear that lead concentrations were not reduced to below the standard by the inhibitor. Because of this fact and problems of algal growth in distribution storage reservoirs, which may have been associated with phosphate addition, the use of ZOP was discontinued.

Alternative 2: pH adjustment with NaOH

Procedure. There was a 6-month interval between the time ZOP use was stopped and pH adjustment was initiated. In May 1977, MDC began adding NaOH to raise the pH to levels between 8 and 9, as shown in Fig. 8.5.

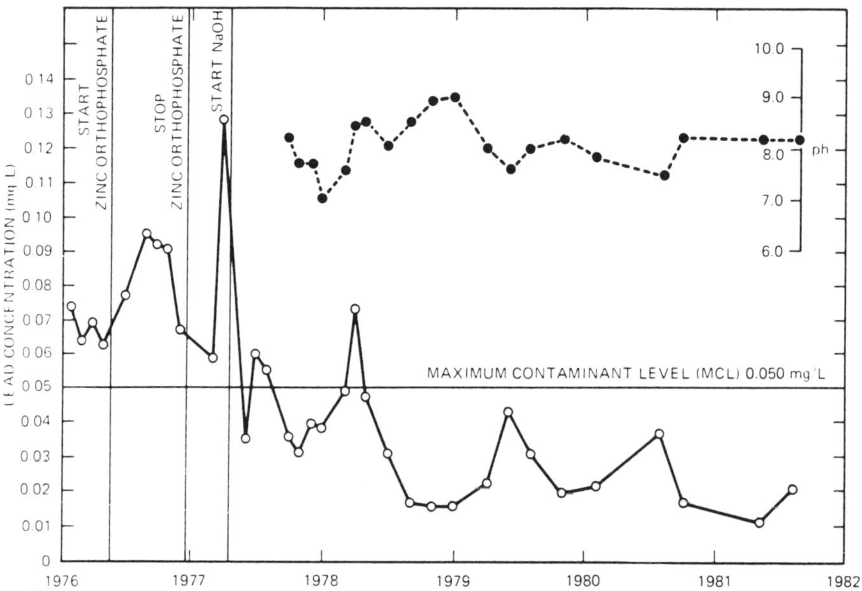

Fig. 8.5. Mean lead levels from samples taken in Boston and Sommerville, Massachusetts, 1976-1981.

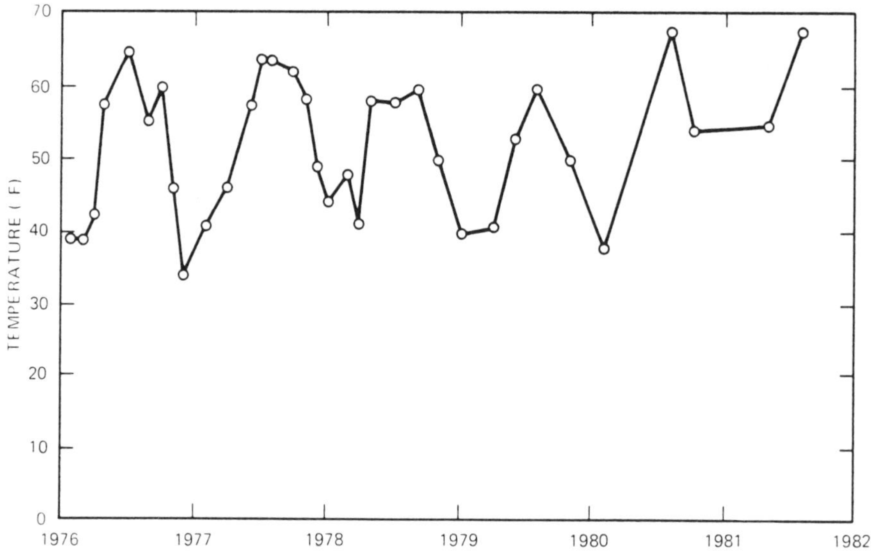

Fig. 8.6. Water temperature, Metropolitan District Commission, Norumbega Reservoir.

Results. As Fig. 8.5 illustrates, using NaOH to adjust pH levels resulted in substantially reduced lead concentrations. There were two brief periods in which average lead concentrations were above 0.05 mg/L due to interruptions in pH adjustment. The first occurred when an underground NaOH line froze during the winter of 1978 because a pump malfunctioned, and the second occurred when a building was struck by lighting in the summer of 1977.

In 1979, a close relationship between average pH values and average lead concentrations, as determined from samples taken at the consumers' taps was observed. From January to June, when the pH dropped from 9 to less than 8, there was a concurrent rise in lead concentrations. From June to December, pH levels increased to more than 8, and drop in lead concentration was observed, leading to the conclusion that there was a causal relationship between pH levels and lead concentrations (i.e., as the pH increased, lead decreased).

Figure 8.7 shows the variation in copper concentrations during the same period of 1976 through 1981. Prior to corrosion control, average copper concentrations ranged as high as 0.35 mg/L, still below the recommended level of 1.0 mg/L. Again, a significant reduction is seen in copper levels after the adjustment of pH using NaOH. Currently, copper concentrations average about 0.05 mg/L.

Figure 8.8 represents average iron concentrations over time. While there is not the dramatic decrease in iron that was seen in copper and lead, note that iron concentrations are at their lowest levels in 5 years. There has been an apparent, gradually downward trend during the past several years, which indicates that pH adjustment has had a positive effect on controlling iron corrosion. With less fluctuation in pH from 1979 to 1981, as compared with previous years, there is apparently less fluctuation in iron concentrations, again with a downward trend.

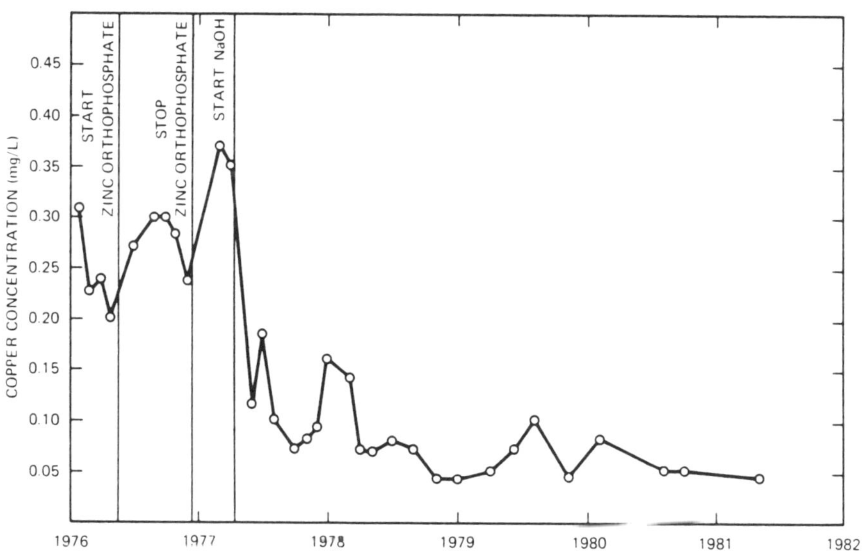

Fig. 8.7. Mean copper levels from samples taken in Boston and Sommerville, Massachusetts, 1976-1981.

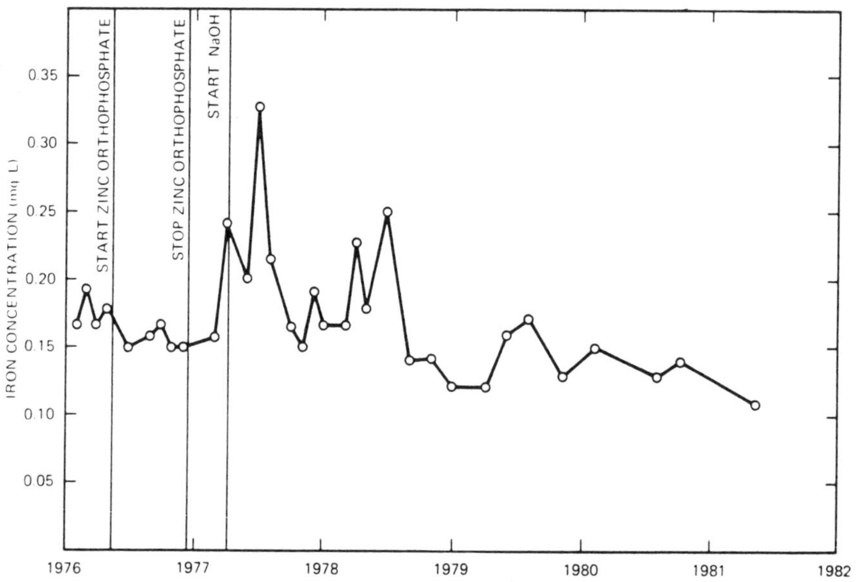

Fig. 8.8. Mean iron levels from samples taken in Boston and Sommerville, Massachusetts, 1976-1981.

Summary and Conclusions

At present, MDC is adding 14 mg/L of 50% NaOH to treat an average daily demand of 301 MGD. Chemical costs for NaOH were $900,000 in 1981, with operating and maintenance expenses of $161,000 for that year. The cost per million gallons treated is $9.64. MDC serves approximately 1,800,000 people in the Boston metropolitan area, which would give a cost per person per year of $0.59.

To summarize, this study shows that pH adjustment using NaOH has effectively reduced both lead and copper corrosion in the Boston area and that a monitoring program is essential to evaluating any proposed corrosion control scheme.

8.6 GALVANIZED PIPE AND THE EFFECTS OF COPPER

This case history differs from the others presented here in that it its not actually one case history but is a composite of incidents from consulting experiences. It is included to illustrate the effects of copper on galvanized pipe and to offer possible remedies.

Background

The waters in these cases were not severely corrosive. None of the waters involved, however, was capable of laying down a protective scale in cold weather. The copper in several of the systems resulted from efforts to control algae in surface water supplies using copper sulfate. The literature reports that concentrations as low as 0.01 mg/L can potentially cause problems.

The corrosion mechanism is as follows: Copper, upon entering a galvanized system, will plate out on the zinc surface. Copper, being the more noble or inactive metal, then becomes the cathode. The zinc (or steel) becomes the anode and goes into solution. This type of problem usually is accompanied by severe tuberculation inside the pipe. Under each tubercle is a pit. In severe cases, the pitting leads to perforation and failure of the pipe.

This problem is not confined to copper that comes in with the water. Hotels, apartments, and some commercial buildings frequently have a central heater which continuously recirculates hot water. Frequently, the heat exchange surfaces (heater coils) are copper, and the plumbing system is galvanized. The problem is the same as described previously.

Possible Remedies

If traces of copper in the water are known or suspected, the builder should use a material other than galvanized pipe in the plumbing system. If that is not possible, the system must be protected *from the outset* by a "scavenger pot." This device is simply a flow-through container which is mounted on the incoming line and provides at least a 1-min empty bed detention time. The unit is charged with a metal higher in the galvanic series than copper so that the copper will plate out on the metal in the scavenger pot and not enter the system. Mossy zinc and magnesium have been used successfully.

In existing systems suffering from this problem, installing a scavenger pot will not cure the problem because the copper is already deposited in the lines. It will merely prevent more copper from aggravating the situation. In such systems, the use of polyphosphate inhibitors has, at times, helped in stifling the cathode reaction. However, caution should be used because if the system is severely tuberculated, the polyphosphate may initially react preferentially with existing corrosion products, resulting in leaks from areas in the system that are severely corroded. These leaks usually manifest themselves within 10 days to 2 weeks after initiation of treatment.

8.7 GREENWOOD, SOUTH CAROLINA

This study, which illustrates the effect of adding ZOP to control corrosion in A-C pipe, was conducted by the CPW, Greenwood, South Carolina, under the sponsorship of EPA. A more detailed account of this study can be found in the EPA report titled "Field Test of Corrosion Control to Protect Asbestos-Cement Pipe" (Grubb 1979).

Background

The water distribution system in Greenwood, South Carolina, contains a great deal of A-C pipe, most of which was installed in the late 1940s and the early 1950s.

The water source for Greenwood is surface water from Lake Greenwood. Prior to this study, treatment consisted of alum coagulation, sedimentation, filtration, pH adjustment with NaOH, and chlorination. Water quality values for raw and finished water are given in Table 8.6. The finished water had an AI of about 10.4 to 10.5, which is considered moderately aggressive.

Table 8.6. Greenwood, South Carolina water quality data

Parameter	Raw water	Finished water
pH, pH units	Variable	8.2–8.3
Alkalinity, mg/L	15	20
Total hardness, mg/L as $CaCO_3$	10	10
Iron, mg/L	0.4	0.1
Free chlorine residual	None	0.75

Initial Investigation and Monitoring Program

At the request of the CPW, EPA tested several Greenwood samples for asbestos fibers. The results of the testing confirmed the presence of asbestos fibers in the water and led the Commissioners to participate in a project to determine whether a chemical could be fed into the water system to coat the pipe sufficiently to prevent further corrosion of the cement and asbestos fibers.

Initial sampling showed that the pH of the water increased as the water passed through the distribution system. Many of the larger mains are concrete or cement-lined pipe, which caused an increase in pH before the water reached the A-C pipe in the system. However, the increase in pH was not sufficient to stabilize the water, and the pH continued to rise when it contacted the A-C pipe.

This study was the first of its type concerned with A-C pipe. At the time this study was started, the apparent corrosion of A-C pipe in use was evaluated almost entirely by electron microscope asbestos fiber counts on water samples from the system. It was later learned that fiber counts can be significantly increased, albeit temporarily, by drilling and tapping the pipe. These temporary increases in asbestos counts can lead to a false evaluation of the general condition of the pipe.

Testing of Control Method

Procedure. Prior to testing, A-C pipe was installed at the two test sites, known as Effie Drive and Canterbury, with a 15-cm (6-in.) section being installed at Effie Drive under low-flow conditions, and a 20-cm (8-in.) section at Canterbury under high-flow conditions.

ZOP was introduced into the system on October 11, 1977, at a feed rate of 3.0 mg/L. The passivation period was terminated on October 14, 1977, short of the intended date of October 15, because of problems encountered by one of the system's largest users, a pharmaceutical company. This firm processes gauze and surgical supplies and used a cotton-filament-wound filter to remove rust or iron particles as well as other particles in the water. The high concentrations of ZOP clogged the filters sufficiently to cause serious problems. The company had to change filters every 2 to 3 days during the initial periods, whereas normally a set of filters would have lasted 6 to 7 months. The feed rate was reduced to 0.5 mg/L in an effort to ease filter problems. Zinc was found at 0.17 mg/L at Effie Drive, the low-flow location, and it was therefore concluded that zinc was present throughout the system.

The problems continued with the filters at the pharmaceutical company, however, and on November 11, 1977, the zinc feed rate was lowered to 0.3 mg/L, resulting in a decrease in the concentration of zinc at the Effie Drive location to less than 0.1 mg/L. This rate was continued until the end of the study, although problems were still encountered with the company's filters.

Samples of water were routinely taken from two locations where A-C pipe was used. Samples were checked for zinc content, pH, alkalinity, and calcium as $CaCO_3$.

Water samples were also forwarded to the EPA laboratory in Cincinnati to determine the number of asbestos fibers present. The two pipe sections were removed and examined for the amount of zinc deposited on their surfaces.

Results. Routine tests, such as pH and alkalinity, showed no significant changes in these parameters during the study period. The AI of the water at the closer, high-flow location ranged from 10.6 to 11.2. At the more distant, low-flow location, the AI varied from 11.2 to 11.9—a range considerably higher than the 10.4 to 10.5 AI values at the treatment plant.

After the initial passivation period, zinc concentration at the closer sampling location averaged about 0.2 mg/L; at the distant location, it was generally less than 0.2 mg/L, averaging 0.1 mg/L. These concentrations were not sufficient to prevent a rise in pH and calcium between the two sampling locations. Most of the calcium increase in the distribution system occurred between these sites, where zinc concentration was lowest.

Two times after A-C pipe had been tapped in the sampling location vicinity, high asbestos fiber counts were observed. Because, as was stated earlier in this section, pipe tapping operations can temporarily raise the fiber count in the distribution system, the use of tapping machines equipped to flush away drilling and tapping debris is recommended.

Electron microscope photographs and energy dispersive X-ray spectra analyses showed coatings of zinc products on the two pipe samples.

The scanning electron microscope (SEM) was used to examine the interior pipe wall of pipe samples removed from both sampling locations. The interior surface of the 20-cm (8-in.) pipe was smooth, and some coating resulting from zinc treatment could be identified. The interior of the 15-cm (6-in.) pipe, used with a lower concentration of zinc (0.1 mg/L), had much larger uncoated areas on the pipe surface.

Regardless of which portion of the pipe was examined, the 20-cm (8-in.) pipe always showed better coverage than the 15-cm (6-in.) pipe. This indicates that the zinc or zinc compounds do adhere to the A-C pipe, and that to be effective, the zinc concentration should be 0.2 mg/L, or higher if possible.

Because zinc precipitates in the system, the amount of ZOP dosed at the treatment plant to obtain the desired zinc residual would have to be determined for each system.

The report concluded that the release of asbestos fibers from A-C pipe can be reduced by maintaining a zinc coating on the pipe material by adding ZOP.

9. Costs of Corrosion Control

This section outlines costs associated with some common corrosion control procedures. Costs are presented for sampling and analysis, rate measurement tests, and various types of equipment and chemicals used in corrosion control. The costs may vary considerably among utilities of different sizes and in different regions of the country and should not be used by any individual utility to estimate the cost of a specific corrosion control program. The data presented here are useful, however, for comparing costs of alternative corrosion control methods.

9.1 MONITORING COSTS

Sampling and Analysis

Sampling and analytical costs to monitor and control corrosion will vary among utilities, depending on the number of parameters analyzed, the number of samples collected, the type of materials used in the system, and the type of control program being monitored by the utility.

To comply with the 1980 NIPDWR amendments, only one or two (if the surface water supplies are used) samples for the following parameters are required:

1. alkalinity, milligrams per liter (mg/L) as calcium carbonate ($CaCO_3$);
2. pH, as pH units;
3. hardness, mg/L as $CaCO_3$;
4. temperature, °C or °F; and
5. TDS, mg/L.

The cost of conducting these analyses is minimal and ranges from less than $20 to $75, depending on whether the utility performs an in-house analysis or sends the samples to an outside laboratory.

Additional sampling and analysis are required to determine if corrosion is occuring and what materials are being corroded, as discussed in Sect. 6. Table 9.1 presents typical costs of analyzing several water quality parameters which affect corrosion rates. Analyzing the samples in-house whenever possible can result in cost savings.

Weight-Loss Measurements

The main costs of coupon or weight-loss methods are

1. the initial purchase and installation of the coupons;
2. labor costs of setting up the test;
3. dismantling and weighing the coupons after a specified time period;
4. the cost of any water quality modifications tested during the test period (such as pH adjustment, reduction of oxygen, pipe lining, or inhibitor treatment).

The costs vary depending on the number of coupons placed in the system, the number of different materials tested, and whether the utility performs the study in-house or hires an outside consultant to conduct the tests. Middlesex Water Company, which conducted in-house weight-loss measurements, reported in 1980 that the cost of their monitoring program is currently about $15 per coupon, excluding chemical costs. A more detailed description of the Middlesex monitoring program is given in Sect. 8.3, Case Histories.

Table 9.1. Cost of typical analytical services for drinking water (1982)

Primary standards		Secondary standards		General	
Parameter	Cost	Parameter	Cost	Parameter	Cost
Arsenic (As)	10	Chloride (Cl)	14	Total hardness ($CaCO_3$)	14
Barium (Ba)	10	Color	10	Total alkalinity ($CaCO_3$)	14
Cadmium (Cd)	10	Copper (Cu)	10	N.C.H. ($CaCO_3$)	2
Chromium (Cr)	10	Corrosivity	45	Bicarbonate (HCO_3)	2
Lead (Pb)	10	Foam Agents (MBAS)	25	Calcium (Ca)	10
Mercury (Hg)	25	Iron (Fe)	10	Magnesium (Mg)	10
Selenium (Se)	10	Maganese (Mn)	10	Carbon dioxide (CO_2)	2
Silver (Ag)	10	Odor	35	Bicarbonate ($CaCO_3$)	2
Nitrate (N)	24	pH	10	Carbonate ($CaCO_3$)	2
Fluoride (F)	14	Sulfate (SO_4)	17	Hydroxide ($CaCO_3$)	2
Turbidity (NTU)	10	Sodium	10	pH_s index	2
	143	TDS	17	RSI	2
		Zinc (Zn)	10	LSI	2
Pesticides	250			H_2S	20
Endrin					
Lindane					
Methoxychlor					
Toxaphene					
2,4–D					
2,4–5 Silvex					
Total cost	393		233		86

Costs given in 1982 dollars.

9.2 CONTROL COSTS

The equipment and chemical costs presented in this section are approximate costs for the purpose of making comparisons. Equipment costs will vary depending on size, quality, features, and construction materials. In addition, many site-specific factors will affect the total system costs. These factors include labor costs, quantity of piping and valves needed, construction materials, housing requirements, bulk storage, unloading/conveyance systems, and site preparation needed.

Chemical costs for water quality modification will vary with location, transportation costs, and volume of chemicals purchased. Water utility personnel are advised in all cases to contact water treatment chemical and equipment suppliers in their areas to determine actual costs of an in-place control system.

Equipment Costs

Lime Feed System Costs. Small lime feed systems [<40 to 50 pounds per hour (lb/h)] usually feed hydrated lime [$CA(OH)_2$], purchased in 100-lb bags. These feed systems generally consist of a storage hopper supplied with a volumetric or gravimetric feeder. The feeder transfers the lime to a dissolving or slurry tank. The lime slurry is then pumped to the point of application by a metering pump.

Larger systems usually feed quicklime, CaO, and require a lime slaker to hydrate the lime and produce a lime slurry. The lime slurry is pumped, or flows by gravity, from the slaker to the point of application.

For stabilization, a lime dose of about 10 mg/L is often adequate. Cost estimates for lime feed systems for several plant sizes are given below. The costs are for equipment sized to feed a dose of 10 mg/L.*

* Updated costs based on costs presented in Environmental Protection Agency publication "Estimating Water Treatment Costs," Vol. 2. By Culp/Wesner/Culp; August 1979.

Plant size[a]	Capital cost[b]	Annual maintenance cost[c]
3 MGD	$20,000	$ 4,000
30 MGD	$75,000	$15,000

[a]MGD = million gallons per day.

[b]Includes manufactured equipment (slaker, storage hopper, bins, pumps, etc.), labor, piping valves, and electrical instrumentation. Housing, bulk storage, and unloading/conveyance system costs are not included.

[c]Includes operation and maintenance labor and electrical power costs.

Sodium hydroxide feed systems. Small systems (<200 lb/d)which feed NaOH generally use dry NaOH. The dry NaOH is delivered in drums and then mixed manually or with a volumetric dry feeder to a 10% solution onsite and is fed with a metering pump. The cost for a small system (1 to 2 MGD) equipped with a volumetric feeder, storage hopper, feed/mixing tank with mixer, and metering pump, including heated indoor storage and appropriate piping and valves, would be approximately $17,000 to $20,000.

For larger systems, NaOH is generally purchased as a 50% liquid solution, containing 6.4 lb of NaOH per gallon. Because 50% liquid NaOH begins to crystallize at 54°F, bulk storage facilities for NaOH must either be located in a heated building or have heating coils in the storage tank. The total capital cost for a bulk liquid NaOH feed system suitable for a 50-MGD plant would be $60,000 to $65,000 (based on updated costs from the EPA report "Estimating Water Treatment Costs," EPA 1979). This includes the cost of indoor bulk storage tanks having fiber-glass-reinforced polyester housing, metering pumps, flow monitoring equipment, electrical instrumentation, piping and valves, and installation labor.

Silicate feed systems. Although sodium silicate is a white powder, it is usually marketed as an opaque solution in 50-gal drums or tank cars.

Many small systems often feed sodium silicate directly from the shipping container to the point of application using a metering pump and polyvinyl chloride (PVC) piping. Larger systems use a bulk storage tank and feed the silicate to the point of application through PVC piping using a metering pump.

The cost for a silicate feed system for a less than 3-MGD plant is approximately $1,000 to $1,300. This cost includes a metering pump to feed directly from the shipping drum to the point of application plus associated PVC piping and valves. For a larger system (50-MGD), feeding from a bulk storage facility, the cost would be in the range of $15,000 to $20,000. This includes the cost of bulk storage, a metering pump, PVC piping, installation labor, and electrical instrumentation.

Phosphate feed systems. Phosphate compounds for corrosion control are available in liquid form but are commonly bought and shipped as dry solids. In systems handling less than 10 MGD, the dry phosphate compound is usually put into solution in a day tank and fed with a chemical metering pump. In systems larger than about 10 MGD, a gravimetric or volumetric feeder which transfers the dry material to a dissolving tank is usually required. A chemical metering pump is used to feed the solution from the tank or from an additional dry tank.

The cost of a phosphate feed system for plants handling up to about 10 MGD using two feed tanks—one equipped with a mixer and dissolving tray—plus a chemical metering pump, PVC piping, valves, and flow meter, is in the range of $1,500 to $2,000.

For larger systems which use a dry-solids feeder and loading hopper in addition to the feed tanks the cost is approximately $12,000 to $15,000. This also includes the cost of a metering pump, PVC piping, valves, installation labor, and flow meter.

Operation and maintenance costs for these systems are moderate. The major cost for both systems is the phosphate. For smaller systems, an operator must periodically prepare a batch of feed solution. This may occur once or several times per day, depending on the system size. Electrical costs are negligible due to the small motor sizes used in the mixer and the metering pump.

Operation requirements for larger systems include periodically charging the hopper with dry phosphate, equipment maintenance, and monitoring. Power costs with the larger systems are more but are still negligible.

Sodium carbonate feed system. Sodium carbonate is sold as a dry white powder in bags or barrels or in bulk (i.e., carloads and truckloads). Its solubility varies with temperature. At 68°F, its solubility is 1.5 lb/gal; at 86°F, its solubility is 2.3 lb/gal. Small systems can feed sodium carbonate by manually making up batch solutions in dissolving tanks and feeding the solution with a metering pump. The cost for a small system such as this, including the tank, metering pump, flow meter, associated PVC piping, valves, and installation labor would be approximately $1,500 to $2,000.

A larger system would require the use of a gravimetric or volumetric feeder to feed the material from a storage hopper into the dissolving tank. Because the material tends to adhere to the sides of the bin, arch, and flood, a hopper agitator is required for the light and powdery grades.

A system of this type for a larger plant (50 MGD) would cost in the range of $12,000 to $15,000, including a vibrator-equipped storage hopper, volumetric feeder, dissolving tank, metering pump, PVC piping, valves, flow meter, and installation labor.

Chemical Costs

Chemical costs for the most common chemicals used in corrosion control are given in Table 9.2. These costs can vary considerably depending on the size and location of the plant, the time of year, and the particular chemical supplier. The costs are not intended to represent actual costs to a utility. Each utility is advised to contact local chemical suppliers to determine the costs for a specific plant. The figures do, however, indicate a cost range which can be useful in considering alternative corrective actions for corrosion control.

Table 9.2. Typical annual chemical costs for corrosion control (1982)
Costs do not include freight

Chemical	Use	Feed rate	Cost per unit ($)	Cost per year 3-MGD plant ($)	Cost per year 50-MGD plant ($)
Quicklime, CaO	pH adjustment	1–20 mg/L 8–170 lb/MG	63/ton bulk	277–5,865	4,500–97,700
Hydrated lime, $Ca(OH)_2$	pH adjustment	1–20 mg/L 8–170 lb/MG	78/ton bag 65/ton bulk	342–7,254 285–6,045	5,700–121,000 4,750–101,000
Caustic soda, NaOH (50% solution)	pH adjustment	1–20 mg/L 12–150 lb/MG	200/ton bulk	1,310–21,900	27,400–456,000
Soda ash, Na_2CO_3	pH adjustment	1–40 mg/L 8–350 lb/MG	16/cwt bag 152/ton bulk	1,402–61,320 666–30,375	23,400–>1,000,000 11,100–506,000
Inorganic phosphates	Inhibitor	3 mg/L 25 lb/MG	65/cwt bag	17,800	297,000
Sodium silicate	Inhibitor	2–8 mg/L 17–67 lb/MG	5.00/cwt tank	930–3,670	15,500–61,200

Source: Various chemical suppliers.

Glossary*

Active—a state in which metal tends to corrode (opposite of passive).

Active metal—a metal ready to corrode, or being corroded.

Additive—a substance added in a small amount, usually to a fluid, for a special purpose—such as to reduce friction, corrosion, etc.

Aeration cell—an oxygen concentration cell; an electrolytic cell resulting from differences in dissolved oxygen at two points.

Aggressive—a property of water which favors the corrosion of its conveying structure.

Aggressive Index (AI)—corrosion index established by the American Water Works Association (AWWA) Standard C-400; established as a criterion for determining the corrosive tendency of the water relative to asbestos-cement pipe; calculated from the pH, calcium hardness (H), and total alkalinity (A) by the formula $AI = pH + \log (AH)$.

Alkalinity—the capacity of a water to neutralize acids; a measure of the buffer capacity of a water. The major portion of alkalinity in natural waters is caused by (1) hydroxide, (2) carbonates, (3) and bicarbonates.

Aerobic—presence of unreacted or free oxygen (0_2).

Anaerobic—an absence of unreacted or free oxygen [oxygen as H_2O Na_2SO_4 (reacted) is not "free"].

Anion—an ion or radical which is attracted to the anode because of the negative charge on the ion or radical (as Cl^-, OH^-).

Anode—(opposite of cathode) the electrode at which oxidation or corrosion occurs. A common anode reaction is:

$$Zn \rightarrow Zn^{++} + 2 \text{ electrons.}$$

Anodic polarization—polarization of anode; i.e., the decrease in the initial anode potential resulting from current flow effects at or near the anode surface. Potential becomes more noble (more positive) because of anodic polarization.

Anodic protection—an appreciable reduction in corrosion by making a metal an anode and maintaining this highly polarized condition with very little current flow.

Aqueous—pertaining to water; an aqueous solution is a water solution.

Bicarbonate alkalinity—that part of the total alkalinity that is due to the bicarbonate ion (HCO_3^-).

Bimetallic corrosion—corrosion resulting from dissimilar metal contact; galvanic corrosion.

Biological corrosion—corrosion that results from a reaction between the pipe material and organisms such as bacteria, algae, and fungi.

Carbonate alkalinity—that part of the total alkalinity due to the carbonate ion ($CO_3^=$).

Cathode (opposite of anode)—the electrode where reduction (and practically no corrosion) occurs. A typical cathode reaction:

$$4 \text{ electrons} + O_2 + 2H_2O \quad 4OH^-$$

Cathodic corrosion—an unusual condition (esp. with Al, Zn, Pb) in which corrosion is accelerated at the cathode because cathodic reaction creates an alkaline condition which is corrosive to certain metals.

* Portions of this glossary were prepared by Anton deS. Brasunas, Professor of Metallurgical Engineering, University of Missouri—Rolla for the NACE Basic Corrosion Course.

Cathodic polarization—polarization of the cathode; a reduction from the initial potential resulting from current flow effects at or near the cathode surface. Potential becomes more active (negative) because of cathodic polarization.

Cathodic protection—reduction or elimination of corrosion by making the metal a cathode by means of an impressed d.c. current or attachment to a sacrificial anode (usually Mg, Al, or Zn).

Cation—A positively charged ion (H^+, Zn^{++}) or radical (as NH_4^+) which migrates toward the cathode.

Cavitation—formation and sudden collapse of vapor bubbles in a liquid; usually resulting from local low pressures—as on the trailing edge of a propeller; this develops momentary high local pressure which can mechanically destroy a portion of a surface on which the bubbles collapse.

Cavitation-corrosion—corrosion damage resulting from cavitation and corrosion: metal corrodes, pressure develops from collapse of the cavity and removes corrosion product, exposing bare metal to repeated corrosion.

Cavitation-damage—deterioration of a surface caused by cavitation (sudden formation and collapse of cavities in a liquid).

Cavitation-erosion—see "Cavitation damage," the preferred term.

Cell—a circuit consisting of an anode and a cathode in electrical contact in a solid or liquid electrolyte. Corrosion generally occurs only at anodic areas.

Concentration cell—a cell involving an electrolyte and two identical electrodes, with the potential resulting from differences in the chemistry of the environments adjacent to the two electrodes.

Concentration polarization—polarization of an electrode caused by concentration changes in the environment adjacent to the metal surface.

Conductivity—a measure of the ability of a solution to carry an electrical current. Conductivity varies both with the number and type of ions the solution carries.

Corrosion—the destruction of a substance, usually a metal, or its properties because of a reaction with its (environment) surroundings.

Corrosion-erosion—corrosion which is increased because of the abrasive action of a moving stream; the presence of suspended particles greatly accelerates abrasive action.

Corrosion fatigue—the combined action of corrosion and fatigue (cycling stress) in causing metal fracture.

Corrosion index—measurement of the corrosivity of a water (e.g., Langelier Index, Ryznar Index, Aggressive Index, etc.).

Corrosion potential—the potential that a corroding metal exhibits under specific conditions of concentration, time, temperature, aeration, velocity, etc.

Corrosion rate—the speed (usually an average) with which corrosion progresses (it may be linear for a while); often expressed as though it were linear, in units of mdd (milligrams per square decimeter per day) for weight change, or mpy (mils per year) for thickness changes.

Couple—a cell developed in an electrolyte resulting from electrical contact between two dissimilar metals.

Crevice corrosion—localized corrosion resulting from the formation of a concentration cell in a crevice formed between a metal and a nonmetal, or between two metal surfaces.

Dealloying—the selective corrosion (removal) of or a metallic constituent from an alloy—usually in the form of ions.

Demineralization—removal of dissolved mineral matter, generally from water.

Depolarization—the elimination or reduction of polarization by physical or chemical means; depolarization results in increased corrosion.

Deposit attack (deposition corrosion)—pitting corrosion resulting from deposits on a metal surface which cause concentration cells.

Dezincification—the parting of zinc from an alloy (in some brasses, zinc is lost, leaving a weak, brittle, porous, copper-rich residue behind).

Distribution lines—those facilities used to carry water from the transmission lines to the service lines, including water mains, distribution reservoirs, elevated storage tanks, booster stations, and valves.

Electrical current—an electric current is caused by the flow of electrons. However, the electric current flows in a direction opposite to the flow of electrons. (This is accepted though seemingly illogical.

Electrochemistry—the result of an electrical and chemical reaction such as when a metal goes into solution as an ion or reacts in water with another element to form a compound resulting in a flow of electrons (electricity).

Electrochemical methods—direct corrosion monitoring method based on the electrochemical nature of corrosion in water. Measures instantaneous corrosion rates, usually in mils per year (mpy).

Electrochemical reaction—a chemical reaction involving the transfer of electrons which involves oxidation (the loss of electrons) and reduction (the gain of electrons).

Electrode—a metal in contact with an electrolyte which serves as a site where an electrical current enters the metal or leaves the metal to enter the solution.

Electrolysis—chemical changes in an electrolyte caused by an electrical current. The use of this term to mean corrosion by stray currents should be discouraged.

Electrolyte—an ionic conductor (usually in aqueous solution).

Electron acceptor—any substance which accepts electrons from some other substance in an electrochemical reaction.

Equilibrium potential—the electrode potential at equilibrium.

Erosion—deterioration of a surface by the abrasive action of moving fluids. This is accelerated by the presence of solid particles or gas bubbles in suspension. When deterioration is further increased by corrosion, the term "erosion-corrosion" is often used.

Fatigue—a process leading to fracture resulting from repeated stress cycles well below the normal tensile strength. Such failures start as tiny cracks which grow to cause total failure.

Filiform corrosion—(see "Underfilm corrosion," the preferred term).

Film—a thin surface layer that may or may not be visible.

Fouling—a term used to describe the covering of submerged surfaces covered by marine growths such as barnacles.

Galvanic—pertaining to an effect caused by the cell—often dissimilar metal contact which results in electrolytic potential.

Galvanic cell—a cell consisting of two dissimilar metals in contact with each other and with a common electrolyte (sometimes refers to two similar metals in contact with each other but with dissimilar electrolytes; differences can be small and more specifically defined as a concentration cell).

Galvanic corrosion—corrosion that is increased because of the current caused by a galvanic cell (sometimes called "couple action").

Galvanic series—a list of metals arranged according to their relative corrosion potentials in some specific environment; sea water is often used.

General corrosion—corrosion in a uniform manner.

Graphitization (graphitic corrosion)—corrosion of gray cast iron in which the metallic constituents are converted to corrosion products, leaving the graphite flakes intact. Graphitization is also used in a metallurgical sense to mean the decomposition of iron carbide to form iron and graphite.

Grain—a portion of a solid metal (usually a fraction of an inch in size) in which the atoms are arranged in an orderly pattern. The irregular junction of two adjacent grains is known as a grain boundary. (Also a unit of weight, 1/7000th of a pound.)

Half cell—a pure metal in contact with a solution of known concentration of its own ion, at a specific temperature develops a potential which is characteristic and reproducible; when coupled with another half cell, an overall potential develops which is the sum of both half cells.

Inert material—a material which is not very reactive, such as a noble metal, plastic, or cement.

Inhibitor—a substance which sharply reduces corrosion, when added to water, acid, or other liquid in small amounts.

Internal corrosion—corrosion that occurs inside a pipe because of the physical, chemical, or biological interactions between the pipe and the water as opposed to forces acting outside the pipe, such as soil, weather, or stress conditions.

Ion—an electrically charged atom (Na^+, Al^{+3}, Cl^-, S^{-2}) or group of atoms known as "radicals" (NH_4^+, SO_4^{-2} PO_4^{-3}).

Ionization—dissociation of ions in an aqueous solution (e.g., $H_2\,CO^3 \rightleftarrows H^+ + HCO_3^-$ or $H_2O \rightleftarrows H^+ + OH^-$).

Langelier Index—a calculated saturation index for calcium carbonate that is useful in predicting scaling behavior of natural water.

Local action—corrosion due to action of local cells, i.e., galvanic cells caused by nonuniformities between two adjacent areas at a metal surface exposed to an electrolyte.

Local cell—a galvanic cell caused by small differences in composition in the metal or the electrolyte.

Metal ion concentration cell—a galvanic cell caused by a difference in metal ion concentration at two locations on the same metal surface.

National Interim Primary Drinking Water Regulations (NIPDWR)—regulations established by EPA (*Federal Register*, Vol. 40, No. 51—March 14, 1975) which set maximum contaminant levels (MCLs) for various parameters in public drinking water systems to protect the public health.

National Secondary Drinking Water Regulations (NSDWR)—regulations established by EPA (*Federal Register*, Vol. 42, No. 62—March 31, 1977) which specify secondary maximum contaminant levels (SMCLs) for 12 parameters which primarily affect the aesthetic qualities relating to the public acceptance of drinking water.

Noble metal—a metal which is not very reactive—as silver, gold, copper—and may be found naturally in metallic form on earth.

Nonuniform corrosion—corrosion that attacks small, localized areas of the pipe. Usually results in less metal loss than uniform corrosion but causes more rapid failure of the pipe due to pits and holes.

Oxidation—loss of electrons, as when a metal goes from the metallic state to the corroded state (opposite of "Reduction"). Thus, when a metal reacts with oxygen, sulfur, etc. to form a compound as oxide, sulfide, etc., it is oxidized.

Oxidizing agent—a chemical or substance that causes a loss of electrons such as causing a metal to go from the metallic state to the corroded state. An electron acceptor.

Oxygen concentration cell—a galvanic cell caused by a difference in oxygen concentration at two points on a metal surface.

Passivator—an inhibitor which changes the potential of a metal appreciably to a more cathodic or noble value (as when chromate is added to water).

Passive—the state of a metal when its behavior is much more noble (resists corrosion) than its position in the Emf series would predict. This is a surface phenomenon.

Passive-active cell—a cell composed of a metal in the passive state and the same metal in the active state.

Passivity—the phenomenon of an active metal becoming passive.

pH—a measure of the acidity or alkalinity of a solution. A value of seven is neutral; low numbers are acid, large numbers are alkaline. Strictly speaking, pH is the negative logarithm of the hydrogen ion concentration.

pH_s—the pH at which a water is saturated with calcium carbonate ($CaCO_3$).

Pitting—highly localized corrosion resulting in deep penetration at only a few spots.

Pitting factor—the depth of the deepest pit divided by the "average penetration" as calculated from weight loss.

Polarization—the shift in electrode potential resulting from the effects of current flow, measured with respect to the "zero-flow" (reversible) potential; i.e., the counter-emf caused by the products formed or concentration changes in the electrolyte.

Protective potential—a term sometimes used in cathodic protection to define the minimum potential required to suppress corrosion. For steel in sea water, this is claimed to be about 0.85 volt as measured against a saturated calomel.

Raman spectroscopy—a direct corrosion monitoring method that reflects an infrared beam off a pipe surface and records the change in frequency of the beam as the Raman spectrum. The spectrum, which is different for all compounds, is compared with Raman spectra of known materials to identify constituents of the corrosion film on the pipe system.

Reduction—gain of electrons, as when copper is electro plated on steel from a copper sulfate solution (opposite of "Oxidation").

Rusting—corrosion of iron or an iron base alloy to form a reddish-brown product which is primarily hydrated ferric oxide.

Saturated solution—a solution that can dissolve no more of a given substance and will not precipitate any of that substance.

Scaling—(1) high-temperature corrosion resulting in formation of thick corrosion product layers. (2) Deposition of insoluble materials on metal surfaces, usually inside water boilers or heat exchanger tubes.

Selective corrosion—the selective corrosion of certain alloying constituents from an alloy (as dezincification) or in an alloy (as internal oxidation)

Solubility—the amount of one substance that will dissolve in another to produce a saturated solution.

Stabilization—the production of a water that is exactly saturated with calcium carbonate ($CaCO_3$).

Stray current corrosion—corrosion that is caused by stray currents from some external source.

Supersaturated solution—a solution that contains more of one substance than needed to be saturated.

Thermogalvanic corrosion—galvanic corrosion resulting from temperature differences at two points.

Tuberculation—localized corrosion at scattered locations resulting in knoblike mounds.

Underfilm corrosion—corrosion which occurs under lacquers and other organic films in the form of randomly distributed hairlines (also called "Filiform corrosion").

Undersaturated solution—a solution that contains less of a substance than needed to saturate it.

Uniform corrosion—corrosion that results in an equal amount of material loss over an entire pipe surface.

X-Ray diffraction—a direct corrosion monitoring method that identifies the scale constituents on a pipe by evaluation of a diffraction pattern.

Weight-loss method—a direct corrosion monitoring method that measures the rate of corrosion by metallic weight loss from a pipe section (or coupon) that has been contacted with a water supply over a period of time.

Additional Source Materials

Additional Source Materials for Chapter 2

Adams, O.H. 1977. "The Safe Drinking Water Act Impacts on State Water Programs," presented at the 5th Annual American Water Works Association Water Quality Technology Conference, Kansas City, Mo., December 1977.

Craun, G.F., and McCabe, L.J. 1975. "Problems Associated with Metals in Drinking Water," *Journal of the American Water Works Association*, November 1975, pp. 593-599.

Hudson, H.E., Jr., and Gilcreas, F.W. 1976. "Health and Economic Aspects of Water Hardness and Corrosiveness," *Journal of the American Water Works Association*, 68(4): 201-204.

Sussman, S. 1978. "Implications of the EPA Proposed National Secondary Drinking Water Regulation on Corrosivity," presented at the American Water Works Association Seminar on Controlling Corrosion Within Water Systems, Atlantic City, N.J., June 1978.

U.S. Environmental Protection Agency. 1975. Primary Drinking Water—Proposed Interim Standards. *Federal Register*, 40(51): 11990-11998

1977a. Drinking Water and Health Recommendations of the National Academy of Sciences. *Federal Register*, 42(132): 35764-35779.

1977b. National Secondary Drinking Water Regulations—Proposed Regulations, *Federal Register*, 42(62): 17143-17147.

1981. National Interim Primary Drinking Water Regulations, *Code of Federal Regulations*, Title 40, Part 141, pp. 309-354.

Additional Source Materials for Chapter 3

AWWA Research Foundation. 1982a. Research News: Corrosion Control, No. 32, Denver, Colo.

1982b. Water Quality Research News. Treatment: Control of Lead Concentrations, No. 30, Denver, Colo.

Buelow, R.W., Millette, J.R., McFarren, E.F., and Symons, J.M. 1979. The Behavior of Asbestos-Cement Pipe Under Various Water Quality Conditions: A Progress Report, U.S. Environmental Protection Agency, Municipal Environmental Research Laboratory, Drinking Water Research Division, Cincinnati, Ohio.

Christman, R.F., and Ghassemi, M. 1966. "Chemical Nature of Organic Color in Water." *Journal of the American Water Works Association*, 58(6): 723-741.

Craun, G.F., and McCabe, L.J. 1975. "Problems Associated with Metals in Drinking Water," *Journal of the American Water Works Association*, November 1975, pp. 593-599.

Davis, M.J., Herndon, B.L., Shea, E.P., and Snyder, M.K. 1979. Occurrence, Economic Implications, and Health Effects Associated with Aggressive Waters in Public Water Supply Systems: Final Report. Midwest Research Institute, Kansas City, Mo.

Gros, W.F.H. 1977. Internal Corrosion in Water Distribution Systems, presented at the 5th Annual American Water Works Association Water Quality Technology Conference, Kansas City, Mo., December 1977.

Houck, D.H. 1981. Structural Performance of Asbestos-Cement Pipe in Corrosive Potable Water Environment, Paper No. 73, presented at the International Corrosion Forum, Toronto Ontario, Canada, April 6-10, 1981.

Karalekas, P.C., Jr. 1980. Water Treatment for Control of Lead Corrosion, presented at the New England Water Works Association Seminar on Corrosion Control in Drinking Water Systems, Randolf, Mass., March 24-25, 1980.

Kruger, J. 1981. Corrosion: Its Character and Consequences. ASTM Standardization News, 9(5): 21-23.

Larson T.E. 1975. Corrosion by Domestic Waters. Prepared for the State of Illinois, State Water Survey Division, Urbana, Ill.

Lassovszky, P., Vogt, C., and Cotruvo, J.A. 1980. Environmental Protection Agency Activities Concerning Corrosion in Municipal Drinking Water Systems, presented at the New England Water Works Association Seminar on Corrosion Control in Drinking Water Systems, Randolf, Mass., March 24-25, 1980.

McFarren, E.F., Buelow, R.W., Thurnou, R.C., Gardels, M., Sorrell, R.K., Snyder, P., and Dressman, R.C. 1977. Water Quality Deterioration in the Distribution System, presented at the 5th Annual American Water Works Association Water Quality Technology Conference, Kansas City, Mo., December 1977.

Rossi, D.L. 1980. Causes of Corrosion, presented at the New England Water Works Association Seminar on Corrosion Control in Drinking Water Systems, Randolf, Mass., March 24-25, 1980.

Sanders, D.O., 1978. Bacterial Growth and Effect, presented at the American Water Works Association Seminar on Controlling Corrosion Within Water Systems, Atlantic City, N.J., June 1978.

Schock, M.R., Logsdon, G.S., and Clark, P.J. 1981. Evaluation and Control of Asbestos-Cement Pipe Corrosion. U.S. Environmental Protection Agency, Municipal Environmental Research Laboratory, Drinking Water Research Division, Cincinnati, Ohio.

Singley, J.E. 1978. Principles of Corrosion, presented at the American Water Works Association Seminar on Controlling Corrosion Within Water Systems, Atlantic City, N.J., June 1978.

Stumm, W., and Morgan, J.J., 1981. Aquatic Chemistry: An Introduction Emphasizing Chemical Equilibria in Natural Waters, Wiley Interscience, New York.

Sylvester, M., and Cornelius, W.K. 1979. Factors Influencing Corrosiveness of Well Water Toward Copper Piping Systems. Prepared by the Johns Hopkins School of Hygiene and Public Health, Baltimore Md., for the U.S. Environmental Protection Agency, Municipal Environmental Research Laboratory, Cincinnati, Ohio.

Taylor, F.B. 1980. Metals—Corrosion or Dissolution, presented at the New England Water Works Association Seminar on Corrosion Control in Drinking Water Systems, Randolf, Mass., March 24-25, 1980.

U.S. Environmental Protection Agency. 1975. Primary Drinking Water—Proposed Interim Standards, *Federal Register*, 40(51): 11990-11998.

1977. Drinking Water and Health—Recommendations of the National Academy of Sciences, *Federal Register*, 42(132): 35764-35779.

Victoreen, H.I. 1978. Microbial Intervention in Corrosion and Discolored Water, presented at the American Water Works Association Seminar on Controlling Corrosion Within Water Systems, Atlantic City, N.J., June 1978.

Von Wolzogen Kuhr, C.A.H., and Van der Vlugt, L.S., 1953. "Aerobic and Anaerobic Iron Corrosion in Water Mains," *Journal of the American Water Works Association*, 45(1). 33-46.

Additional Source Materials for Chapter 4

Anonymous. 1981. 1981 Water Main Pipe Survey: Cost, Not Material, Shifting Pipe Choice, *American City and County*, June 1981, pp. 41-44.

AWWA Staff Report. 1960. A Survey of Operating Data for Water Works in 1960—Staff Report, American Water Works Association, Inc., New York.

AWWA Standards Committee on Plastic Pipe. 1971. "Plastic Pipe and the Water Utility—Committee Report," *Journal of the American Water Works Association*, 63(6): 352-354.

AWWA Task Group. 1960. "Cold-Water Corrosion of Copper Tubing," *Journal of the American Water Works Association*, 52(8): 1033-1040.

Bottles, D.G. 1970. "Use of Plastic Pipe," *Journal of the American Water Works Association*, 62(1): 55-58.

Buelow, R.W., Millette, J.R., McFarren, E.F., and Symons, J.M. 1979. The Behavior of Asbestos-Cement Pipe Under Various Water Quality Conditions: A Progress Report, U.S. Environmental Protection Agency, Municipal Environmental Research Laboratory, Drinking Water Research Division, Cincinnati, Ohio.

Davis, M.J., Herndon, B.L., Shea, E.P., and Snyder, M.K. 1979. Occurrence, Economic Implications, and Health Effects Associated with Aggressive Waters in Public Water Supply Systems: Final Report. Midwest Research Institute, Kansas City, Mo.

Dressendorfer, P.V., and Halff, A.H. 1972. Large Water Mains: Experience and Practice of Three Large Users, *Journal of the American Water Works Association*, 64(7): 435-440.

Fitzgerald, J.H., III. 1968. Corrosion as a Primary Cause of Cast-Iron Main Breaks, *Journal of the American Water Works Association*, 60(8): 882-897.

Gros, W.F.H. 1977. Internal Corrosion in Water Distribution Systems, presented at the 5th Annual American Water Works Association Water Quality Technology Conference, Kansas City, Mo., December 1977.

Higgins, M.J. 1980. Ductile Iron Pipe Corrosion, presented at the New England Water Works Association Seminar on Corrosion Control in Drinking Water Systems, Randolf, Mass., March 24-25, 1980.

Houck, D.H. 1981. Structural Performance of Asbestos-Cement Pipe in Corrosive Potable Water Environment, Paper No. 73, presented at the International Corrosion Forum, Toronto, Ontario, Canada, April 6-10, 1981.

Hucks, R.T., Jr. 1972. Designing PVC Pipe for Water-Distribution Systems, *Journal of the American Water Works Association*, 64(7): 443-447.

Hudson, W.D. 1966. Studies in Distribution System Capacity in Seven Cities, *Journal of the American Water Works Association*, 58(2): 157-164.

Karalekas, P.C., Jr. 1980. Water Treatment for Control of Lead Corrosion, presented at the New England Water Works Association Seminar on Corrosion Control in Drinking Water Systems, Randolf, Mass., March 24-25, 1980.

Larson T.E. 1975. Corrosion by Domestic Waters. Prepared for the State of Illinois, State Water Survey Division, Urbana, Ill.

Lassovszky, P., Vogt, C., and Cotruvo, J.A. 1980. Environmental Protection Agency Activities Concerning Corrosion in Municipal Drinking Water Systems, presented at the New England Water Works Association Seminar on Corrosion Control in Drinking Water Systems, Randolf, Mass., March 24-25, 1980.

Leckie, H.P., and Uhlig, H.H. 1966. Environmental Factors Affecting the Critical Potential for Pitting in 18-8 Stainless Steel, *Journal of the Electrochemical Society*, 113(12): 1262-1267.

McCabe, L.J., Symons, J.M., Lee, R.D., and Robeck, G.G. 1970. Survey of Community Water Supply Systems, *Journal of the American Water Works Association*, 62(11): 670-687.

McFarren, E.F., Buelow, R.W., Thurnou, R.C., Gardels, M., Sorrell, R.K., Snyder, P., and Dressman, R.C. 1977. Water Quality Deterioration in the Distribution System, presented at the 5th Annual American Water Works Association Water Quality Technology Conference, Kansas City, Mo., December 1977.

Miller, W.T. 1965. Durability of Cement—Mortar Linings in Cast-Iron Pipe. *Journal of the American Water Works Association*, 57(6): 773-782.

National Association of Corrosion Engineers (NACE). 1980. Prevention and Control of Water-Caused Problems in Building Potable Water Systems, TPC Publication No. 7, Houston, Tex.

Nesbitt, W.D. 1980. PVC Water Pipe in Corrosive Environments, presented at the New England Water Works Association Seminar on Corrosion Control in Drinking Water Systems, Randolf, Mass., March 24-25, 1980.

Patterson, J.W., and O'Brien, J.E. 1979. Control of Lead Corrosion. *Journal of the American Water Works Association*, May 1979, pp. 264-271.

Scott, J.B., and Caesar, A.E. 1975. Survey of Water Main Pipe in U.S. Utilities Over 2,500 Population, prepared by *The American City Magazine*, Morgan-Grampian Publishing Co., Pittsfield, Mass.

Seidel, H.F., and Cleasby, J.L. 1966. A Statistical Analysis of Water Works Data for 1960, *Journal of the American Water Works Association*, 58(12):1507-1527.

Streicher, L. 1956. Effects of Water Quality on Various Metals, *Journal of the American Water Works Association*, 48(3): 219-238.

SumX Corporation. 1982. Final Report—Corrosion in Potable Water Systems. Prepared for the U.S. Environmental Protection Agency, Science and Technology Branch, Washington, D.C., Austin, Texas.

Taylor, F.B. 1980. Metals—Corrosion or Dissolution, presented at the New England Water Works Association Seminar on Corrosion Control in Drinking Water Systems, Randolf, Mass., March 24-25, 1980.

Timblin, L.O., Jr., Selander, C.E., and Causey, F.E. 1972. Progress Report on the Evaluation of RPM Pipe, *Journal of the American Water Works Association*, 64(7): 449-456.

Additional Source Materials for Chapter 5

Bennett, W.F., Holler, A.C., and Hurst, W.D. 1977. An Unusual Form of Corrosion, *Journal of the American Water Works Association*, 69(1):26-30.

Davis, M.J., Herndon, B.L., Shea, E.P., and Snyder, M.K. 1979. Occurrence, Economic Implications, and Health Effects Associated with Aggressive Waters in Public Water Supply Systems: Final Report. Midwest Research Institute, Kansas City, Mo.

Larson T.E. 1966. Deterioration of Water Quality in Distribution Systems, *Journal of the American Water Works Association*, 58(10): 1307-1316.

1975. Corrosion by Domestic Waters. Prepared for the State of Illinois, State Water Survey Division, Urbana, Ill.

Lassovszky, P., Vogt, C., and Cotruvo, J.A. 1980. Environmental Protection Agency Activities Concerning Corrosion in Municipal Drinking Water Systems, presented at the New England Water Works Association Seminar on Corrosion Control in Drinking Water Systems, Randolf, Mass., March 24-25, 1980.

Millette, J.R., Hammonds, A.F., Pansing, M.F., Hansen, E.C., and Clark, P.J. 1980. Aggressive Water: Assessing the Extent of the Problem, *Journal of the American Water Works Association*, 72(5): 262-266.

Morris, R.E., Jr. 1967. Principal Causes and Remedies of Water Main Breaks, *Journal of the American Water Works Association*, 59(7): 782-798.

National Association of Corrosion Engineers (NACE). 1980. Prevention and Control of Water-Caused Problems in Building Potable Water Systems, TPC Publication No. 7, Houston, Tex.

Rambow, C.A., and Holmgren, R.S., Jr. 1966. Technical and Legal Aspects of Copper Tube Corrosion, *Journal of the American Water Works Association*, 58(3): 347-353.

Singley, J.E. 1978. Principles of Corrosion, presented at the American Water Works Association Seminar on Controlling Corrosion Within Water Systems, Atlantic City, N.J., June 1978.

Additional Source Materials for Chapter 6

Adams, O.H. 1977. The Safe Drinking Water Act Impacts on State Water Programs, presented at the 5th Annual American Water Works Association Water Quality Technology Conference, Kansas City, Mo., December 1977.

American Water Works Association Water Quality Committee, Pacific Northwest Section. 1980. *Manual for Determining Internal Corrosion Potential in Water Supply Systems*, Final Draft.

American Water Works Association Committee on Corrosion and Deposition. 1978. Current Corrosion Experiences in Large Utilities: Results of a Committee Survey, presented at the American Water Works Association Annual Conference and Exposition, Atlantic City, N.J., June 1978.

American Water Works Association Research Foundation. 1982. Research News: Corrosion Control, No. 32, Denver, Colo.

American Water Works Association Research Foundation. Water Quality Research News. 1982. Treatment: Control of Lead Concentrations, No. 30., Denver, Colo.

Benedict, R.L., Opincar, V.E. 1980. Planning to Mitigate External Corrosion—Case Study, presented at the New England Water Works Association Seminar on Corrosion Control in Drinking Water Systems, Randolf, Mass., March 24-25, 1980.

Buelow, R.W., Millette, J.R., McFarren, E.F., and Symons, J.M. 1979. The Behavior of Asbestos-Cement Pipe Under Various Water Quality Conditions: A Progress Report, U.S. Environmental Protection Agency, Municipal Environmental Research Laboratory, Drinking Water Research Division, Cincinnati, Ohio.

Byars, H.G., and Gallop, B.R. 1975. An Approach to the Reporting and Evaluation of Corrosion Coupon Exposure Results, *Materials Performance*, 14(11): 9.

Casey, J.R., and Eakins, W. 1980. Impact of Cleaning and Lining Cast-Iron Pipe Corrosion— Case History of Lynn, MA, presented at the New England Water Works Association Seminar on Corrosion Control in Drinking Water Systems, Randolf, Mass., March 24-25, 1980.

Costello, J.J. 1978. Lime Use for Corrosion Control, presented at the American Water Works Association Seminar on Controlling Corrosion Within Water Systems, Atlantic City, N.J., June 1978.

Davis, M.J., Herndon, B.L., Shea, E.P., and Snyder, M.K. 1979. Occurrence, Economic Implications, and Health Effects Associated with Aggressive Waters in Public Water Supply Systems: Final Report. Midwest Research Institute, Kansas City, Mo.

Dye, J.F. 1958. Correlation of the Two Principal Methods of Calculating the Three Kinds of Alkalinity, *Journal of the American Water Works Association*, 50(6): 800-820.

Graves, D.J., and Jorden, E.C. 1980. The Use of Indices to Describe and Control Corrosion, presented at the New England Water Works Association Seminar on Corrosion Control in Drinking Water Systems, Randolf, Mass., March 24-25, 1980.

Gros, W.F.H. 1977. Internal Corrosion in Water Distribution Systems, presented at the 5th Annual American Water Works Association Water Quality Technology Conference, Kansas City, Mo., December 1977.

Grubb, C.E. 1979. Field Test of Corrosion Control to Protect Asbestos-Cement Pipe. Prepared for the U.S. Environmental Protection Agency, Municipal Environmental Research Laboratory, Drinking Water Research Division, Cincinnati. Greenwood, S.C.

Houck, D.H. 1981. Structural Performance of Asbestos-Cement Pipe in Corrosive Potable Water Environment, Paper No. 73, presented at the International Corrosion Forum, Toronto, Ontario, Canada, April 6-10, 1981.

Larson T.E. 1975. Corrosion by Domestic Waters. Prepared for the State of Illinois, State Water Survey Division, Urbana, Ill.

Lassovszky, P., Vogt, C., and Cotruvo, J.A. 1980. Environmental Protection Agency Activities Concerning Corrosion in Municipal Drinking Water Systems, presented at the New England Water Works Association Seminar on Corrosion Control in Drinking Water Systems, Randolf, Mass., March 24-25, 1980.

Logsdon, G.S., and Millette, J.R. (n.d.) Monitoring for Corrosion of A/C Pipe, U.S. Environmental Protection Agency, Drinking Water Research Division, Cincinnati, Ohio.

McCauley, R.F. 1960. Controlled Deposition of Protective Calcite Coatings in Water Mains, *Journal of the American Water Works Association*, 52(11):1386-1396.

McClanahan, M.A., and Mancy, K.H. 1974a. Comparison of Corrosion—Rate Measurements on Fresh vs. Previously Polarized Samples, *Journal of the American Water Works Association*, 66(8): 461-466.

1974b. Effect of pH on Quality of Calcium Carbonate Film Deposited from Moderately Hard and Hard Water, *Journal of the American Water Works Association*, 66(1): 49-53.

McClelland, N.I., and Mancy, K.H. 1972. Water Quality Monitoring in Distribution Systems: A Progress Report, *Journal of the American Water Works Association*, 64(12): 795-803.

McFarren, E.F., Buelow, R.W., Thornau, R.C., Gardels, M., Sorrell, R.K., Snyder, P., and Dressman, R.C. 1977. Water Quality Deterioration in the Distribution System, presented at the 5th Annual American Water Works Association Water Quality Technology Conference, Kansas City, Mo., December 1977.

McFarren, E.F., Thornau, R.W., Gardels, M., Sorrell, R.K., Snyder, P., and Dressman, R.C. 1977. Water Quality Deterioration in the Distribution System. U.S. Environmental Protection Agency, Municipal Environmental Research Laboratory, Water Supply Research Division, Cincinnati, Ohio.

Medlar, S. 1980. Evaluating Steam Corrosion, presented at the New England Water Works Association Seminar on Corrosion Control in Drinking Water Systems, Randolf, Mass., March 24-25, 1980.

Nalco Chemical Co. (n.d.) Evaluating Corrosion in Steam and Water Systems, Oak Brook, Ill. Water Treatment Chemicals TF 37.

National Association of Corrosion Engineers (NACE). 1980. Prevention and Control of Water-Caused Problems in Building Potable Water Systems, TPC Publication No. 7, Houston, Tex.

Patterson, J.W., and O'Brien, J.E. 1979. Control of Lead Corrosion. *Journal of the American Water Works Association*, 71(5): 264-271.

Petrolite Corporation, Petreco Division (n.d.) *Technical Manual*: M-3010 *Corrosion Rate Instrument*, Houston, Tex.

Pourbaix, M. 1969. Recent Applications of Electrode Potential Measurements in the Thermodynamics and Kinetics of Corrosion of Metal. *Corrosion*, National Association of Corrosion Engineers, 25(6): 267.

1972. Theoretical and Experimental Considerations in Corrosion Testing. *Corrosion Science*, Vol. 12, pp. 161-191.

Ritter, J.A. 1977. Establishing a Corrosion Monitoring Program, presented at the 5th Annual American Water Works Association Water Quality Technology Conference, Kansas City, Mo., December 1977.

Schock, M.R., Logsdon, G.S., and Clark, P.J. 1981. Evaluation and Control of Asbestos-Cement Pipe Corrosion. U.S. Environmental Protection Agency, Municipal Enviromental Research Laboratory, Drinking Water Research Division, Cincinnati, Ohio.

Schock, M.R., Mueller, W., and Buelow, R.W. 1979. Laboratory Technique for Measurement of pH for Corrosion Control Studies and Waters not in Equilibrium with the Atmosphere. (Draft). U.S. Environmental Protection Agency, Physical and Chemical Contaminant Removal Branch, Cincinnati, Ohio.

Shull, K.E. 1980. An Experimental Approach to Corrosion Control, *Journal of the American Water Works Association*, May 1980, pp. 280-285.

Singley, J.E. 1981. The Search for a Corrosion Index. *Journal of the American Water Works Association*, 73(11): 579.

Singley, J.E., and Lee, T.H. 1982. Development and Use of an Apparatus for Study of Corrosion of Pipe Sections, Gainesville, Fla.

Sussman, S. 1978. "Implications of the EPA Proposed National Secondary Drinking Water Regulation on Corrosivity," presented at the American Water Works Association Seminar on Controlling Corrosion Within Water Systems, Atlantic City, N.J., June 1978.

System Water Quality Committee, California-Nevada Section, American Water Works Association. 1978. *Procedure Manual for Handling Water Quality Complaints*.

Thibeau, R.J., Brown, C.W., Goldfarb, A.Z., and Heidersbach, R.H. 1980. Raman and Infrared Spectroscopy of Aqueous Corrosion Films on Lead in 0.1 M Sulfate Solutions, *Journal of the Electrochemical Society*, 127(9): 1913-1918.

Trussell, R.R., and Russell, L.L. 1977. The Langelier Index, presented at the 5th Annual American Water Works Association Water Quality Technology Conference, Kansas City, Mo., December 1977.

University of Rhode Island, Department of Chemistry. 1980. Quarterly Report, First Quarter, In Situ Analysis of Corrosive and Passive Surfaces by Laser-Excited Raman Spectroscopy, Kingston, R.I.

U.S. Environmental Protection Agency. 1975. Primary Drinking Water—Proposed Interim Standards. *Federal Register*, 40(51): 11990-11998

1977. National Secondary Drinking Water Regulations—Proposed Regulations, *Federal Register*, 42(62): 17143-17147.

1981. National Interim Primary Drinking Water Regulations, *Code of Federal Regulations*, Title 40, Part 141, pp. 309-354.

U.S. Environmental Protection Agency, Office of Drinking Water. 1979. National Secondary Drinking Water Regulations, EPA-570/9-76-000, Washington, D.C.

U.S. Environmental Protection Agency, Office of Drinking Water, Criteria and Standards Division. 1980. Statement of Basis and Purpose for Amendments to the National Interim Primary Drinking Water Regulations, Washington, D.C.

Voyles, C.F. 1978. Stabilizing Southern California Waters, presented at the American Water Works Seminar on Controlling Corrosion Within Water Systems, Atlantic City, N.J., June 1978.

Young, W.T. 1978. Instrumentation for Evaluating Corrosion, presented at the American Water Works Association Seminar on Controlling Corrosion Within Water Systems, Atlantic City, N.J., June 1978.

Additional Source Materials for Chapter 7

American Water Works Association Research Foundation. 1982. Research News: Corrosion Control, No. 32, Denver, Colo.

American Water Works Association Research Foundation. Water Quality Research News. 1982. Treatment: Control of Lead Concentrations, No. 30, Denver, Colo.

American Water Works Association Task Group 2690-P. 1960. Cold-Water Corrosion of Copper Tubing, *Journal of the American Water Works Association*, 52(8): 1033-1040.

Bailey, T.L. 1980. Corrosion Control Experiences at Durham, North Carolina, presented at the New England Water Works Association Seminar on Corrosion Control in Drinking Water Systems, Randolf, Mass., March 24-25, 1980.

Benedict, R.L., Opincar, V.E. 1980. Planning to Mitigate External Corrosion—Case Study, presented at the New England Water Works Association Seminar on Corrosion Control in Drinking Water Systems, Randolf, Mass., March 24-25, 1980.

Buelow, R.W., Millette, J.R., McFarren, E.F., and Symons, J.M. 1979. The Behavior of Asbestos-Cement Pipe Under Various Water Quality Conditions: A Progress Report, U.S. Environmental Protection Agency, Municipal Environmental Research Laboratory, Drinking Water Research Division, Cincinnati, Ohio.

Casey, J.R., and Eakins, W. 1980. Impact of Cleaning and Lining Cast-Iron Pipe Corrosion— Case History of Lynn, MA, presented at the New England Water Works Association Seminar on Corrosion Control in Drinking Water Systems, Randolf, Mass., March 24-25, 1980.

Christman, R.F., and Ghassemi, M. 1966. Chemical Nature of Organic Color in Water, *Journal of the American Water Works Association*, 58(6): 723-741.

Costello, J.J. 1978. Lime Use for Corrosion Control, presented at the American Water Works Association Seminar on Controlling Corrosion Within Water Systems, Atlantic City, N.J., June 1978.

Davis, M.J., Herndon, B.L., Shea, E.P., and Snyder, M.K. 1979. Occurrence, Economic Implications, and Health Effects Associated with Aggressive Waters in Public Water Supply Systems: Final Report. Midwest Research Institute, Kansas City, Mo.

Ghosh, M.M. 1973. Chemical Conditioning to Control Water-Quality Failure in Distribution Systems, *Journal of the American Water Works Association*, 65(5): 348-355.

Grady, R.P. 1980. Corrosion Control Experience in Portland, Maine, presented at the New England Water Works Association Seminar on Corrosion Control in Drinking Water Systems, Randolf, Mass., March 24-25, 1980.

Gros, W.F.H. 1977. Internal Corrosion in Water Distribution Systems, presented at the 5th Annual American Water Works Association Water Quality Technology Conference, Kansas City, Mo., December 1977.

Haskew, G.M. 1978. Use of Zinc Orthophosphate Corrosion Inhibitor— Plant Practice, presented at the American Water Works Association Seminar on Controlling Corrosion Within Water Systems, Atlantic City, N.J., June 1978.

Houck, D.H. 1981. Structural Performance of Asbestos-Cement Pipe in Corrosive Potable Water Environment. Paper No. 73. Presented at the International Corrosion Forum, Toronto, Ontario, Canada, April 6-10, 1981.

Hullinger, D.L. 1975. A Study of Heavy Metals in Illinois Impoundments, *Journal of the American Water Works Association*, 67(10): 572-576.

Karalekas, P.C., Jr. 1980. Water Treatment for Control of Lead Corrosion, presented at the New England Water Works Association Seminar on Corrosion Control in Drinking Water Systems, Randolf, Mass., March 24-25, 1980.

Lane, R.W., Larson, T.E., and Schilsky, S.W. 1977. The Effect of pH on the Silicate Treatment of Hot Water in Galvanized Piping, *Journal of the American Water Works Association*, 69(8): 457-461.

Larson T.E. 1975. Corrosion by Domestic Waters. Prepared for the State of Illinois, State Water Survey Division, Urbana, Ill.

Lassovszky, P., Vogt, C., and Cotruvo, J.A. 1980. Environmental Protection Agency Activities Concerning Corrosion in Municipal Drinking Water Systems, presented at the New England Water Works Association Seminar on Corrosion Control in Drinking Water Systems, Randolf, Mass., March 24-25, 1980.

McCauley, R.F. 1960. Controlled Deposition of Protective Calcite Coatings in Water Mains, *Journal of the American Water Works Association*, 52(11):1386-1396.

McFarren, E.F., Buelow, R.W., Thornau, R.C., Gardels, M., Sorrell, R.K., Snyder, P., and Dressman, R.C. 1977. Water Quality Deterioration in the Distribution System, presented at the 5th Annual American Water Works Association Water Quality Technology Conference, Kansas City, Mo., December 1977.

Medlar, S. 1980. Evaluating Steam Corrosion, presented at the New England Water Works Association Seminar on Corrosion Control in Drinking Water Systems, Randolf, Mass., March 24-25, 1980.

Merrill, D.T., and Sanks, R.L. 1977. Corrosion Control by Deposition of $CaCO_3$ Films: Part 1, A Practical Approach for Plant Operators, *Journal of the American Water Works Association*, 69(11): 592-599.

National Association of Corrosion Engineers (NACE). 1980. Prevention and Control of Water-Caused Problems in Building Potable Water Systems, TPC Publication No. 7, Houston, Tex.

Paris, D.B. 1980. Corrosion Control in Manchester, New Hampshire, presented at the New England Water Works Association Seminar on Corrosion Control in Drinking Water Systems, Randolf, Mass., March 24-25, 1980.

Paterson, J.A. 1978. Corrosion Inhibitors and Coatings, presented at the American Water Works Association Seminar on Controlling Corrosion Within Water Systems, Atlantic City, N.J., June 1978.

Patterson, J.W., and O'Brien, J.E. 1979. Control of Lead Corrosion. *Journal of the American Water Works Association*, 71(5): 264-271.

Sanders, D.O. 1978. Bacterial Growth and Effect, presented at the American Water Works Association Seminar on Controlling Corrosion Within Water Systems, Atlantic City, N.J., June 1978.

Schock, M.R., Logsdon, G.S., and Clark, P.J. 1981. Evaluation and Control of Asbestos-Cement Pipe Corrosion. U.S. Environmental Protection Agency, Municipal Enviromental Research Laboratory, Drinking Water Research Division, Cincinnati, Ohio.

Shull, K.E. 1980. An Experimental Approach to Corrosion Control, *Journal of the American Water Works Association*, May 1980, pp. 280-285.

Stevens, R.L., and Dice, J.C. 1981. Chlorination—A Proven Means of Disinfecting Mains, *OpFlow*, 7(10): 1.

U.S. Environmental Protection Agency, Office of Drinking Water, Criteria and Standards Division. 1980. Statement of Basis and Purpose for Amendments to the National Interim Primary Drinking Water Regulations, Washington, D.C.

Victoreen, H.T. 1978. Microbial Intervention in Corrosion and Discolored Water, presented at the American Water Works Association Seminar on Controlling Corrosion Within Water Systems, Atlantic City, N.J., June 1978.

Voyles, C.F. 1978. Stabilizing Southern California Waters, presented at the American Water Works Seminar on Controlling Corrosion Within Water Systems, Atlantic City, N.J., June 1978.

Yapijakis, C. 1977. Controlling Corrosion in Distribution Systems, *Water and Sewage Works*, 124(4): 96.

Additional Source Materials for Chapter 8

Davis, M.J,, Herndon, B.L., Shea, E.P., and Snyder, M.K. 1979. Occurrence, Economic Implications, and Health Effects Associated with Aggressive Waters in Public Water Supply Systems: Final Report. Midwest Research Institute, Kansas City, Mo.

Grubb, C.E. 1979. Field Test of Corrosion Control to Protect Asbestos-Cement Pipe. Prepared for the U.S. Environmental Protection Agency, Municipal Environmental Research Laboratory, Drinking Water Research Division, Cincinnati. Greenwood, S.C.

Karalekas, P.C., Jr., Ryan, C.R., and Taylor, F.B. 1982. Control of Lead Pipe Corrosion in the Boston Metropolitan Area, presented at the Annual Conference of the American Water Works Association, Miami Beach, Fla., May 16-20, 1982.

Logsdon, G.S. 1981. Project Summary: Field Test of Corrosion Control to Protect Asbestos-Cement Pipe, U.S. Environmental Protection Agency, EPA-600/S2-81-023, Cincinnati.

Additional Source Materials for Chapter 9

Bailey, T.L. 1980. Corrosion Control Experiences at Durham, North Carolina, presented at the New England Water Works Association Seminar on Corrosion Control in Drinking Water Systems, Randolf, Mass., March 24-25, 1980.

Benedict, R.L., Opincar, V.E. 1980. Planning to Mitigate External Corrosion—Case Study, presented at the New England Water Works Association Seminar on Corrosion Control in Drinking Water Systems, Randolf, Mass., March 24-25, 1980.

Courchene, J.E. 1978. Seattle Internal Corrosion Control Plan—Summary Report, presented at the American Water Works Association Seminar on Controlling Corrosion Within Water Systems, Atlantic City, N.J., June 1978.

Grady, R.P. 1980. Corrosion Control Experience in Portland, Maine, presented at the New England Water Works Association Seminar on Corrosion Control in Drinking Water Systems, Randolf, Mass., March 24-25, 1980.

Nelson, J.A. 1978. One Utility's Approach to Solving Copper Corrosion, presented at the American Water Works Association Seminar on Controlling Corrosion Within Water Systems, Atlantic City, N.J., June 1978.

Paris, D.B. 1980. Corrosion Control in Manchester, New Hampshire, presented at the New England Water Works Association Seminar on Corrosion Control in Drinking Water Systems, Randolf, Mass., March 24-25, 1980.

Ryder, R.A. 1980. The Costs of Internal Corrosion in Water Systems. *Journal of the American Water Works Association*, 72(5): 267-279.

Part II

Review of Monitoring, Detection, Prevention and Control Techniques

The information in Part II is from *Corrosion in Potable Water Systems* by David W. DeBerry, James R. Kidwell and David A. Malish of SumX Corporation for the U.S. Environmental Protection Agency, February 1982.

1. Introduction

BACKGROUND

The Safe Drinking Water Act of 1974 (PL 93-523) was passed by Congress to safeguard public drinking water supplies and to protect public health. This act essentially states that National Drinking Water Regulations will be established and enforced for all public drinking water supplies. As a result, the Environmental Protection Agency (EPA) proposed National Interim Primary Drinking Water Regulations (NIPDWR) which became effective June 24, 1977. These interim regulations have established maximum allowable concentrations of various contaminants in drinking water supplies and require that water suppliers sample and analyze the water on a regular basis. Contaminants identified by the NIPDWR for limitations in drinking water supplies include bacteria, turbidity, radioactivity, trihalomethanes, ten inorganic chemicals, and six organic pesticides. These regulations will be reviewed at least once every three years and can be amended at any time.

Corrosion control regulations are addressed directly in the National Secondary Drinking Water Regulations which state that potable waters should be non-corrosive for protection of the public welfare. These Secondary Regulations are federal guidelines only, but may be adopted and enforced by individual states.

The EPA has determined that corrosive materials in the water works industry can pose a serious threat to public health and, on August 27, 1980, issued amendments to the NIPDWR which specifically outline monitoring for corrosion related parameters. These regulations require public water systems to identify the presence of specific materials of construction within the distribution systems and to monitor and report corrosivity characteristics including pH, alkalinity, hardness, total dissolved solids, and the Langelier Index. Additional regulations will be developed and enforced by EPA that set forth requirements for systems distributing corrosive waters to increase monitoring for corrosion byproducts such as lead and cadmium.

Attempts to develop regulations controlling corrosion in the water works industry is controversial owing to the complexity of the corrosion problem. Major problems include the lack of legal definition for corrosivity, the lack of a generally acceptable method for measuring corrosivity, and the lack of corrosion control methods which are effective throughout the entire distribution system as well as compatible with other potable water supply objectives. The problem is further complicated by the requirement to regulate corrosion at the "free flowing outlet of the ultimate consumer of a public water system." However, in defining maximum contaminant levels, the NIPDWR excludes contaminants added to the water under circumstances controlled by the user, except those resulting from corrosion of piping and plumbing caused by the water quality. This regulation implies that the water supplier must be cognizant of the pipe materials in the service lines and within the household as well as the distribution system.

Corrosion in potable water systems may be caused by either inherent factors or design, construction, or operational deficiencies. Most materials used in potable water systems are susceptible to corrosion in waters containing oxygen. This inherent susceptibility is often decreased by formation of protective coatings on the bare material surface. The coatings are formed by precipitation of substances such as calcium carbonate from solution or growth of insulating corrosion reaction product films on the material surface. Thus, the properties of the material and the composition of the water are interactive in the occurrence or inhibition of inherent corrosion. Corrosion susceptibility is also influenced by many other factors including temperature, local flow rate, pH, alkalinity, carbon dioxide, dissolved solids, and minor chemical constituents of either the water or corroding material.

Corrosion problems may also be caused by poor choice of materials, coupling of dissimilar metals, improper installation practices, incorrect design allowing unnecessary stagnant areas or crevices, release of unstable waters, or addition of additives. While in principle subject to correction, it is necessary to assume that many of these defects will persist for some time in present systems.

Potable water distribution and plumbing systems may contain a variety of materials. The environmental conditions noted above can have effects on individual materials that differ both in type and degree. It is possible that alleviation of corrosion problems for one material or type of installation could create problems with another or result in deterioration of overall water quality.

Before corrosion control measures can be implemented, sufficient evidence must be available which detects the presence of corrosion. Additionally, the location and cause of the corrosion occurrence must be identified. This information can be obtained through the initiation of a comprehensive monitoring program of potential corrosion byproducts. However, under existing regulations, water utilities are required to monitor only once per year for facilities using surface water sources and only once every three years for facilities using groundwater sources. The only major corrosion products required to be monitored are lead and cadmium. Thus, not only is the current monitoring schedule insufficient to reliably detect corrosion, much less to isolate the location and possible causes, but some corrosion recognized by the consumer, in terms of economics and aesthetics, may not be recognized by the water supplier.

The secondary drinking water regulations state that potable waters should be non-corrosive. No attempt has been made to assign a numerical value for a maximum contaminant level (MCL) for corrosion. A major problem in attempting to produce a "non-corrosive" water is the lack of a generally applicable measure of water corrosivity. Several corrosive or "aggressive" indices have been developed to rate and evaluate the corrosivity of potable waters. These indices include the Langelier Index, the aggressive index, the Ryznar Stability Index, the Larson Index, the Casil Index, the Riddick Index, and the driving force index. Predictions using these indices are often found to be in disagreement with field experience.

These discrepancies are often a result of the differences between theoretical deviations and actual conditions. For instance, indices predicting $CaCO_3$ deposition do not incorporate factors for metastable conditions nor do they reflect that the protective capacity of the deposit is a function of the physical and chemical conditions existing during its formation.

Development of a general index is also difficult because of the multiple roles of chemical species in potable water. The most common and abundant aggressive ions in potable waters are usually considered to be chloride and sulfate, although they may aid formation of more protective calcium carbonate scale. Also, although oxygen provides the main driving force for many types of corrosion in potable water, there is evidence that a certain minimum solution oxygen level is necessary for the initial formation of protective calcium carbonate films on steel. Flow velocities are also important for the formation of protective layers and there is some evidence that low levels of natural organic or silicate inhibitors may be influential.

Direct tests have been performed using pipe sections, metal coupons, and water quality analyses to determine corrosivity. These tests must be performed over a fairly long period of time and may be prohibitively expensive for small communities. The results of these tests are also often inconsistent with observations of field equipment and often are of limited use in isolating a specific problem.

The inconsistency between practical and theoretical measurements and field observations can be attributed to the large variety of mechanisms by which corrosion can occur. All factors which can influence or cause corrosion are not necessarily identified using indices or direct field tests. For example, pitting corrosion can be initiated from oxygen being trapped and not evenly distributed in areas within the distribution system. Corrosion from erosion or impingement caused by excessive velocities at certain areas or occasional sand discharges can remove protective films and accelerate corrosion. Increased velocities often result from improper design of home or commercial plumbing facilities.

Another problem exists with the wide range of materials and installation practices used in the water works industry. A specific water quality may be non-corrosive to some materials but corrosive to others. Most older distribution systems were constructed of cast iron pipe. Materials used more recently for distribution systems include welded steel, ductile iron, precast concrete, and asbestos cement. Asbestos cement is used most extensively today because of its excellent physical properties. However, asbestos cement pipe is limited to 30 inches in diameter and cannot be used if large hydrostatic pressures are required. Steel pipes are often used for larger diameter transmission mains and may be protected with a cement mortar coating. In alkaline conditions, steel or cast iron pipes are sometimes lined with coal-tar enamel or an asphaltic layer over the cement lining for protection.

Materials used for the construction of facilities which cannot be protected with coatings such as valves, small pipes, and pumps include lead, copper, zinc, aluminum, and alloys such as brass, bronze, and stainless steels. Many municipal plumbing codes are not restrictive and a wide variety and mix of material types and installation practices are used for home plumbing. This practice can be conducive to galvanic corrosion. Domestic corrosion can be also be aggravated by the use of home water softening units, inferior grade hot water heaters, and incorrectly sized pipes.

OBJECTIVES

The purpose of this investigation is to collect, review, evaluate, and present existing information to determine whether a sufficient data base is available to develop corrosion control regulations for the water works industry as required by the Safe Drinking Water Act. To accomplish this objective, an exhaustive literature search was completed which included a review of the various materials used in the water works industry and their corrosion characteristics, corrosion monitoring and detection techniques, and corrosion prevention and control strategies. Materials which are addressed are iron-based materials, copper-based materials, lead-based materials, aluminum, asbestos cement, concrete, and plastics. Results of laboratory and field research on each material as related to corrosion in the water works industry are extensively reviewed and data is presented as appropriate. Major emphasis is placed on assessing the conditions of service and water quality characteristics in potable water systems on the corrosion or deterioration of each material.

The review of corrosion monitoring and detection techniques addresses the various methodologies used to identify and evaluate corrosive waters. This discussion includes a review of the search for corrosion indices using water quality monitoring data as well as direct monitoring using coupon exposure tests. The limitations of each of the techniques are presented.

Available corrosion prevention and control techniques are also evaluated and presented. These techniques include adjustments in water quality characteristics, additions of corrosion inhibitors, and the application of various pipe coatings. Additionally, case histories of corrosion control programs are presented for examples.

Finally, the information and data presented in these reviews are compiled and presented in table or matricies form as a summary. These tables provide an overall view of the nature of the corrosion problems in the water works industry and can be used as a data base for the initial consideration of corrosion control regulations.

2. Corrosion and Water Chemistry Background

GENERAL ASPECTS OF CORROSION AND LEACHING IN POTABLE WATER

In general, corrosion refers to the degradation of a metal by electrochemical or chemical reaction with its environment or by physical wearing away. Leaching implies the removal of a soluble constituent, not necessarily metallic, by the action of a percolating liquid. While more often used in relation to ground water systems, leaching has also been used to describe some corrosive dissolutions. This report is concerned with both the physical and chemical degradation of any material used in potable water systems that impart additional species to the water. The nature of corrosion and degradation of materials varies to a large extent with the specific material and environment. Detailed discussions of these interactions with domestic water environments are given in Section 4. These discussions are primarily based on the corrosion and water works literature. Although this literature is voluminous, it is also rather chaotic in the sense that reported results are often not accompanied by sufficient background data or are anecdotal or subject to multiple conclusions. This situation is not surprising in view of the highly complex physical-chemical nature of corrosion processes in general and especially in large-scale field installations. Much of the pertinent literature is oriented to engineering implications of corrosion; there is relatively little information on leaching of constituents of materials with regard to health implications. A number of modes of corrosion or degradation may exist. This report is primarily concerned with those processes which occur on surfaces contacting the potable water and which impart substances to the water. This excludes external corrosion and stress corrosion cracking; the latter is rare in domestic water environments in any case.

The fundamentals of corrosion are given in several textbooks. (See bibliography for this section.) The book by Fontana and Greene is recent and lucid. The text by Butler and Ison has a stronger orientation to corrosion in natural waters and the chemistry of natural waters. The 1948 edition of The Corrosion Handbook by Uhlig compiles a very large amount of data obtained prior to its publication; the more recent work by Uhlig is a useful introductory text. The other references listed are useful for a somewhat different outlook or additional detail on the broad subject of corrosion.

TYPES OF CORROSION

Corrosion types may be classified as either uniform or localized. Uniform corrosion is the loss of a more or less equal amount of material over the surface of a pipe or other structure. It may proceed directly to metal ions which go into solution or by way of a solid reaction product such as metal oxide, hydroxide, carbonate, or other compound. In this case the amount of material imparted to solution may become limited by the solubility of the reaction product compound or by its dissolution kinetics. There is also the possibility of periodic sloughing off of particles or chunks of the reaction product either due to erosion or stresses built up during growth of the layer on the metal. Uniform corrosion is not often a major concern in domestic waters from an engineering or maintenance standpoint. It is a more likely concern from a water quality view since a fairly low uniform corrosion rate, spread over considerable area, can impart more impurity to a given amount of water than a few deep pits.

Localized, or non-uniform, corrosion results in relatively rapid attack and penetration on small areas of metal surface while the remainder of the surface is not affected. Such attack can affect the structural or hydraulic integrity of equipment. Pitting of iron can develop the tuberculation which decreases the flow capacity of pipes. For these reasons, localized corrosion is often an engineering or maintenance consideration. Since attack can be rapid and may result in selective leaching of a metal from an alloy, localized corrosion may also have environmental repercussions. It is noteworthy that either changing to slightly more resistant materials or modifying the parameters in a corrosive medium to reduce the uniform corrosion rate may produce conditions in which localized corrosion is the predominant mode.

Localized corrosion types of interest in potable water systems include pitting, galvanic, concentration-cell, and selective-removal corrosions. Pitting is a general term that refers to the formation of a pit where local anodic conditions exist relative to a nearby cathodic area. Tuberculation occurs when oxides of corrosion products are deposited over or adjacent to the pit. Galvanic corrosion results when two metals of different solution potentials contact each other. The anodic metal will corrode, affording "protection" to the cathodic metal. Concentration-cell corrosion occurs when localized differences in the potential of a single metal exist. Conditions creating this environment could include differences in acidity, cation or anion concentrations, dissolved oxygen, or even temperature fluctuations. Crevice corrosion is a form of concentration-cell corrosion where the rate of oxygen reaching the metal surface is controlled by diffusion in a confined area. Selective-removal corrosion would include dezincification, the removal of zinc from brass, or graphitization, the removal of the iron silicon metal alloy from cast iron, leaving graphite.

CORROSION INDICES

The following presents a brief general discussion of historical attempts to develop a corrosion index. While these indices developed have a potential use, none can be considered as a single tool for formulating a responsible corrosion control program.

In 1912, J. Tillmans proposed the carbonate saturation theory of pipe protection. Since then, several theoretical and empirical approaches have been made to determine a single parameter that would indicate the ability of a given water to protect or corrode the carrying pipes.

In 1936, W. F. Langelier developed the Langelier Saturation Index (9). The SI is based on the theoretical tendency of a water to deposit or dissolve calcium carbonate. The index is derived from the solubility product of calcium carbonate, the dissociation constant of water, the second dissociation constant of carbonic acid ($H^+ + CO_3^= \rightleftarrows HCO_3^-$), and a stoichiometric equilibrium between alkalinity and protons versus the carbonate, bicarbonate, and hydroxyl species. By restricting the applicable pH to 6.5-9.5, a saturation pH, pH_s, is determined as:

$$pH_s = (pK_2' - pK_s') + pCa + pAlk$$

where $pX = \log\left(\frac{1}{X}\right)$ and

Ca = calcium ion concentration in moles/liter

Alk = total alkalinity as equivalents/liter

K_2' = second dissociation constant of H_2CO_3, corrected for ionic strength and temperature

K_s' = solubility product of $CaCO_3$, corrected for ionic strength and temperature

The saturation index is: $SI = pH - pH_s$.

It is a logarithm of the ratio of the hydrogen ion concentration that the water must have if saturated with calcium carbonate to the actual hydrogen ion concentration. A negative value indicates an undersaturation of $CaCO_3$ and, hence, a tendency to dissolve anv existing $CaCO_3$ coating. A positive value indicates oversaturation, a tendency to precipitate $CaCO_3$ and to form a protective layer.

Correlation attempts between the calculated SI and observed corrosion effects have often, but not always, been in agreement. In general, a corrosion free system exists when the SI is greater than -0.5 for cold water and 0.0 for hot water, provided the water is of moderate hardness and alkalinity (7), where the Langelier presumption of calcite being the $CaCO_3$ phase is less apt to be discrepant from impure calcite or precipitation inhibition. The SI is similarly not suitable for use in soft, saline waters where a low buffer capacity and ionic species such as chlorides may disrupt the $CaCO_3$ equilibrium conditions. Parameters that must be evaluated to calculate the SI are methyl orange alkalinity, pH, temperature, total dissolved solids, and the calcium ion concentration.

When using the SI, it is imperative to remember that this parameter states a difference between actual pH and a pH at which $CaCO_3$ equilibrium is theoretically achieved for that water. It says nothing about the driving force or tendency for the $CaCO_3$ to dissolve or crystallize in the given water. Differences may result from temperature effects, specific uncharged or ionic constituents in the water, or crystal growth inhibition. Increased temperatures will decrease the solubility of calcium carbonate as well as increase reaction velocities, and consequently establish local equilibrium situations much faster than at normal temperatures. There may also exist a difference in pH between the $CaCO_3$ saturation point and the point at which crystal growth actually begins. This difference, called the metastable region, may increase in the presence of other dissolved ions, especially those that form slightly soluble salts with calcium or carbonates such as magnesium or sulfate, causing a poor correlation between the calculated SI and actual conditions of solubility.

A commonly used refinement of the SI that includes temperature and ionic strength corrections is

$$pH_S = A + B - \text{Log}\ (Ca^{++}) - \log\ (\text{alkalinity})$$

where A and B are constants derived from the following tables:

Water Temperature C	A	mg/l Total Filterable Residue	B
0	2.60	0	9.70
4	2.50	100	9.77
8	2.40	200	9.83
12	2.30	400	9.86
16	2.20	800	9.89
20	2.10	1,000	9.90
25	2.00		
30	1.90		
40	1.70		
50	1.55		
60	1.40		
70	1.25		
80	1.15		

TABLE 1. Correction Parameters Used With The Refined SI

Other corrosion indicators include the Ryznar, Larson, Driving Force, Casil, Aggressive, and Riddick indices. The Ryznar Stability Index is defined as

$$RI = 2pH_s - pH$$

with pH_s and pH defined as before. A value of seven or greater indicates an aggressive water while 6 or less indicates a tendency to form scale (17). This index may be used with moderate to hard waters, but is not applicable to soft or saline waters for the reasons previously cited.

The Larson index attempts to measure the aggressive nature of specific ions and is defined as:

$$LI = \frac{Cl + SO_4}{Alk}$$

where Cl and SO_4 are the chloride or halogen concentrations and sulfate concentrations, repectively, and Alk is total alkalinity. All three are expressed in mg/l of equivalent $CaCO_3$. When this ratio of reactive anions to alkalinity is greater than 0.5, the possibility of corrosive action exists. Unlike the SI, this index does not refer to the solubility of $CaCO_3$ but rather to the faster rates of corrosion of metals because of conductivity effects. It is not applicable to water that is soft or has a low dissolved solids concentration.

The driving force index is defined as:

$$DFI = (Ca^{++}) \times (CO_3^{=})/K_s' \times 10^{10}$$

$$DFI = 10^{SI}$$

where the calcium and carbonate concentrations are expressed in mg/l as $CaCO_3$, and K_s' is the $CaCO_3$ solubility product, corrected for ionic strength and temperature. This index is a ratio of the actual ion product to that which would exist during equilibrium conditions. Values greater than 1.0 indicate a tendency for deposition of $CaCO_3$ while values lower than 1.0 indicate a $CaCO_3$ dissolution condition.

The Casil Index is a modification to the calcium carbonate solubility indices that accounts for the effect of other parameters for soft waters. It is defined as

$$CI = Ca + Mg + HSiO_3 - \frac{Anions}{2}$$

where each concentration is expressed in milliequivalents per liter. Negative values are considered indicators of very corrosive water, values between 0 and 0.1 indicate slightly corrosive waters, and values above 0.1 indicate noncorrosive conditions.

The aggressive index was formulated to determine the quality of water that can be transported through asbestos-cement pipe without adverse structural effects. It, however, does not incorporate temperature or TDS effects nor does it indicate the tendency of the pipe to release fibers or allow $Ca(OH)_2$ leaching.

Aggressive Index = pH + log (AH) where

A = total alkalinity as mg/l $CaCO_3$

H = calcium hardness as mg/l $CaCO_3$

According to the AI, values greater than 12 define a nonaggressive water; values less than 10 define a highly aggressive water; and values between 10 and 12 define a moderately aggressive water.

The Riddick Corrosion Index is an empirically based formula that weighted several corrosion-influencing factors including dissolved oxygen, chloride ion concentration, noncarbonate hardness, and silica. The Riddick Index is

$$RCI = \frac{75}{Alk} [CO_2 + \frac{1}{2} (Hardness - Alk) + Cl^- + 2N] \left(\frac{10}{SiO_2}\right) \left(\frac{D.O. + 2}{Sat.D.O.}\right) (12)$$

where CO_2 is expressed as mg/l $CaCO_3$

Hardness is expressed as mg/l $CaCO_3$

Cl^- is the chloride ion concentration as mg/l

N is the nitrate ion concentration as mg/l

D.O. is the dissolved oxygen as mg/l

Sat.D.O. is the saturated oxygen value as mg/l.

The results are interpreted as

0 - 5	extremely noncorrosive
6 - 25	noncorrosive
26 - 50	moderately corrosive
51 - 75	corrosive
76 -100	very corrosive
> 100	extremely corrosive.

The values obtained correlated well with the soft waters in the eastern part of the U.S., but not to the harder waters found in the middle states.

A more recent attempt to index the corrosivity of waters resulted from a combination of the ratio:

$$\frac{(Ca^{++})\ (HCO^-)^2}{(CO_2)}$$

which represents the $CaCO_3$ precipitation equilibrium in the reaction:

$$Ca^{++} + 2HCO_3^= \rightleftharpoons CaCO_3(s) + CO_2(g) + H_2O$$

and the Larson Index. When corrected for low hardness cases, the result is:

$$Y = AH + B[Cl^-] + [SO_4^=] \exp(-\frac{1}{AH}) + C$$

where $A = 3.5 \times 10^{-4}$

$B = 0.34$

$C = 19.0$

$$H = \frac{(Ca^{++})\ (HCO_3^-)^2}{(CO_2)}$$

$[Cl^-]$, $[SO_4^=]$, $[Ca^{++}]$, $[CO_2]$ are expressed in ppm

$[HCO_3^-]$ is expressed in ppm as $CaCO_3$

A correlation between this index and the scale formed by waters of that constituency is shown in Table 7. The scale was quantified by impedance measurements and only three samples were analyzed. However if substantiated, this index would indicate that the chloride and sulfate concentrations, while conventionally regarded as corrosive factors, may actually assist in the crystal growth of calcium carbonate and resultant pipe protection.

A summation of the indices is presented below.

TABLE 2. CORROSION INDICES
(numbers in parenthesis refer to corrosion indices bibliography, symbols are explained in the test)

Langelier Saturation Index (9, 10)

$$S.I. = pH - \{(pK_2' - pK_s') + pCa + pAlk\}$$

Ryznar Stability Index (17)

$$R.I. = 2pH_s - pH$$

Larson Index (11, 12)

$$L.I. = \frac{Cl + SO_4}{Alk}$$

Driving Force Index (14)

$$DFI = (Ca^{++}) \times (CO_3^{=})/K_s' \times 10^{10}$$

Casil Index (13)

$$C.I. = Ca + Mg + HSiO_3 - \frac{Anions}{2}$$

Aggressive Index (15)

$$A.I. = pH + Log[AH]$$

Riddick Corrosion Index (16)

$$R.C.I. = \frac{75}{Alk}\left[CO_2 + \frac{1}{2}(Hardness - Alk) + Cl + 2N\right]\left(\frac{10}{SiO_2}\right)\left(\frac{D.O. + 2}{Sat.D.O.}\right)(12)$$

Feigenbaum, Gal-or, Yahalom combination (8)

$$Y = AH + B([Cl^-] + [SO_4^{=}])\exp\left(-\frac{1}{AH}\right) + C$$

GENERAL CORROSION BIBLIOGRAPHY

1. Butler, G. and H. C. Ison, Corrosion and Its Prevention in Waters, Reinhold, New York (1966).

2. Fontana, M. G., and N. D. Greene, Corrosion Engineering, McGraw-Hill, New York (1978).

3. Larson, T. E., Corrosion by Domestic Waters, Illinois State Water Survey, Urbana, Bulletin 59 (1975).

4. Speller, F. N., Corrosion Causes and Prevention, McGraw-Hill, New York, (1951).

5. Uhlig, H. H. (ed.), The Corrosion Handbook, John Wiley & Sons, New Yo (1948).

6. Uhlig, H. H., Corrosion and Corrosion Control, John Wiley, New Yor (1963).

CORROSION INDICES BIBLIOGRAPHY

7. DeMartini, F. E., "Corrosion and the Langelier Calcium Carbonate Saturation Index," JAWWA, Vol. 30, No. 1, pp 85-111.

8. Feigenbaum, C., L. Gal-or, and J. Yahalom, "Microstructure and Chemical Composition of Natural Scale Layers," Corrosion, Vol. 34, No. 2, pp 65-70, 1978.

9. Langelier, W. F., "The Analytical Control of Anti-Corrosion Water Treatment," JAWWA, Vo. 28, No. 10, pp. 1500-1521, 1936.

10. Langelier, W. F., "Chemical Equilibria in Water Treatment," JAWWA, Vol. 38, No. 2, pp 169-179, 1946.

11. Larson, T. E., and F. W. Sollo, "Loss in Water Main Carrying Capacity," JAWWA, Vol. 59, p 1564, 1967.

12. Larson, T. E., "Corrosion by Domestic Waters," Bulletin 59, Illinois State Water Survey, Urbana, 1975.

13. Loschiavo, G. P., "Experiences in Conditioning Corrosive Army Water Supplies in New England," Corrosion, Vol. 4, pp 1-14, 1948.

14. McCauley, R. F., "Controlled Deposition of Protective Calcite Coatings in Water Mains," JAWWA, Vol. 52, 1960.

15. Millette, J. R., A. F. Hammonds, M. F. Pansing, E. C. Hanson, and P. J. Clark, "Aggressive Water: Assessing the Extent of the Problem," JAWWA, Vol. 72, No. 5, 1980.

16. Riddick, T. M., "The Mechanism of Corrosion of Water Pipes," Water Works and Sewerage, p 133, 1944.

17. Ryzner, J. W., "A New Index for Determining Amount of Calcium Carbonate Scale Formed by Water," JAWWA, Vol. 36, 1944.

3. Materials Used in the Water Works Industry

A variety of materials are used by the water works industry for the construction of facilities for treatment, storage, and distribution of potable water supplies. The majority of materials are used for pipes and piping and for water storage or pressure tanks. For many small installations, no treatment facilities exist and the water utility facilities consist of only pumps, pipelines, and storage/pressure tanks.

The various materials used by the water works industry are identified and briefly described in this section. Emphasis is placed on those materials used for pipes and piping and for water storage as any corrosion control regulations or utility programs will be primarily governed by the performance of these facilities and materials. Materials used for the construction of water treatment facilities are essentially the same as those used for pipelines and storage tanks and, therefore, are not neglected from this presentation.

PIPES AND PIPING

Pipes used in the water works industry are categorized under four classifications, excluding household plumbing. These four classifications are transmission lines, distribution mains, service lines, and in-plant systems.

Transmission lines are those pipes used to transport water from the water resource to the treatment facilities or finished water from the treatment facilities to a community distribution system. These pipes can be significantly large and occasionally a tunnel is required if the maximum size pipe available is insufficient for design. Transmission lines are usually designed for gravity flow to avoid pumping costs and to reduce line pressures. Design flow velocities should not exceed 5 fps, but sometimes range from 12 to 15 fps. Factors which should be considered when selecting a particular material for a transmission line are corrosion resistance, structural qualities, hydraulic characteristics, installation and field conditions, and economics.

Distribution systems are those facilities used to carry water from the transmission lines and distribute it throughout a community. The distribution system includes a network of pipelines or mains, distribution reservoirs, elevated storage tanks, booster stations, and valves. Components of the distribution system include arterial mains, distribution mains, and a valve system. Arterial mains, sometimes called trunk mains or feeders, are used to connect transmission lines to the distribution lines. Arterials are normally placed in a loop arrangement to avoid dead ends. Distribution lines are connected to the arterial loop forming a grid system. These lines are used to serve communities or commercial areas and hook up to individual service lines.

Materials commonly used for transmission lines and distribution mains are asbestos cement, cast iron and ductile iron, concrete, plastic, steel, and wrought iron (2). The advantages and disadvantages of the use of these materials are presented in Table 3. Aluminum is also used for pipelines, but to a lesser extent (1). Plastic and wrought iron are more commonly used in service lines and in-house plumbing systems (4).

Service lines are small diameter pipes that connect the consumer to the distribution main. The selection of a particular material for a service line is influenced by required size, durability, water characteristics, corrosion resistance, material availability, ease of installation, and economics. These criteria as well as corrosive tendencies of the waters in the specific area are usually reflected in the local plumbing codes. Because of its excellent physical characteristics, lead was the earliest material used for service lines. However, the use of lead is now being questioned because of its cost and its tendency to dissolve in soft waters of low pH. Copper is now more frequently selected for service lines and approximately 50 percent of the water utilities in the U.S. use copper exclusively. However, plastic pipe is becoming more popular (4).

Minimum size service lines range from 3/4 to 1.0 inches in diameter. For larger residences with numerous baths, minimum size service lines will range from 1-1/4 to 1-1/2 inches in diameter. Approximately one-half of all service lines in the U.S. are owned by utilities and the other one-half are owned by the customers. However, approximately two-thirds of all service lines are installed by utilities (4).

The most commonly used piping materials for service lines are asbestos cement, brass, cast iron, copper, galvanized iron, lead, plastics, steel, and wrought iron. The hydraulic flow characteristic of all these piping materials is good when initially installed. These flow characterictics generally remain good for asbestos cement, copper, lead, and plastic. These materials are listed and briefly characterized in Table 4.

Flexible materials used for service lines are usually connected directly to the corporation cock on the main and to the stop valve within the household. Nonflexible materials require the use of a "gooseneck" connection to the corporation cock and possibly some type of flexible connection to the household plumbing system. Goosenecks are available in lead, copper (if permitted by local plumbing codes), and flexible plastic (4).

Every type of piping material previously discussed is used for in-plant piping systems. Other materials used include glass and rubber. Glass and rubber are not usually used for conveying potable water within the plant, but rather for other in-plant operational functions. Pipe materials used for

TABLE 3. SEVERAL MATERIALS USED FOR TRANSMISSION AND DISTRIBUTION LINES (4)

Materials	Available Size Diam. (in.)	Advantages	Disadvantages
Asbestos Cement	4-36	Corrosion resistant; good flow characteristics; light weight; easy handling; low maintenance.	Low flexural strength in small sizes; more subject to impact damage; difficult to locate underground.
Cast Iron (cement-lined)	2-48	Durable and strong; good corrosion resistance; easily tapped; flow characteristics good.	Subject to electrolysis and attack from acid and alkaline soils; heavy to handle.
Ductile Iron (cement-lined)	4-54	Durable, strong, high flexural strength; lighter weight than cast iron; greater carrying capacity for same external diameter; fracture resistant; easily tapped.	Similar to cast iron.
Concrete (reinforced)	12-168	Durable with low maintenance; good corrosion resistance; flow characteristics good; resists backfill and external loads.	May deteriorate in alkaline soil, if cement type is improper, or in acid soil if not protected.
Concrete (prestressed)	16-120	Durable, low maintenance; good corrosion resistance; good flow characteristics; resists backfill and external loads.	Same as above.

(Continued)

TABLE 3 (Continued)

Materials	Available Size Diam. (in.)	Advantages	Disadvantages
Steel	4-120	Light weight and easily installed; high tensile strength; low cost; good hydraulically when lined; adapted to locations where some movement may occur.	Subject to electrolysis; external corrosion in acid or alkaline soil; poor corrision resistance unless properly lined, coated, and wrapped; low resistance to external pressure in larger sizes; air-vacuum valves imperative for large sizes; subject to tuberculation when unlined.
Plastics			
ABS	1/2-12	Smooth interior surface minimizes pumping losses; chemically inert; corrosion resistant, non-reactive with water; light weight.	Jointing sometimes difficult; tendency to creep; brittle at low temperatures.
PE	1/2-6		
PVC	1/2-16		
Wrought Iron	1/4-30	Tough; ductile; malleable; weldable and corrosion resistant.	Generally rough surface; hot dip galvanizing or non-metallic coating generally required.

TABLE 4. MATERIALS USED FOR SERVICE LINES (4)

Material	Size range (in.)	Comments
Asbestos-Cement	3,4,6	Corrosion resistant; not available below 3-in.; slip coupling joints.
Brass	1/2 to 6	Long life under normal conditions; corrodes in acid soils; uses threaded coupling joints; requires gooseneck connection.
Cast iron	2,3,4,6	Corrosion resistant when lined and coated; not available in small diameters; rigidity and short length require joints and gooseneck connection.
Copper	1/2 to 6	Direct connection to mains; corrosion resistant; dissolves in soft water with high CO_2 content.
Galvanized iron	1/2 to 3	Not highly resistant to corrosion; requires threaded joints and gooseneck connection.
Lead	3/8 to 2	Direct connection mains; corrosion resistant except in soft waters with high CO_2; some tendency to creep or crack unless properly formulated.
Plastics		
ABS	1/2 to 6	
PE	1/2 to 2	
PVC	1/2 to 6	
Steel	1/2 to 6	Available in three grades; strong, extra-strong, double extra-strong; not resistant to corrosion unless cement-lined.
Wrought iron	1/2 to 6	Same comments apply as for steel.

in-plant systems for transporting potable waters have the same design and are manufactured by the same process as previously discussed.

The extent of use of various materials for piping by water utilities serving more than 2500 persons in the U.S. as of 1975 and for 1975 was surveyed and compiled by Scott and Caesar and is summarized in Table 5 (3). Approximately 75 percent of all water main piping in the U.S. as of 1975 was cast iron and it accounted for approximately 46 percent of total pipe installed in 1975. Asbestos cement pipe and steel pipe accounted for approximately 13 and 6 percent, respectively, of all water mains in place by 1975. The use of steel pipe appears to be declining as it accounted for only approximately 3.4 percent of all water pipes installed in 1975.

STORAGE TANKS

Materials used for the construction of water storage tanks are wood, fiberglass, concrete, and steel. Aluminum is sometimes used for construction of storage tank roofs. The type of material selected is generally determined by the required capacity, the specific use, and economics. Concrete is considered economical for large storage tanks with capacities ranging between 1.25 and 5 million gallons. Concrete tanks are estimated to constitute 15 percent of all new water tanks (2).

Steel tanks are considered more economical for tank capacities smaller than 1.25 million gallons, and they are more adaptable for elevated use when natural relief or topography is not available (2). However, steel tanks are often lined with a protective coating, such as a coal tar, to minimize corrosion. Consequently, the maintenance costs of steel tanks is generally higher than that for concrete tanks (2).

Currently there are approximately one-half million steel water storage tanks in the U.S. ranging in capacity from 50,000 gallons to 10 million gallons. It is estimated that over 1000 new steel water tanks are constructed each year (2).

Wooden tanks generally have capacities ranging in size from 25,000 to 50,000 gallons, but can be as large as 250,000 gallons. Because these tanks have limited capacities and are prone to leak, their use is limited (2).

Fiberglass-reinforced plastic tanks are also used for storage of potable water, but to a limited extent owing to problems encountered in field construction. Because of these construction problems, most of these tanks are limited in capacity to that which can be shop fabricated and transported. Fiberglass tanks have been shop fabricated up to capacities of nearly 50,000 gallons (2).

Steel storage tanks are usually lined with a water-impervious coating. Traditionally, these coatings have been coal tar based enamels, but recent difficulties in controlling fumes during application and reports that toxic

TABLE 5. PIPE USED IN U.S. WATER SUPPLY DISTRIBUTION SYSTEMS BY UTILITIES SERVING OVER 2,500 PERSONS (3)

TYPE OF PIPE	MILEAGE IN PLACE (beginning of 1975)	% OF TOTAL	MILEAGE INSTALLED (1975)	PERCENT OF PIPE MILEAGE IN PLACE AT BEGINNING OF 1975 BY DIAMETER			
				Under 6"	6"-12"	13"-24"	Over 24"
Cast Iron	481,816	75.29	6,847	15.7	76.7	6.6	1.0
Asbestos Cement	83,871	13.11	3,743	9.7	86.3	4.0	0.1
Steel	37,852	5.91	505	53.4	29.5	10.7	6.4
Reinforced Concrete	10,083	1.58	517	0.2	4.1	43.4	52.3
Plastic	6,981	1.09	1,826	62.0	37.3	0.6	0.1
Ductile	7,498	1.17	1,388	*	*	*	*
Galvanized Wrought Iron	2,364	0.37	3	*	*	*	*
Wood	1,246	0.19	*	*	*	*	*
RCP Steel Cylinder	652	0.10	8	*	*	*	*
Black Galvanized Iron	431	0.07	*	*	*	*	*
Copper	312	0.05	*	*	*	*	*
All others and unidentified	6,879	1.07	96				

Source: Scott and Caesar, 1975 (3)

*Not Specified

substances may be introduced to the waters by their use have resulted in the use of epoxy and vinyl paints (2). Various estimates report that 50-90% of all new water tanks are lined with vinyl and 10-50% are lined with epoxy (2).

REFERENCES

1. Booth, F. F., Murray, G. A. W. and H. P. Godard, "Corrosion Behavior of Aluminum in Fresh Waters with Special Reference to Pipeline," Br. Corros. J., Vol. 1, No. 2, 1965, pp. 80-86.

2. Goldfarb, A. S., Konz, J., and Pamela Walker, "Coal Tar Based Materials and Their Alternatives," Interior Coatings in Potable Water Tanks and Pipelines, The Mitre Corp., Mitre Technical Report MTR-7803, U.S. EPA Contract No. 86-01-4635, January 1979.

3. Scott, J. B. and Adelaide E. Caesar, Survey of Water Main Pipe in U.S. Utilities Over 2500 Population, Morgan Grampian Publishing Co., Pittsfield, Massachusetts, 1975.

4. Symons, G. E., Ph.D., "Water Systems, Pipes and Piping, Part 1/Piping Systems Design," Water and Wastes Engineering, Manual of Practice Number Two, Vol. 4, No. 5, May 1976, pp. M3-M50.

4. Corrosion Characteristics of Materials Used in the Water Works Industry

The corrosive behavior of specific materials when subject to the environmental conditions of potable water systems is presented in this section. This information is compiled primarily from published results of laboratory and field research. In general, most studies reviewed are consistent and in agreement in identifying the conditions of service and water quality characteristics which initiate and maintain the corrosion or deterioration of a specific material. However, specific data presented by various investigators is sometimes inconsistent or in disagreement. This inconsistency usually results f m variations in the conditions of testing and/or reporting. It is also noted that the literature often fails to fully describe or present the details of the testing procedures which are often critical for assessing test results.

The corrosion behavior of each material is discussed independently and the presentation format for each material is dictated by the information available in the literature. Emphasis is placed on presenting numerical results of various corrosion testing and monitoring as this data serves as the basis for considering specific corrosion control alternatives.

IRON-BASED MATERIALS

Iron-based materials are among the most common piping materials. They are also subject to a variety of corrosion mechanisms that may occur in potable water systems. This subsection discusses the various iron-bearing metals that may be encountered.

Corrosion of Iron

The corrosion behavior of steel and cast iron materials in potable water environments is highly complex. Many factors can be involved and are often interrelated. The effects of several factors can vary from beneficial to conducive to greater corrosion, depending on the specific situation. It is often difficult to say what the main factor controlling the corrosion of steel is, due to these subtle relationships. The following discussion outlines the basic corrosion mechanism of iron and then discusses these contributing factors.

The corrosion of iron and steel in waters is basically electrochemical in nature. The actual metal loss is due to an oxidation of iron atoms on the metal surface to give ferrous ions which can go into solution and electrons which stay with the metal:

$$Fe \rightarrow Fe^{2+} + 2e^- \qquad (1)$$

In order for this process to proceed, the electrons must be taken up by a reduction process which can take place on another part of the surface. Usually this complementary process is the reduction of dissolved oxygen (Eq. 2) or the reduction of hydrogen ions or related species (Eq. 3).

$$O_2 + 2H_2O + 4e^- \rightarrow 4OH^- \qquad (2)$$

$$2H^+ + 2e^- \rightarrow H_2 \qquad (3)$$

The oxidation and reduction reactions are parallel-coupled events which must proceed at identical rates. The overall corrosion rate is limited by the slower of the two coupled reactions. The oxidation reaction (Eq. 1) is rapid in most media. The rates of the reduction reactions are limited in natural waters by reactant concentrations, sluggish electrochemical kinetics, or a combination of factors. Thus the overall uniform corrosion rate is normally controlled by the rate of the reduction reaction, as amplified below.

For corrosion to occur, the difference between the electrochemical potentials for the oxidation reaction (occurring at anodic sites) and reduction reactions (occurring at cathodic sites) must be such that the overall free energy change drives the reactions as written. This potential difference ultimately appears as a driving force which can be viewed as being divided between the two reactions in such a way that the inherently slower reaction receives the larger share of driving force. For iron corrosion with hydrogen ion or water reduction, this overall driving force is relatively small and decreases with increasing pH. On the other hand, the driving force for iron corrosion with oxygen reduction is very large. The reduction of oxygen is a complex electrode process which is inherently quite slow. The detailed mechanism is not well known. In spite of the low inherent rate, the driving force is so large that the reduction can be fast, and transport of O_2 to the iron surface often becomes the rate limiting process.

The metal loss will be uniform or general over the surface as long as the oxidation and reduction sites constantly shift in location and the fractional coverage of sites is roughly the same. If an oxidation site becomes small, fixed, and surrounded by a much larger reduction area, then localized corrosion such as crevice corrosion or pitting can result. This localization of an oxidation site can be caused by a variety of factors such as local breakdown of a protective oxide film, presence of a crevice, a break in a deposit on the metal, and so on. Uniform corrosion is favored by a clean metal surface and ample supply of cathodic reactant. Thus in acidic solutions where oxide films are not stable and the concentration of hydrogen ions is high, steel generally corrodes uniformly. Localized corrosion is often favored by conditions which reduce the rate of uniform corrosion. Factors influencing the type of corrosion are also related to the effects of oxygen concentration, pH, flow rate, temperature and electrolyte concentration, which in turn do not operate independently. These factors are taken up separately below, but their interdependence should be kept in mind.

It is tempting to attribute a majority of the factors involved in corrosion of iron in aerated near neutral pH waters to control by oxygen transport (cathodic control). Stumm strongly suggests, however, that at long exposure duration, formation of protective films on iron and other events may make the rate of the anodic iron dissolution process an important factor (97). Although cathodic control appears to explain a large number of environmental effects, the latter possibility should also be kept under consideration.

Effect of Dissolved Oxygen--

Dissolved oxygen plays a key role in corrosion or iron in natural waters, but its effects can be conflicting and partially dependent on other environmental factors. In near neutral pH waters at ambient temperature, dissolved oxygen provides the reduction reaction (Eq. 2) which sustains the corrosion of iron. However, oxygen also plays a role in formation of semi-protective iron oxide films on the metal, and the more protective films are formed at higher oxygen concentrations. The presence of oxygen also appears to be necessary for formation of protective layers on steel by calcium carbonate deposition (7). Once these films are formed, however, oxygen provides the main driving force for initiation of pitting (leading to tuberculation) or other forms of localized corrosion.

The first part of this discussion is for conditions in the absence of calcium carbonate or other external inhibiting species. At sufficiently high pH values, iron oxide or hydroxide layers can be formed. The first of these is probably ferrous hydroxide which can be formed by an overall reaction such as Eq. 4.

$$Fe + H_2O + 1/2\ O_2 \rightarrow Fe(OH)_2 \qquad (4)$$

This solid is often found next to the metal surface and can act as a diffusion barrier to oxygen. Further oxidation of this product yields hydrous ferric oxide which comprises most of ordinary rust. An intermediate oxidation stage, $Fe_3O_4 \cdot nH_2O$ often forms as a layer between the ferric and ferrous compounds.

In the absence of dissolved oxygen, the corrosion rates for both pure iron and steel becomes negligible in near neutral pH water at room temperature. Corrosion rates may be high when the metal is first exposed to air-saturated water, but the iron oxide films formed over a period of a few days act as a barrier to diffusion of oxygen to the surface and a steady state corrosion rate is obtained. This steady state rate is proportional to oxygen concentration, as shown in Figure 1,since the oxygen diffusion rate is proportional to its concentration. An oxygen concentration of about 6 mg/l corresponds to air-saturated water. At still higher oxygen concentrations, the uniform corrosion rate of mild steel may decrease abruptly, as shown in Figure 2. This effect is apparently due to passivation of the iron which involves either the oxidation of the normal ferrous hydroxide layer to one which has better protective properties or the formation of a thin chemisorbed oxygen layer on the metal surface. More oxygen is required in waters

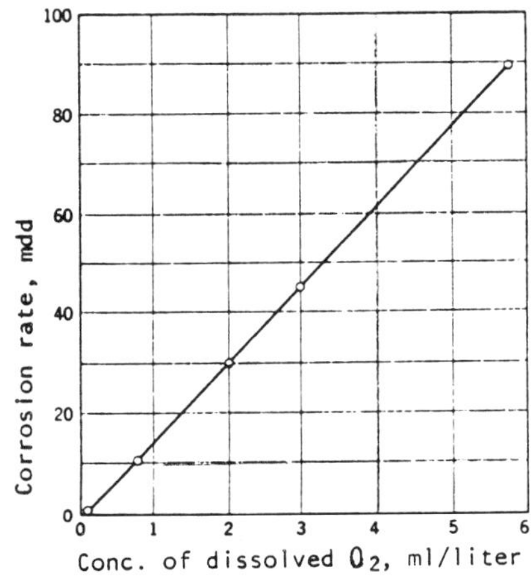

Figure 1. Effect of O_2 concentration on corrosion of mild steel in slowly moving water containing 165 ppm $CaCL_2$, 48-hour test, 25°C (107).

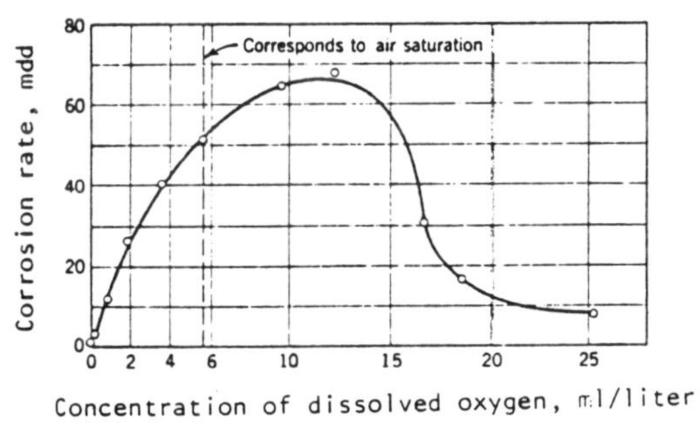

Figure 2. Effect of O_2 concentration on corrosion of mild steel in slowly moving distilled water, 48-hour test, 25°C (107).

containing chloride ions and passivation of mild steel connot occur if the chloride concentration is high enough. The breakdown of passivity in the presence of moderate amounts of chlorides (more than about 20 ppm, see below) is often accompanied by severe pitting or crevice corrosion.

The effect of solution flow rate on mass transport of oxygen to the corroding surface can also give rise to diverse effects. At moderate oxygen concentrations and flow rates, increasing the flow rate increases the corrosion rate due to the increase in amount of oxygen transported to the surface. At higher flow rates, the surface oxygen concentration can become high enough to cause passivation, provided the chloride content is not too high. These effects are shown in Figure 3. Still higher flow rates, over 15 ft/sec, can greatly accelerate corrosion by erosion of the protective films, combined with fast transport of oxygen to the surface by turbulent flow. At the other extreme, stagnant conditions are usually most conducive to pitting and other forms of localized corrosion. The threshhold for chloride effects on passivity is ill defined. As little as 20 ppm chloride may cause breakdown and pitting but the threshhold may be higher depending on solution composition (vida infra).

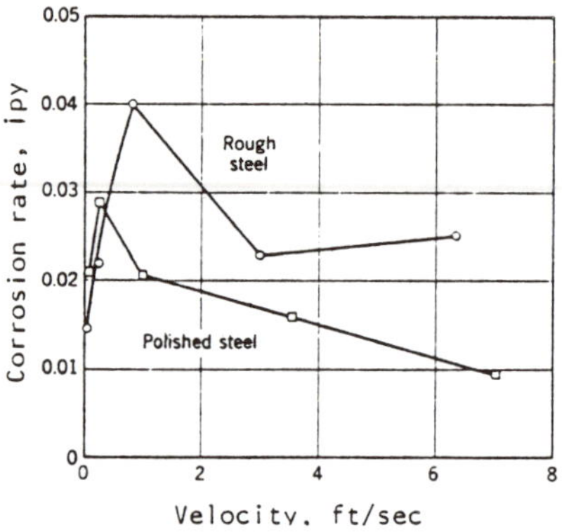

Figure 3. Effect of velocity on corrosion of mild steel tubes containing Cambridge water, 21°C, 48-hour tests (87).

Effect of pH--

Conflicting results and conclusions have been obtained for the effect of pH on corrosion of steel and cast iron in potable water environments. This is probably because a number of effects are possible in the range of interest, from about pH 6 to pH 10, and both the mode of corrosion as well as degree of corrosion and structure of surface films may be affected. In addition, all attempted correlations are done with respect to the bulk pH of the liquid phase, while the actual pH at the material surface can be affected by the corrosion reactions, surface iron hydroxide layers, other precipitated layers, other chemical species in solution, and solution flow rate.

In the moderately acidic region, corrosion is generally uniform and the rate increases rapidly with decreasing pH. The onset of this region is about pH 4 to 5 if no other acidic species other than hydrogen ions are present, and from about pH 5 to 6 if acidic species such as dissolved carbon dioxide are present in appreciable quantities. Corrosion is accelerated in these regions because hydrogen evolution can occur utilizing fairly concentrated reagents (hydrogen ions or protonated anions). In addition, the oxide films which act as diffusion barriers are dissolved. The latter effect also offers greater access of oxygen to the metal surface which can be more important than hydrogen evolution at the higher pH end of this range. Both hydrogen evolution and oxygen reduction tend to raise the pH at the metal surface and stabilize oxide films. The presence of acidic species such as dissolved CO_2 tends to mitigate this effect since the total acidity of such a system is greater than that of a completely dissociated acid at the same pH.

In the near neutral pH range of most natural waters, oxygen transport can in some cases dominate changes in pH, although in this region many complex pH effects have been noted. The simplest type of behavior for corrosion of steel by aerated water at 22°C is shown in Figure 4 (112). For this case, in which either NaOH or HCl were added to aerated water to adjust the pH, the corrosion rate is constant from about pH 4 to 10. This is reasonable behavior if the rate depends only on diffusion of oxygen to the metal surface through a diffusion barrier of ferrous hydroxide or hydrated ferrous oxide which is continuously renewed by the corrosion process. Also, in this bulk solution pH region, the iron surface is always in contact with the saturated ferrous hydroxide solution which should maintain a pH of about 9.5. Similar results were obtained at 40°C except the constant corrosion rate plateau (18 mpy) extends from pH 4.5 to 8.5. When the pH is adjusted with CO_2 instead of HCl, the rapid increase in corrosion rate occurs at pH 5.4 instead of 4.1 (112).

Later studies have occasionally shown a maximum in corrosion rate occurring in the pH range from about 6 to 9. Examples are given by Eliassen, et al, along with controlled corrosion tests using synthetic "average" water (84 ppm hardness, 57 ppm methyl orange alkalinity, and a pH after aeration of 8.05, pH_s = 8.28) (25). These tests were done for steel pipe as a function of flow rate and duration of exposure. A maximum in the corrosion rates was observed in the pH range 6.5 to 7.5 for short-term tests (7-14 days exposure) at flow velocities of 0.33 and 1.0 fps. At zero velocity a much lower corrosion rate was observed which was constant from pH 4 to 11. The authors note that their short-term results are quite similar to others in the literature which show corrosion maxima with pH (25). However, their "long-term" tests (30-40 days) do not show maxima and tend to agree with the early results of Whitman (112). These results are illustrated in Figure 5. These authors also observed very severe pitting at a pH of 6.5, slight pitting at pH 8.0 to 10.0 and no pitting at pH 4.0 to 5.5. They generally agree with the model of pH effects proposed by Whitman, but point out that variations may be caused by changes in mode of corrosion and nature of corrosion products formed as a

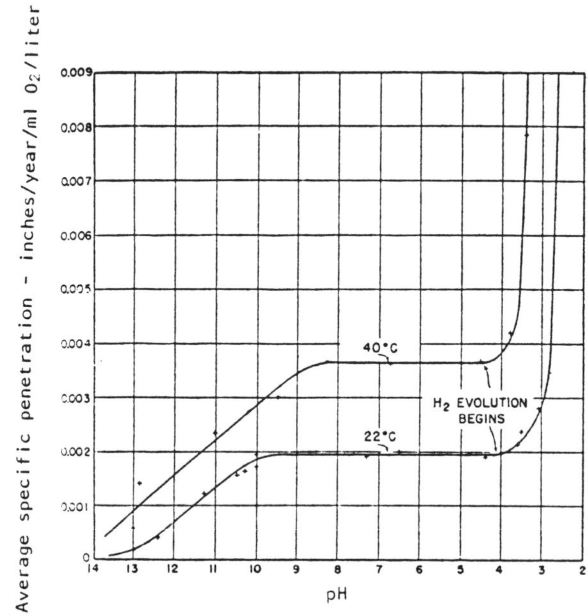

Figure 4. Effect of pH on corrosion of mild steel (112).

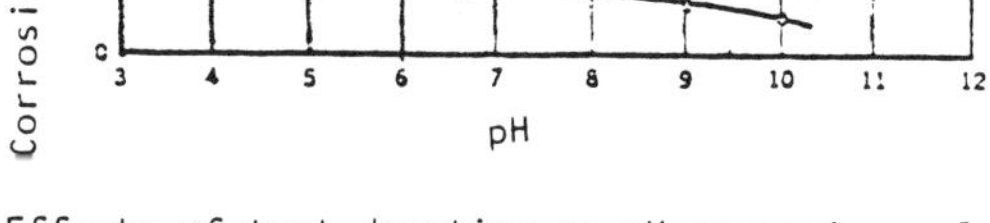

Figure 5. Effects of test duration on pH-corrosion relationship (25).
A. 7-14 day tests.
B. 30-40 day period.

function of pH and also time. The corrosion products they observed were gelatinous and loosely attached to the metal in acid waters, but hard, raised, and well attached at pH values greater than 8.0.

Some additional examples of more complex pH behavior have been given by Larson (57, 59, and 60) and by Fontana (32). Larson's results were obtained with waters containing bicarbonate and chloride. In one case, at a Cl^-/alk mole ratio of 0.4, a sharp maxima at a pH of 8.0 is seen which increases with duration of exposure, in contrast to Eliassen's results (60). These results are shown in Figure 6. In another case with a higher chloride content, the corrosion rates increase by a factor of about 3 in going from pH 7 to pH 8.5 (at a Cl^-/alk ratio of 1.0). Fontana obtained results for a sample maintained in a high flow rate location (39 ft/sec) in distilled water at 50°C. A sharp peak in corrosion rate is observed at pH 8 which is about 10 times that observed at pH 6 or pH 10. The effect was attributed to an atypical reaction product scale of granular Fe_3O_4 which formed at pH 8. In regions of low attack the products were $Fe(OH)_2$ and $Fe(OH)_3$ which provided better barriers to diffusion of oxygen and ions. The results are said to be supported by power plant experience showing greater attack at pH 8 as compared to slightly lower values (32).

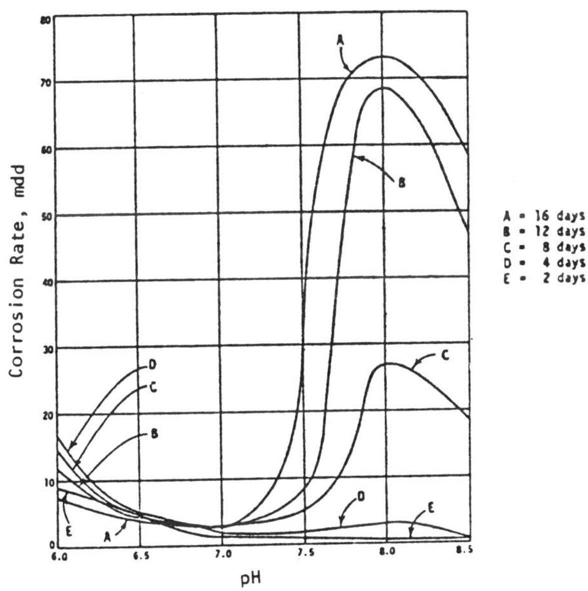

Figure 6. Effect of pH on corrosion rate at chloride-alkalinity ratio of 0.4; duration: A, 16 days; B, 12 days; C, 8 days; D, 4 days; E. 2 days (60).

An increase in corrosion rate with increasing pH (from pH 6.9 to 8.6) has also been reported for ground cast iron samples in a variety of natural and synthetic waters after 50 days exposure (97). The author attributes this to a decrease in buffer capacity of the electrolyte (for a given alkalinity) with increasing pH and to the pH dependence of relative local anode/cathode areas. These views are discussed in the calcium carbonate protection section below.

Some relatively recent electrochemical results may give additional insight on the pH behavior. The fundamental corrosion reactions on iron have been mainly studied in deaerated solutions at a pH less than 5 under conditions where no solid reaction products are formed. The oxidation reaction (Eq. 1) appears to proceed through intermediates involving hydroxyl ions in such a way that its individual rate increases with increasing pH from about pH 1 to pH 5 (46). Hydroxyl ion catalysis at near neutral pH might help explain the disparate experimental results in natural waters.

Effect of Dissolved Salts--

Dissolved salts (for example, ions such as Na^+, Cl^-, $SO_4^=$, Ca^{++}, Mg^{++}, HCO_3^-, $CO_3^=$, etc.) can have a number of effects, some of which depend only on the general effect of ions, while others depend on the individual chemical species. Examples of the first class of effects are given below.

- Increasing salt concentration increases solution conductivity which can have several effects.
- Increasing salt concentration generally decreases the equilibrium concentration of dissolved oxygen and CO_2.

Contrary to early predilections, solution conductivity itself has little effect on most modes of uniform or localized corrosion. This is because the local anodes and cathodes are so close together that the resistance offered by the solution is much less than equivalent electrochemical reaction rate resistances. Solution conductivity can, however, affect the range over which the effects of attack due to galvanic coupling of dissimilar metals is extended. Attack can extend for example, from about 1 cm from the joint in soft waters to more than 10 cm in water with significant dissolved salt content. Galvanic corrosion is discussed in a later section.

Changes in solution conductivity can have more subtle and significant effects. Uhlig proposes that moderate increases in conductivity (by dissolved NaCl) from that of very soft or distilled water can lead to increased corrosion rates due to formation of a less protective $Fe(OH)_2$ film (107). The corrosion of iron in aerated solution as a function of NaCl concentration is shown in Figure 7. The film is less protective since it is formed further away from the surface than in less conductive waters. This in turn is due to coupling of local anodic (source of Fe^{2+}) and cathodic (source of OH^-) areas at greater distances in the more conductive solution. The initial increase in corrosion rate might also have to do with some specific effects of chloride and a number of postulates on these effects have been offered (31).

Specific effects of chloride are discussed below. There is little direct evidence for either the $Fe(OH)_2$ film or chloride effect arguments.

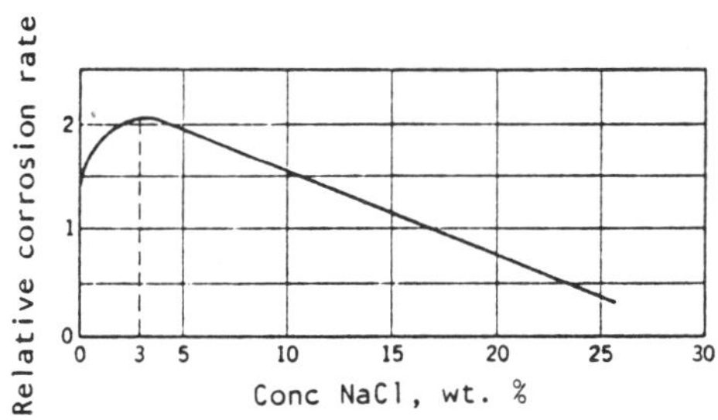

Figure 7. Effect of NaCl concentration on corrosion of iron in aerated solutions, room temperature (107).

It has recently been proposed that an increase in solution conductivity may have the effect of actually promoting a more protective coating on iron in the presence of calcium carbonate film forming precursors (28). Correlation between the electrical impedances and protectiveness of scales and a quantity related to the conductivity of the solution was found. Conductive salts facilitate protective film formation, provided that sufficient temporary hardness exists. This effect is dicussed in more detail under the calcium carbonate film section below.

The other overall effect of dissolved salts, that of decreasing the solubility of dissolved gases, is also illustrated in Figure 7. Since the dissolved oxygen content is the main controlling factor under the conditions, "salting out" of oxygen causes a decrease in corrosion rate. This effect only becomes pronounced at higher salt concentrations and is most likely unimportant in natural fresh waters.

Chloride is the most deleterious individual ionic species normally occurring in natural waters. Much of this appears to be due to the ability of chloride to promote pitting by penetration or local destruction of otherwise protective iron oxide films. It is quite difficult for a uniform, true passive iron oxide film to be formed in the presence of chloride ions. Foley lists five possible specific roles for chloride in iron corrosion (31). It is difficult, however, to quantify the relation of corrosivity to chloride concentration in natural waters. The treshold concentration of chloride, above which pitting of iron is possible, is said to be about 10 ppm (31). The role of sulfate is more nebulous, especially since sulfate does not appear to have the film piercing properties that chloride displays. Larson has done extensive studies of the effect of these salts on corrosion of steel under conditions simulating potable water environments (57, 58, 59, 60). He concludes that the corrosivity of air-saturated domestic waters depends on the following factors (60).

- The proportion of corrosive agents (chloride and sulfate ions) to inhibitive agents (bicarbonate, carbonate, hydroxide, and calcium ions).
- The concentration and degree of effectiveness of the corrosive and inhibitive species.
- The velocity of flow which affects the rates of diffusion of both types of species to the surface.

He also notes that the relative effectiveness of the species involved is not definitive, but may be influenced by one or more of the others. The interdependence of the relevant factors is also emphasized by the observation that "intermediate proportions" of corrosive to inhibitive species, which result in incomplete protection (to uniform corrosion) at a particular solution flow rate, are conducive to pitting and/or tuberculation. Representative results obtained by Larson are shown in Figure 8 (60).

Effect of Dissolved Carbon Dioxide--

The effect of dissolved carbon dioxide depends in large part on solution pH since this determines the relative amounts of "carbonic acid" (hydrated CO_2), bicarbonate, and carbonate present. The carbonic acid form is aggressive towards iron since it can serve as a relatively concentrated reactant for hydrogen evolution at a relatively high pH (23). This effect is probably not very important in aerated natural waters above a pH of about 6. Carbonic acid can also act to dissolve calcium carbonate/hydrated iron oxide films and thus remove protective diffusion barriers.

Bicarbonate is the predominant CO_2 species from about pH 7 to 10. Larson has classified bicarbonate as a mild, but effective inhibitor of steel in aerated natural waters, in the absence of calcium (60). This was a general experimental result but was not explained physically. Recent work by Davies and Burstein in concentrated bicarbonate solutions, borate buffered at pH 8.8, indicates that the anodic dissolution of iron is accelerated by bicarbonate due to formation of the complex $Fe(CO_3)_2^{2-}$ (22). Formation of solid $FeCO_3$ along with $Fe(OH)_2$ is also indicated under some conditions. Pitting is attributed to the heterogeneous nature of the surface reaction products and formation of the complex. This bulk solution is quite different from the bulk bicarbonate content of natural waters, but the results may be relevant to localized corrosion. The environment under a calcium carbonate scale on iron or in a developing pit or tubercule could in some instances have a high effective bicarbonate concentration and aid in initiation or growth of localized corrosion. The effect on uniform corrosion on relatively bare surfaces would still be expected to be minimal. Except as implied by the foregoing, carbonate ions in normal domestic water are expected to be essentially neutral except as they can act beneficially to form $CaCO_3$ films or decrease acidity.

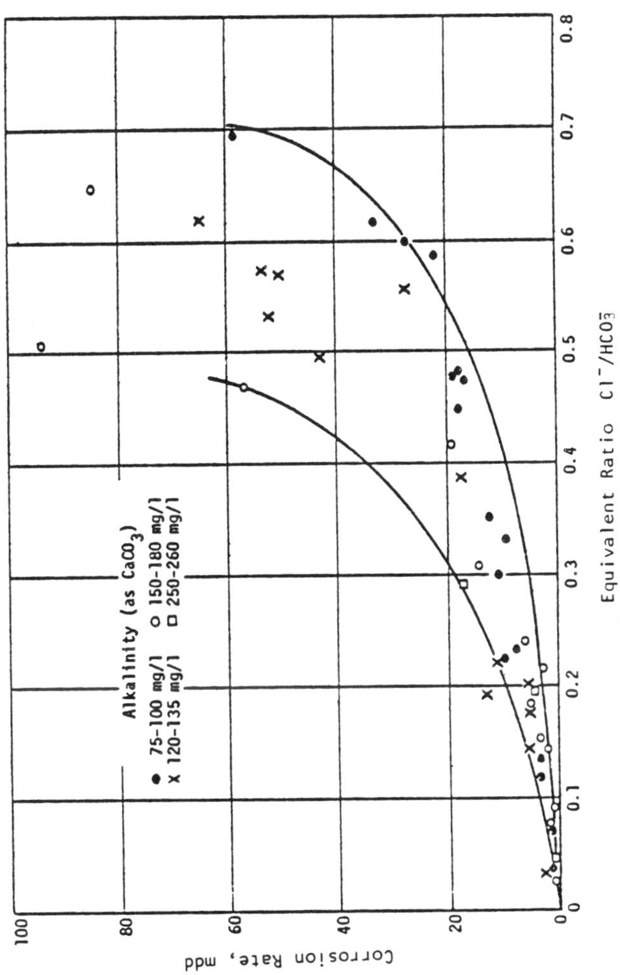

Figure 8. Effect of chloride-bicarbonate salts ratio on corrosion of mild steel (60).

Effect of Calcium--

Much of the protection, both natural and man-made, for iron-based materials in potable water environments is attributed to the formation of a calcium carbonate film or scale on the surface of the material. Presumably, this film provides a diffusion barrier to oxygen, thus further limiting the oxygen reduction rate which is usually rate controlling in natural aerated waters. Various indices have been proposed to approximate the tendency of calcium carbonate to deposit or to dissolve in natural waters, as discussed in Section 2. The actual mechanism of protection, however, is much more complicated than simple deposition of a layer of $CaCO_3$. The saturation indices are often useful as guides, but are too indirect to be applied indiscriminately. Langelier was one of the early proponents of applying a $CaCO_3$ saturation index to corrosion control and described both an index and its correlation with results obtained in New York City pipe corrosion tests (53). He also presented refinements and reviewed early applications (54). The methods do not provide a quantitative measure of the amount or rate of corrosion or $CaCO_3$ deposition. By 1954, Larson noted that water works practice indicated that the saturation index does not necessarily show corrosivity.

Inhibition by calcium carbonate appears to be intimately connected with the corrosion reactions on iron or steel. In a review of treatment methods for desalination product water, Bopp and Reed emphasize that sufficient dissolved oxygen (they quote a minimum of 5 ppm) is needed for the proper formation of protective $CaCO_3$ (+ iron oxide) films (7). Untreated product water will rapidly attack iron and other normal materials of construction of municipal systems. McCauley studied the properties of $CaCO_3$ coatings formed on cast iron under different conditions (67). In general, development of an adherent layer required the initial deposit of dense material well bonded to the metal. Even if this initial layer was very thin, a tenacious coating could be developed. The films formed in static tests produced poorly bonded mixture of calcium, ferrous carbonate and iron oxide in a porous layer of rust. Adherent, durable layers were usually formed under high flow rate conditions on corroded samples. The presence of colloidal $CaCO_3$ was beneficial. These adherent layers developed were largely hydrous ferric oxide (in the form of limonite) with 5 to 40 percent calcite ($CaCO_3$). Siderite ($FeCO_3$) was usually observed close to the metal surface in ridges (67).

Larson reports that calcium, independent of saturation index, is a mild but effective corrosion inhibitor of machined cast iron at least in the presence of sufficient alkalinity (60). The corrosion rate depended on calcium concentration, practically independent of flow velocity from 0.08 to 0.85 fps, pH, minor variations in chloride to alkalinity ratios, and presence or absence of chlorine, chloramine, and silica. It was found that a certain length of time was needed for the effectiveness of calcium to become apparent. The effect was explained on the basis of the corrosion reactions providing an environment at the surface conducive to the formation of $CaCO_3$, even though the bulk water is below saturation with respect to $CaCO_3$.

Stumm measured corrosion rates of cast iron under relatively well characterized conditions (97). Results are shown in Table 6. According to his analysis, neither the $CaCO_3$ saturation equilibrium, nor the relative amount of $CaCO_3$ deposited at the electrode surface are significant parameters of

TABLE 6. CHEMICAL COMPOSITION AND CORROSION CHARACTERISTICS OF WATERS INVESTIGATED (97)

Water Number	Type[a]	Temperature in degree C	Analysis pH	as mg per L $CaCO_3$ Ca+2	Mg+2	Alkalinity	mg per L Cl^-	SO_4-2	Si	$CaCO_3$ Deposition, in mg per cm^2 0	20 Days	50 Days	Saturation Index, in pH units	Buffer Capacity, mg per pH	Corrosion Rate, mg/dm^2 day Initial	5 days	20 days	50 days	100 days
1	Sy	10	7.1	180	0	250	14	20	-	11.0	0.35	0.43	-0.40	1.92	182	37.6	5.2	1.1	-
2	Gr	9.8	7.6	185	69	245	3	20	7	11.0	3.5	3.6	+0.07	0.074	165	45	7.7	2.0	0.5
3	Sy	10	7.1	250	0	250	14	20	-	11.0	0.24	0.31	-0.25	1.92	182	54	10.4	2.3	-
4	Gr	12	7.5	158	37	180	12	-	-	2-4	-	-	-0.20	0.81	106	54	17.6	4.6	-
5	Gr	12	7.4	203	55	228	4	-	-	9	-	0.8	-0.03	1.01	150	57.3	13.0	2.9	-
6	Su	4.7	7.7	124	24	136	2	16	2	11.5	2.6	2.7	-0.40	0.33	80	53	21.8	7.7	1.6
7	Su	6.4	8.0	99	12	103	2	-	2.5	12.2	6.5	9.2	-0.20	0.14	100	59	20.8	6.6	1.8
8	Sy	10	7.0	0	0	250	14	20	-	11.0	0	0	-	2.30	125	67.9	22.2	6.4	-
9	Sy	10	8.2	38	0	83	-	20	-	11	2.4	4.6	-0.40	0.08	111	79.4	38.5	14.2	-
10	Sy	3-5	7.0	40	8	12	-	-	-	12.5	-	-	-2.5	0.11	77	65.0	42.6	22.8	10
11	Sy	10	8.3	20	0	100	14	20	-	11.0	2.5	4.8	-0.50	0.00	83.2	68.9	43.2	21.3	-
12	Sy	10	8.6	38	0	209	-	20	-	11.0	6.1	12.0	+0.30	0.22	156	96	36.7	12	-
13	Su	10	8.4	65	17	47	16	32	-	11	0.03	0.10	-1.6	0.524	105	70.3	48.9	22.9	-
14	Su	10-18	8.5	56	17	47	16	32	-	3-11	2.6	5.3	±0.0	0.045	105	67.1	58.2	25.8	10.4
15[b]	Gr	12	7.1	278	37	285	12	-	-	5-6	-	-	-0.15	2.20	142	105.5	52.4	21.5	-
16	Sy	10	8.0	0	0	250	-	20	-	11	0	0	-	0.28	208	146	63	24.4	-
17	Sy	10	8.1	0	0	180	12	20	-	11	0	0	-	0.17	102	130	71.2	29.2	-

a: Sy = synthetic, Su = surface, Gr = ground water

b: High nitrate and occurrence of iron bacteria

corrosion inhibition. He used ground cast iron samples in a number of natural and synthetic waters and exposures over 50 days. The deposition of $CaCO_3$ is primarily controlled by the electrochemical changes at the surface and thus is associated with the corrosion reactions and accompanying pH changes. He also speculates that the buffer capacity of the solution exerts a considerable influence (greater buffer capacity, i.e., alkalinity, being less corrosive) and that the anode/cathode relative area is important and pH dependent. The relative size of the local anode areas supposedly increases with increasing pH. Deposition of $CaCO_3$ is stimulated by elevated pH of local cathode areas but acts to reduce the anode area fraction (97). These considerations make $CaCO_3$ deposition more effective at a pH of about 7 than at higher pH values, and also more effectively applied to well buffered waters.

Patterson contends that effective $CaCO_3$ protection can only be provided when the water contains an alkalinity of at least 50 mg/L (as $CaCO_3$), and an equal amount of calcium (expressed as equivalent $CaCO_3$) (75). Using these minimum values, the pH required to maintain the $CaCO_3$ coating is much higher than the pH calculated using most saturation indices. The $CaCO_3$ layer deposited at a high pH has often been found to be less effective than that formed at moderate pH. Excessively high pH values may promote pitting and tuberculation.

Recent work by Feigenbaum and co-workers stresses the structure of natural calcium/iron scales (27). Fifteen natural scale layers formed in potable water systems carrying waters of various compositions were examined by scanning electron microscopy, x-ray diffraction, and microanalysis. The specimens studied showed a distinct outer zone (adjacent to the scale/water interface) and inner zone (adjacent to the metal/scale interface). The outer zone is relatively compact and consists of contiguous crystals mainly of calcite ($CaCO_3$). The inner zone is considerably more porous and comprised of geatlite [$\alpha FeO(OH)$], siderite ($FeCO_3$), and magnetite (Fe_3O_4) that favor a needle-like and granular porous structure. A steep gradient in Fe and Ca concentrations was found in the bulk scale. Depth of the gradient in the scale varied from scale to scale and appeared to play a role in protectiveness (27). In a later study, these workers proposed a model based on the structure and porosity of the scales they had observed and made AC impedance measurements on scaled specimens to associate with the diffusion resistances used in the model (28). Correlations were developed between the individual impedances of the 15 natural scales and their crystalline phase composition and water composition. A new criterion for the tendency of protective scale deposition was proposed and compared to others. Results of the correlation of scale impedance (spatial compactness) and water quality factors are shown in Table 7. Further comparison of scale resistance with long-term corrosion experience indicated good correlation with the γ value. According to this criterion, provided sufficient temporary hardness exists, the presence of chlorides and sulfates can improve the protective properties of scale (28).

TABLE 7. RESULTS OF CORRELATION ANALYSIS (28)

Number	Combinations	Correlation Coefficient	Standard Deviation
1	$\frac{[Ca^{++}]\,[HCO_3^-]^2}{[CO_2]}$	0.71	52
2	Langelier index	0.34	70
3	$\frac{[Alkalinity]}{[Cl^-] + [SO_4^=]}$	0.49	223
4	$Y = AH + B\,([CL^-] + [SO_4^=])\exp(-1/AH) + C$	0.92	32
	where $A = 3.5 \times 10^{-4}$, $B = 0.34$, $C = 19.0$, and $H = \frac{[Ca^{++}]\,[HCO_3^-]^2}{[CO_2]}$		

Effect of Flow Rate and Temperature--

Examples of the diverse and often opposing effects of solution flow rate on corrosion of iron have been noted in the previous sections of this discussion. The extremes of flow rate can produce serious corrosion: stagnant situations promoting pitting and tuberculation, and very high flow rates causing widespread metal losses due to erosion-corrosion. In the intermediate range, the effect of flow rate on corrosion rate has been modeled (apparently for conditions where velocity dependent $CaCO_3$ deposition or high oxygen passivation do not occur) (66). The equations are based on a double resistance model in which one resistance is significantly time dependent. An adequate representation of new data obtained at 150°F and available literature data was obtained using the semi-empirical correlation and as a function of Re number and a dimensionless diffusion group (66).

The effect of temperature on corrosion of iron in natural water is also highly complex. It has received very little independent study. Temperature changes can affect all of the aqueous equilbria, diffusion rates, deposition rates and electrochemical reaction rates. In relatively simple systems such as when the iron corrosion rate is controlled by diffusion of oxygen through the reaction product film, the rate increases as the increase in oxygen diffusion rate increases with temperature. In this case, the corrosion rate doubles with every 30°C rise in temperature up to about 80°C. Above 80°C, in open systems, the corrosion rate decreases sharply due to the marked decrease in solubility of oxygen with increasing temperature (107).

Effects of Other Species in Solution--

This section gives a brief discussion of the effects of free chlorine, chloramine, nitrate, humic acids, and sulfide on the corrosion of iron in natural waters. Variation of species such as sodium ion, potassium ion, or magnesium ion is not expected to have appreciable effects on corrosion rates.

The effect of free Cl_2 concentration (mg/L) is shown in Figure 9 where they are superimposed on data obtained with no Cl_2 present (60). These results were obtained for mild steel in aerated water of about 120 to 135 mg/L alkalinity, about 30 mg/L NaCl, at pH 7 and 8 and at low flow rates. It can be seen that the corrosion rate is increased in the presence of free chlorine concentrations greater than 0.4 mg/L. As shown, chloramine actually acts as a mild inhibitor at low concentrations. The threshold concentration of free chlorine for accelerated corrosion may be a function of the chloride to alkalinity ratio, but this was not investigated. Chlorine can act as an oxidizing agent which is "stronger" than oxygen in neutral solutions.

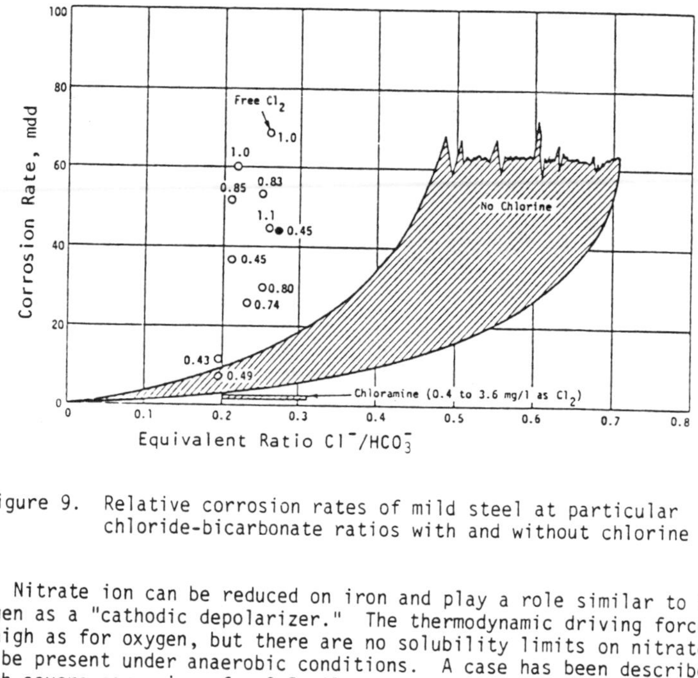

Figure 9. Relative corrosion rates of mild steel at particular chloride-bicarbonate ratios with and without chlorine (60).

Nitrate ion can be reduced on iron and play a role similar to that of oxygen as a "cathodic depolarizer." The thermodynamic driving force is not as high as for oxygen, but there are no solubility limits on nitrate and it can be present under anaerobic conditions. A case has been described in which severe corrosion of a 2.5 mile steel main carrying anaerobic well water was caused primarily by 4-7 ppm (as N) nitrate (12). A detectable decrease in nitrate concentration and corresponding increase in nitrite, ammonia and hydroxyl ion (products of nitrate reduction) and dissolved iron was found as water passed through the main. Increasing the pH from 6.4 to 8.0 completely arrested the corrosion both in the presence and absence of chlorine. Nitrate can under some conditions act as a passivating agent for iron, but this is an undependable type of inhibition.

The effect of humic acids on the corrosion of black steel pipes in natural waters has recently been reported (86). These compounds were found to inhibit corrosion for a range of hardness, flow rate, and chloride values. The authors interpret this as being due to the inhibition by the humic material of the oxidation of the siderite ($FeCO_3$) product layer. They attribute considerable protective properties to siderite layers. It also seems possible that large organic molecules such as these could also act as direct adsorption inhibitors or lead to the formation of reaction product layers whose structure is more protective, regardless of composition.

Hydrogen sulfide or other sulfide species should not be present in any properly maintained water system. In spite of this, cases do arise where water containing sulfides is conveyed to consumers usually from small water suppliers using underground sources (111a). The presence of sulfides is almost always objectionable to the consumer. In addition, sulfide waters can be quite corrosive, attacking iron and steel to form "black water" and also attacking copper, copper alloys, and galvanized piping, even in the absence of oxygen. The mode of attack by sulfide is often complex and its effects may either begin immediately or not be apparent for months only to become suddenly severe. Much of the corrosive action of sulfide may be due to the partial replacement of oxide or hydroxide films on iron or copper by metal sulfide films which either disrupt the normal protective nature of the film or initiate galvanic corrosion. Wells has discussed methods for removal of hydrogen sulfide and sulfides from water in detail (111a).

Comparison of Cast Iron and Mild Steel--

Cast irons are ferrous alloys containing more than 1.7 percent carbon. Gray fracture due to the presence of free graphite is seen in normal slowly-cooled cast form. This graphite causes the brittleness of cast iron and is the important metallurgical difference from mild steel. From a corrosion standpoint, the main differences are:

- a surface skin of iron oxide, silicates, and alumina silicates which is formed on cast iron during production.
- the existence of graphite sites which occur at 0.04 mm intervals on ground cast iron surfaces (57).

graphitic corrosion of cast iron is possible.

The exterior skin can increase corrosion resistance of cast iron relative to mild steel, but this layer is often partially removed by grinding, especially prior to the application of linings. Grinding exposes the graphite sites, and these can stimulate corrosion relative to steel during initial exposure by galvanic attack. There seems to be little difference between corrosion rates of ground cast iron and steel at long durations. Under some conditions a selective leaching of iron (due to the galvanic cell formed by graphite and iron) can occur ultimately leaving a porous mass consisting of graphite, voids, and rust. This is usually a slow process.

Corrosion of Galvanized Iron

Galvanized (zinc coated) steel is an example of a coating used as a cathodic protection device. The zinc coating is put on the steel not because it is corrosion resistant, but because it is not. The zinc corrodes preferentially and protects the steel, acting as a sacrificial anode. Electrodeposited zinc coatings are more ductile than hot-dipped coatings and usually quite pure. Hot-dipped coatings form a brittle alloy layer of zinc and iron at the coating interface. Corrosion rates of the two coatings are comparable except that hot-dipped coatings, compared to rolled zinc and probably electrodeposited Zn, tend to pit less in hot or cold water. This difference suggests that either specific potentials of the intermetallic compounds favor uniform corrosion, or that the incidental iron content of hot-dipped zinc is beneficial. In this connection, it is reported that Zn alloyed with either 5 or 8 percent Fe pits less than pure Zn in water (107). Zinc used for hot-dip galvanizing may contain 0.01 to 0.1% cadmium and up to 1% lead as impurities (73).

Effect of Water Quality Parameters--

In aqueous environments at room temperature the overall corrosion rate of zinc in short-term tests is lowest within the pH range 7 to 12. In acid or very alkaline environments, major attack is by H_2 evolution. Above about pH 12.5, zinc reacts rapidly to form soluble zincates by the following reaction.

$$Zn + OH^- + H_2O \rightarrow HZnO_2^- + H_2$$

In general, both zinc and cadmium react readily with nonoxidizing acids to release hydrogen and give divalent ions. Cadmium, however, is relatively stable in bases since cadmiate ions, if formed, are much less stable than zincate ions. The effect of pH on corrosion of Cd is shown in Figure 10. In the intermediate pH range of main interest here, the main cathodic reaction in aerated waters is probably reduction of oxygen. The corrosion rate of zinc in distilled water increases with oxygen concentration and with CO_2 from air saturation (105). Nonuniform aeration of the surface can cause differential concentration cells and localized corrosion of zinc. The corrosion rate of zinc increases with temperature as discussed below. In general, corrosion in actual use is greater in soft waters than hard waters (52,108). Chlorine additions, in the amounts normally used for health protection of water supplies, do not increase the corrosion of zinc in water (2).

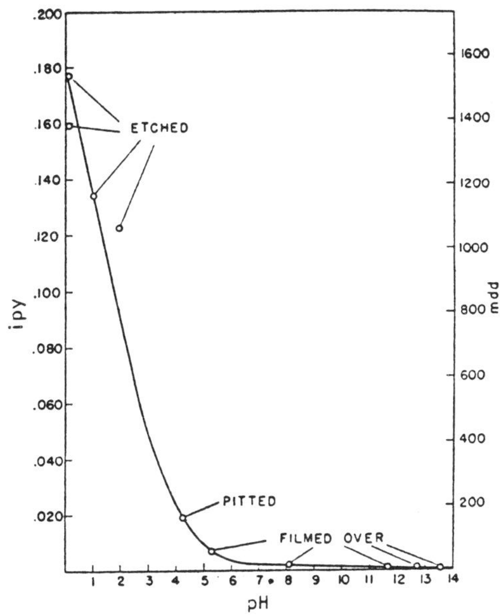

Figure 10. Corrosion of Cadmium vs. pH in continuously flowing, uniformly agitated and aerated solutions of HCl or NaOH (108).

Material: 5 x 10 x 0.63 cm (2 x 4 x 1/4") cast cadmium.

Temperature: 24 ± 0.5°C (74 ± 1°F).

Time: 7 days for pH below 2; 41 days for pH above 2.

Wagner has summarized results from field and laboratory tests on the effect of water quality parameters on corrosion of galvanized steel tubes (109). He shows a definite correlation between corrosion rate and pH, at least for the zinc phase of the coating and with steady flow of water (at 0.5 m/s). These results, shown in Figure 11, indicate that corrosion rate increases rapidly with a decrease in pH in the pH range 7 to 8. This effect is said to exist in spite of other water quality parameters. According to Wagner, there is negligible effect of buffer capacity and saturation index on the corrosion rate of galvanized steel tubes, although the composition of the deposits are altered. Corrosion rate does vary with time, first decreasing as zinc corrosion products grow. Once formed, the coating gives a constant (but pH-dependent) rate as long as the metallic zinc phase is present. Once the Zn/Fe alloy phase is reached, the rate decreases again but reaches another constant value which is also pH dependent. Effects of additives and organic acids are also discussed (109).

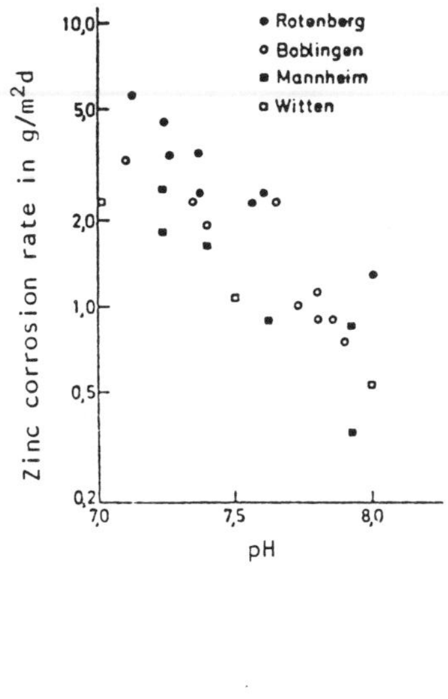

Figure 11. Effect of pH on corrosion on galvanized steel tubes (109).

One of the important environmental factors for galvanized steel or iron is the dissolved copper content of the water. A corrosion study of galvanized steel and galvanized wrought iron pipes in 25 selected domestic waters for two years has been reported (70). Computer correlation of corrosion grades with a large number of factors such as chloride, pH, saturation index, alkalinity, hardness, flow rate, etc. was attempted. The only definite correlation for corrosion of galvanized pipe with water quality was for dissolved copper concentration. Chloride concentration was a possible accelerating factor. In general, the remainder of the results were difficult to interpret. Attack was observed only on the zinc, not iron. There was no evidence that high alkalinity (above 100 ppm) or silica had inhibitory effects. Several case histories illustrating the copper effect are given by Cruse (19). His results, showing the correlation of copper found in corrosion products with maximum pitting rates in potable water, are shown in Figure 12.

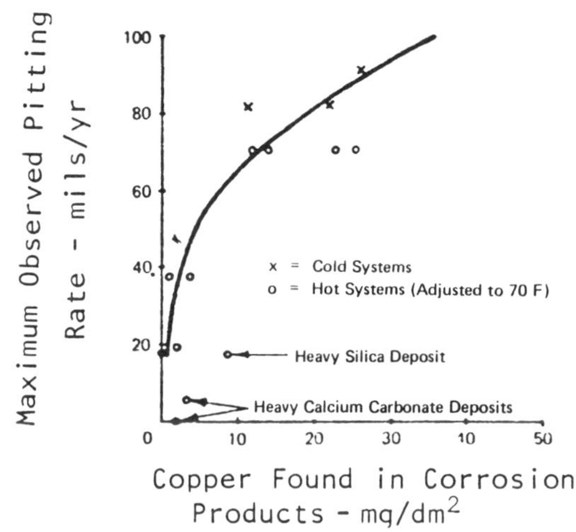

Stagnant Conditions--
The increase in concentration zinc, along with cadmium and lead, from hot-galvanized pipes in domestic drinking water systems has recently been discussed (10). Under stagnant conditions, concentrations of 5 to 10 ppm zinc may be obtained after 8 to 40 hour exposure to new galvanized pipes. This behavior is almost independent of water composition. Most of the corrosion products are not dissolved but present as an "oxidic colloid" or finely divided solid. The content of corrosion products in stagnant water could be decreased by treating the water with phosphates or silicates. Their research results indicate stagnation water in new pipes can also contain cadmium and lead products but in rather lower proportion to zinc than their concentration in the galvanizing layer. The authors are apparently referring to tests using piping corresponding to a German standard in which the hot-galvanized coating consists of about 97 percent Zn and 3 percent Fe with maximum limits of 0.8 percent Pb and 0.01 percent Cd.

A field investigation dating from 1973 was also undertaken for new plumbing in a three story building (10). All samples were taken from one tap on the top floor. The water composition is shown in Table 8; initial lead and cadmium contents probably coming from "old pipework." Zinc, cadmium, and lead contents in stagnation waters are shown in Table 9. The scatter in individual measurements is attributed to the corrosion products being mainly present in the form of suspended solids. Taking the initial values of Cd and Pb to be 2 μg/L Cd and 25 μg/L Pb, the concentration increases due to building plumbing and relative amounts shown in Table 10 were obtained. Random samples from the pipe materials in this plumbing system showed a composition of 97 percent Zn, 0.7 percent Pb, and 0.06 percent Cd for the galvanizing layer. This gives ratios of Zn/Cd = 1600 and Zn/Pb = 140 which, according to the authors, indicate some retention of Pb and Cd in the pipe covering layers (10).

TABLE 8. WATER COMPOSITION (10)

Parameter	Content (millimole/L, except as noted)
pH	7.0 - 7.3 pH Units
Alkalinity (to pH 4.3)	3.6 - 4.5
Acidity (to pH 8.2)	0.85 - 1.3
Chloride	1.6 - 2.3
Sulfate	0.93 - 1.33
Nitrate	0.2 - 0.3
Cu^{2+}	< 0.01 mg/L
Cd	1 - 3 μg/L
Pb	20 - 30 μg/L
O_2	3 - 7 mg/L

TABLE 9. ZINC, CADMIUM AND LEAD CONTENTS IN STAGNATION WATER (1-3 DAYS) (10)

TIME OF INVESTIGATION:	1973	1974	1975	1976	1977
Cadmium Contents in μg/L					
Number of Determinations	0	17	9	10	3
Range of Variation	--	2/8	5/10	1/11.6	4.8/10
Mean Values	--	4.6	6.0	5.4	7.6
Lead Contents in μg/L					
Number of Determinations	0	12	9	10	3
Range of Variation	--	20/120	69/90	47/96	25/50
Mean Values	--	56	80	67	39
Zinc Contents, mg/L					
Number of Determinations	12	23	9	10	3
Range of Variation	2.1/9.9	1.9/9.4	5.6/7.1	3.7/6.7	2.1/4.4
Mean Values	6.9	6.7	6.3	5.6	3.5

TABLE 10. CADMIUM AND LEAD CONTENT OF STAGNANT WATER (1-3 DAYS) (10)

Period	Mean Value of Cd (μg/L)	Mean Value of Pb (μg/L)	Zn/Cd (mass	Zn/Pb ratio)
1974	2.6	31	2567	216
1975	4.6	55	1370	115
1976	3.1	45	2100	142

In flowing water the zinc content remained about 8 to 15 ppm for each year from 1973 through 1977, apparently due to extraction from solid zinc compound covering layers. No increase in Pb or Cd (within experimental precision) was noted in flowing water (10).

Comparative tests of galvanized iron in waters having a pH value between 7.5 and 9.5 and containing calcium bicarbonate, but having a very low content of sulphates, chlorides, and nitrates, show that the attack on zinc was soon mediated. But in waters low in calcium bicarbonate or containing appreciable quantities of sulphates, chlorides, or nitrates, zinc suffers pitting attack. Zinc tends to form an insoluble basic zinc carbonate layer on the attacked areas which limits their size. Other findings were that zinc electrochemically protects iron when it is clean, but when coated with a resistant layer of corrosion products the maximum distance at which it still gives protection decreases. Zinc protects the alloy layer of a galvanized coating where it has become laid bare, but the alloy layer will not afford electrochemical protection to bared iron. The calcium carbonate which is deposited on alloy or on iron in the course of electrochemical protection is itself protective, and the ultimate success of the electrochemical protection of any exposed area appears to depend on the building up of the protective effect of this chalk layer at a greater rate than that at which protection is lost by dissolution of zinc from the area adjacent. The thickness of the zinc layer, the calcium bicarbonate content, pH value, and conductivity of the water appeared to be the deciding factors (48).

Hot Water Corrosion--

A number of studies have been performed on the corrosion of zinc in hot water tanks (39, 47). Some of the findings may be summarized as follows:

1. That low alkalinity water containing all or nearly all calcium and magnesium normal carbonates, at a temperature of 150°F, is more corrosive to galvanized metal than water containing no normal carbonates, even though dissolved oxygen is appreciable and some free CO_2 is present.

2. That normal calcium and magnesium carbonates deposit an uneven scale, from 1/32 to 1/16 of an inch in thickness, which is somewhat adherent, whereas the bicarbonates of calcium and magnesium

produce a much thinner scale, probably basic carbonate of zinc, which adheres much better to the metal.

3. That corrosion of zinc depends more upon the alkalinity and pH of the water than upon the dissolved oxygen present.

4. That the attack of normal carbonate water on the galvanized tank, which brings about failure, starts at the pin-holes or the weak spots in the galvanizing. The water soon penetrates through such pin-holes and then attacks the iron. These pits keep spreading and growing deeper until the tank fails.

5. That the products of corrosion fill the pin-holes, thus acting as a protective film.

The effect of temperature on weight losses of zinc was established by using distilled water with aeration and agitation. Weight losses at 150°F were found to be six times greater than at 50°F. This increase in corrosion was found to be caused by a change in the nature of the film from an adherent gelatinous state to a nonadherent granular state. The nonadherent film was found to exist between 131° to 167°F, which is within the range of use in water tanks (39).

In many aerated hot waters, reversal of polarity between Zn and Fe occurs at temperatures of about 60°C (140°F) or above (107). This leads to Zn having the characteristics of a noble coating instead of a sacrificial coating, and hence a galvanized coating under these circumstances induces pitting of the base steel. A 15-year service test on piping carrying Baltimore water at a mean temperature of 46°C (115°F) and maximum of 80°C (176°F) confirmed that pitting of galvanized pipe was 1.2 to 2 times deeper than in black iron pipe (ungalvanized) of the same type, corresponding to shorter life of the galvanized pipe. In cold water, however, pits in galvanized pipe were only 0.4 to 0.7 times as deep as those in black iron pipe, indicating in this case a beneficial effect of galvanizing. It was found that waters high in carbonates and nitrates favor the reversal in polarity, whereas those high in chlorides and sulfates decrease the reversal tendency (107).

The cause of this reversal is apparently related to the formation of porous $Zn(OH)_2$ or basic Zn salts, which are insulators, under those conditions for which Zn is anodic to Fe, but to formation of ZnO, instead, under conditions where the reverse polarity occurs. The latter compound conducts electronically, being a semiconductor. It can therefore perform in aerated waters as an O_2 electrode whose potential, like mill scale on steel, is noble to both Zn and Fe. Accordingly, in deaerated hot or cold waters in which an O_2 electrode does not function because O_2 is absent, Zn is always anodic to Fe, but this is not necessarily true in aerated waters. Apparently, the presence of HCO_3^- and NO_3^- aided by elevated temperature stimulates formation of ZnO, whereas Cl^- and SO_4^{--} favor formation of hydrated reaction products instead.

At room temperature, in water or dilute NaCl, the current output of zinc as anode decreases gradually because of insulating corrosion products which form on its surface. In one series of tests, the current between a couple of Zn and Fe decreased to zero after 60 to 80 days and a slight reversal of polarity was reported. This trend is less pronounced with high purity zinc on which insulating coatings have less tendency to form.

Stainless Steels

According to Reedy, stainless steels used by the water works industry can be divided into the three groups given below (82).

1. Chromium steels (containing 12 percent or more Cr) designated by the American Iron and Steel Institute (AISI) as Type 400 series (best known examples are Types 410 and 430).

2. The 18-8 chromium-nickel stainless steels, designated AISI Type 300 series, of which Types 301, 302, and 304 are frequently used.

3. AISI Type 316 stainless steel with a nominal composition of 18% Cr, 12% Ni, and 2.5% Mo.

Stainless steels are frequently used where protective coatings are not satisfactory or cannot be used such as in corrosive areas of water treatment plants and in components such as pumps, valves, meters, venturis and pressure regulators. Extensive use of stainless steel as a cladding material for uptake and downtake shafts and for control and distribution chambers in a New York City water tunnel has been reported (36). This project will use seven million pounds of stainless steel clad.

The resistance of stainless steels to uniform corrosion is generally good, but overall corrosion stability depends on maintenance of a passive state. Passivity in the sense used here is described below. Maintenance of the corrosion resistant state depends on both the particular type of stainless steel employed and on the environmental conditions.

Passivity--

The corrosion resistance of stainless steels depends primarily on the ability of the material to achieve and maintain a passive state in an aggressive environment. The ability to achieve a passive state determines the reduction in uniform corrosion rate from that of an active metal. The ability to maintain the integrity of the passive state determines the resistance of the material to local attack caused by the chemical environment, eroding particles, or stress. This passive state is characterized by a certain electrode potential or limited potential range over which the dissolution rate changes from a relatively high to a relatively low value. The electrode potentials associated with the onset of passivity and the maximum current density needed to reach the passive state from the active state are a function of the metal composition, chemical composition of the environment, and temperature.

The detailed mechanism of formation and exact nature of the passive surface components have been a subject of controversy for many years. It is probable that primary protection is offered by a very thin (on the order of 10 to 100 angstrom, amorphous, pore free metal oxide film. The film should be an electronic conductor. Thicker, more porous crystalline layers formed on top of the thin passive layer may offer substantial barriers for chemical or mechanical attack on the basic film. However, the primary corrosion resistance is dependent on the initial establishment of the passive layer and its continued stability. The type of passivity defined here should be distinguished from that attributed only to the formation of sparingly soluble reaction products on a metal surface. The latter process, which can be predicted from Pourbaix diagrams, does not necessarily imply an established passive state and the absence of corrosion. It is essential that the reaction products be formed directly on the metal surface and as a direct consequence of the anodic reation; complex processes such as chemisorption may also be intimately involved.

Localized attack occurs when momentary passive film breakdown, in the presence of chloride, exposes a small part of the metal surface which is surrounded by a large film-covered cathodic area. The cathodic area can drive anodic metal dissolution at the bare spot at a high rate (high current density). Once started, hydrolysis of metal ions from the anodic reaction causes the pH in the incipient pit or crevice to decrease which in turn discourages film repair and augments attack. Growth involves migration of chloride into the area which can also augment corrosion. Much of the greater resistance of the more resistant materials is probably due to their greater rate of film repair so that pit initiation is much less likely.

Type of Corrosion and Effect of Alloy Composition--

Passivity in relatively mild environments is obtained by alloying iron with at least 12 percent chromium. The stainless steels formed with more than 12 percent Cr as the only major added component, typified by the Type 400 materials, generally show relatively low uniform corrosion rates in typical potable waters. They are highly susceptible, however, to severe localized corrosion (pitting, crevice, and weld related attack) in waters containing chloride and oxygen (37, 82, 96). Additional alloying is needed to obtain resistance to localized corrosion, but even the more resistant alloys are occasionally susceptible to local breakdown of passivity. Type 400 materials are not recommended for submerged service in potable water systems (82). The stainless steel most favorably used for such service is probably Type 304, which with 18 percent Cr and 8 percent Ni, has much better localized corrosion resistance. Addition of about 2 percent molybdenum (and more nickel) to Type 316 generally gives higher resistance to chloride induced pitting and crevice corrosion than Type 304. While uniform corrosion rates are low for Type 300 materials in potable water environments (36, 37, 96), their application must be well understood to avoid localized attack. Good welding procedures and design to eliminate crevices are particularly important. More resistant types than Type 316 are known, but they are mainly used in extremely corrosive process industries.

Environmental Effects on Corrosion of Stainless Steels--

The two most important chemical factors in potable water systems are chloride and oxygen. Oxygen plays a conflicting role in the corrosion of stainless steels. Oxygen is necessary for the maintenance of the passive state for many stainless steels, but it also provides the driving force for local disruption of the passive film in the presence of chloride to give localized corrosion. The susceptibility of a given material to localized attack increases with chloride concentration. Sulfate and perhaps bicarbonate appear to act as pitting inhibitors and their concentration relative to chloride is a relevant factor (61). The solution pH in the neutral region has relatively little effect on pitting of 300 series stainless steels (M. J. Johnson in Reference 3).

Temperature can have a major effect on the localized corrosion of stainless steels. Susceptibility to pitting increases with increasing temperature for both Types 304 and 316 and the change is fairly marked for moderate (15-30°C) temperature increases above 25°C. Scale and other deposits decrease the corrosion resistance of stainless steels since they provide an opportunity for establishment of differential aeration cells and for crevice corrosion. Stagnant areas can be deleterious for similar reasons. High flow rates can usually be tolerated, in the absence of eroding debris or entrained solids, in the liquid phase. Careful design is needed to ensure satisfactory performance of these materials.

Results in Potable Water--

While stainless steels have been tested under a variety of conditions in the chemical process industry and laboratory, relatively few direct tests in potable water systems have been reported. Results have been reported for Type 410 and Type 316 in Southern California waters including treated Colorado River water (96). Type 410 was stable in four aqueduct and well waters, but was very severely corroded by pitting in both treated and untreated Colorado River water. Since pH and dissolved O_2 were nearly the same in most cases, the reason for increased corrosivity of the Colorado River water may have been its high chloride content (85 ppm) relative to the other waters (16-28 ppm). Type 316 was completely undamaged in any of the waters (96).

Extensive tests of a variety of stainless steels in New York City Reservoir water have been reported (36,37). A typical average analysis of this water is shown in Table 11. Results obtained during exposure in a semi-stagnant area in a shaft 14 feet above the main flow are shown in Table 12. Type 416 was not submerged. The pitting rates for the 300 series materials and 17-4 PH (a cast stainless steel) are all quite small. Weight loss corrosion rates did not exceed 0.01 mpy (36). Similar results were later reported at longer exposures (up to 15 years) for samples submerged in jars containing periodically renewed samples of the same water (37). A submerged Type 416 sample suffered severe crevice corrosion in these tests. Other results were essentially the same; no attack was noted on 300 series or 17-4 PH materials.

CORROSION OF COPPER IN POTABLE WATER SYSTEMS

Copper is a highly regarded material for use in potable water service lines. It is flexible, easy to join and install, has a low resistance to flow, and is considered to be fairly resistant to the corrosive action of most waters. Corrosion can occur under certain conditions, however, causing a number of water quality problems. Most of these problems have been

TABLE 11. TYPICAL AVERAGE ANALYSIS OF NEW YORK CITY RESERVOIR WATER (Catskill-Delaware System--1969) (37)

Component	Concentration (mg/L) (except as noted)
Alkalinity (as $CaCO_3$)	5 - 14
Calcium	3 - 6.3
CO_2	1.8 - 4.0
Chloride	4.0 - 7.0
Copper	0.01 - 0.10
O_2	13.7 - 14.3
Hardness (as $CaCO_3$)	19 - 23
Sulfate	6.0 - 14.0
Silica (as SiO_2)	0.5 - 2.5
Total Solids	40 - 54
pH	6.5 - 7.5
Specific Conductivity (μmho)	60 - 73

TABLE 12. STAINLESS STEELS--PITTING IN RESERVOIR WATER (36)

Alloy	Test Duration (years)	Maximum Pit Depth (mpy)
304	2	0.25
304L	9	0.23
316	9	0.11
316L	9	0.11
317	9	0.34
17-4 PH	2	0.05
416[1]	4	7.20

[1]Exposure at 100 percent humidity and 23°C.

NOTE: Each item represents average of 3 or more specimens.

delineated by an AWWA task group report and are summarized as follows (16). Small concentrations of copper cause the formation of blue or blue-green stains on procelain fixtures. Concentrations greater than 1.0 ppm react with soap to produce insoluble green "curds." A bitter, unpalatable taste results from copper when it is present in concentrations greater than 1.0-1.5 ppm. Traces of copper can accelerate the corrosion of galvanized hot-water tanks and cause pitting of aluminum pots and pans. Traces of copper are objectionable in many industries, i.e., in those involved in the canning of foods and in those using metallic-plating baths. Small amounts of copper in irrigation water are toxic to sugar beets and barley grown in nutrient solution. A concentration of 2 ppm or more is believed to be toxic to tomatoes. Copper is toxic to fish in concentrations of 0.25-1.0 ppm. Pitting corrosion may lead to the failure of copper pipe as a result of pin-hole leaks.

Corrosion of copper in potable waters has been the subject of numerous studies and papers. There are many factors which influence the corrosion process. The interdependence and frequent lack of independent control over these factors has led to a rather chaotic literature. However, favorable conditions for the corrosion stability of copper must predominate in most cases of potable water use based on its overall record. From World War II to 1972 over 6 million miles of copper plumbing tube was put into service (15). Relatively few cases of actual failure have occurred.

General Considerations

Although many modes of corrosion of copper have been distinguished, this discussion will classify the types of corrosion of interest here into only two types: general or uniform corrosion and localized corrosion. Velocity-related corrosion will be discussed under localized corrosion. These modes encompass the most prevalent ones from a water quality standpoint. Their basis is interrelated to a certain degree. It often appears that conditions which suppress uniform corrosion can give way to localized corrosion.

In the total absence of oxygen, copper is thermodynamically incapable of corrosion in potable water environemnts. The usual presence of at least some oxygen is the driving force for corrosion of copper in these environments, but it is not usually the determining factor. It is doubtful that total exclusion or removal of oxygen from potable water systems is desirable or economically feasible. Such efforts could in fact be deleterious due to the formation of sulfide, ammonia, or related compounds by bacterial action which could be harmful to copper or other materials, especially if they are eventually exposed to waters containing oxygen or other oxidants.

Given the presence of oxygen and possibility of corrosion, the actual occurrence of copper corrosion is governed by the presence and stability of inorganic compound films on copper. Probably the most noteworthy of these from a corrosion standpoint is cuprous oxide (Cu_2O) formed by initial corrosion of copper. As will be noted, the mode of formation is very important. Films of cuprous oxide formed by annealing or other manufacturing steps may be detrimental to the stability of the basic metal.

The protective, metal-solution-grown layer of cuprous oxide may be very thin and subject to further attack by solution species and physical erosion. The cuprous oxide layer is often formed slowly and is tenuous in comparison to the type of passive films which offer much of the corrosion protection to metals such as iron, nickel, cobalt, and chromium. As such, it is fortunate that other, thicker, films of insoluble copper compounds are often formed over the primary film and can offer mechanical protection from erosion and an additional barrier to diffusion by aggressive chemical species. Films formed by deposition of calcium carbonate and similar compounds can offer similar, but still somewhat unnatural, forms of protection. But the most basic protection is offered by the delicate, and often slowly grown, thin layers of cuprous oxide. For this reason, conditions occurring during the initial service exposure of copper are very important. Moderate flow rate and chemically mild environments favor formation of small grained, compact and protective layers of Cu_2O. Rapid deposition of disordered Cu_2O under harsher environments may be deleterious since the first layer is not very protective. At high flow rates in aggressive environments, growth may be delayed or very slow. Growth of localized layers of materials such as copper sulfides may be very harmful due to their galvanic influence on adjacent areas of metal.

The dynamic nature of these processes must also be kept in mind. Materials in service are often subjected to stresses which demand more or less continuous maintenance of a protection mechanism.

Uniform Corrosion of Copper--

From experience, the uniform corrosion rate of copper is usually quite low in potable water systems. This observation is based on the long life of the majority of tubing in service. From a health standpoint, however, low corrosion rates over a large uniform surface area can add a significant amount of impurity to a relatively small volume of water. Therefore, although low uniform corrosion rates may be acceptable based on corrosion lifetimes, they may be significant from a water quality viewpoint.

Selected literature results on uniform corrosion of copper in potable waters are given below. Emphasis is placed on references which are illustrative of the literature as a whole. It is noteworthy that the maximum reported copper concentration in standing water of reasonable pH is about 5 ppm and that this value is reported quite often. This could indicate that copper concentration is ultimately limited by solubility of a reaction product.

Effect of O_2

The simplest overall electrochemical uniform corrosion mechanism for copper, requiring simultaneous oxidation and reduction at the material/solution interface, can be represented by the parallel reactions Eqs. 5 and 6.

$$Cu \rightarrow Cu^{2+} + 2e^- \qquad (5)$$

$$1/2O_2 + 2e^- + 2H^+ \rightarrow H_2O \qquad (6)$$

There is evidence that, at least in acidic solution, O_2 is also involved in reaction with intermediate Cu^+ as in the following more elementary steps including the "chemical" step (Eq. 8).

$$Cu \rightarrow Cu^+ + e^- \quad (7)$$

$$Cu^+ + O_2 + 2H^+ \rightarrow Cu^{2+} + H_2O_2 \quad (8)$$

As previously mentioned, in the total absence of oxygen copper is thermodynamically incapable of corrosion in potable water environments. This was substantiated by Schafer's study of corrosion of copper pipes in potable water service in New Zealand (89) where a negligible amount of corrosion was reported in the absence of oxygen. However, the amount of oxygen needed for corrosion to occur may be small. Tronstad and Veimo (105, 106) determined that varying the oxygen content of the tap water between 2.7 ppm and 25 ppm had a relatively small effect on the final dissolved copper concentration. With 100 ppm added $NaHCO_3$ (pH 7.0) the copper cencentration increased about 50% for the 10-fold oxygen increases, while with 500 ppm $NaHCO_2$ (pH 7.2) the copper content increased by about a factor of 3 over the same oxygen range. The authors attributed this to oxidation of dissolved cuprous compounds.

Effects of pH

The effects of pH on copper corrosion have been studied in direct and indirect tests. In their work with water standing in copper tubes, Tronstad and Veimo (105, 106) measured the copper concentrations in water after 24 hour exposure of tubing to tap water alone, and with various additions of oxygen, carbon dioxide, sodium bicarbonate, sodium hydroxide, and calcium oxide. The tap water composition and pertinent experimental details are given in Table 13. All experiments were done at 18°C in the closed tubes.

This tap water can be classified as soft, having a fairly low pH, moderate amount of aggressive CO_2, low alkalinity, and moderately high oxygen content. In the tap water alone, the copper concentration reached a maximum of 0.6 ppm after about 15 hours and then slowly declined. Addition of sodium bicarbonate raised the pH but also consistently raised the amount of copper dissolved over a 24-hour period. For example, addition of 0.4 g/ℓ $NaHCO_3$ produced a pH of 7.2 and dissolved copper of 1.9 ppm. Addition of calcium bicarbonate had a similar effect. Addition of carbon dioxide lowered the pH and increased both the copper content and rate at which a maximum was reached. For example, 9 ppm CO_2 gave a pH of 5.65 and copper content of 3.9 ppm after 24 hours; with 45 ppm CO_2, a copper content of 4 ppm was reached after only 2 hours.

Addition of sodium hydroxide or lime to the tap water caused a decrease in the copper concentration observed after 24 hours until a pH between 8 and 9 was reached. Above a pH of about 10, the copper concentration increased once again. The lime additions resulted in somewhat lower copper concentrations, at the same pH, than sodium hydroxide additions.

TABLE 13. CONDITIONS OF COPPER DISSOLUTION EXPERIMENTS OF TRONSTAD & VEIMO (105)

Tap water composition:	
Total solids	47.5 ppm
Ash	28.5 ppm
Chlorides as $C\ell^-$	9.5 ppm
Iron	0.07 ppm
Combined CO_2	8.0 ppm
Free (aggressive) CO_2	4.0 ppm
pH	6.3
$KMnO_4$ consumed	65 $m\ell/\ell$ of 0.01 N $KMnO_4$
NH_3, NO_x, phosphates	Not detected
Properties of copper tubing:	
Purity: 99.8 to 99.9%	
Treatment: annealed, bent into loops, descaled in 2 N NH_4OH, washed with distilled water	
Specimen Size:	
Volume: 99 cm^3	
Internal surface area: 785 cm^2	
Internal diameter: 0.5 cm	
Length: 500 cm	

Experimental studies under simulated domestic use conditions were done to determine the effect of pH on copper corrosion and were reported by an AWWA Task Group (16). The water composition was not specified, but apparently it was aerated and contained CO_2. In the first set of experiments, water at an adjusted pH value was allowed to flow through 60 feet of new 0.75 inch copper tubing at a rate of 0.067 gpm (0.05 fps). Samples were collected at the end of 1 hour and again after 1.25 hours, the copper concentration determined, and the two values averaged. The observed copper concentrations as a function of pH are shown in Figure 13.

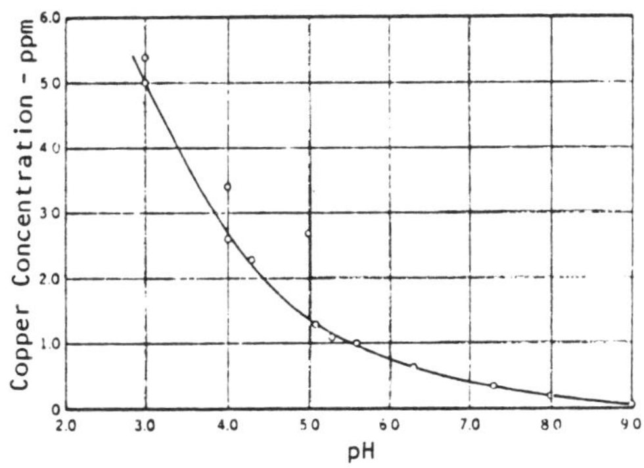

Figure 13. Effect of pH on corrosion of copper (16).

In the second type of experiment, water at the desired pH was passed through new copper tubing of the same dimensions at a flow rate of 0.5 gpm (0.37 fps) for 1 hour. The flow was stopped and the water allowed to stand in the tubing for 16 hours (to simulate overnight conditions). The flow was then started again with water at the initial pH and rate. Water samples were collected immediately and at various time intervals and analyzed for copper. Results as a function of time and pH are shown in Figure 14. The exponential decay suggests a simple rinsing effect of the dissolved copper solution formed during the stand.

The Task Group concluded from these two sets of tests and others that the carbon dioxide content of a water (indirectly measured by its pH) has a very significant effect on the corrosion solubility of copper (16). In addition, raising the pH to a value above 7.0 "greatly minimizes" this action.

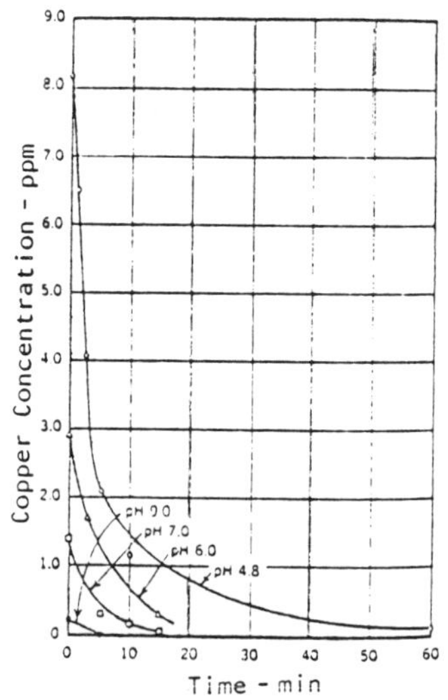

Figure 14. Effect of time and pH on corrosion of copper (16).

Effect of Free CO_2--

The free CO_2 or carbonic acid content of natural waters has often been reported to play a significant role in the uniform corrosion of copper. This could be due to its acting as a source of hydrogen ions at the surface since the free CO_2 concentration can be many times the bulk hydrogen ion concentration. Free carbon dioxide may also act to dissolve or thin the protective films on copper, thus enabling higher corrosion rates. Both mechanisms could occur in practical systems. The amount of CO_2 necessary to cause increased corrosion in potable waters probably depends on many factors. A concentration of 10 ppm is often quoted as a rule-of-thumb "threshold" concentration for anticipating or diagnosing CO_2 corrosion problems.

A review of early results on corrosion of copper and brass pipe and the relations to health is given by Hale (40). His general conclusions are that:

- Soft waters containing very little free carbonic acid will show little corrosion of copper or brass pipe.
- Soft waters containing considerable free carbonic acid (10 ppm and up) may corrode copper and dezincify yellow pipes.

Several examples were given for soft, high CO_2 waters in which standing copper concentrations of 5 ppm were obtained and up to 3.9 ppm with water running continuously. Examples are also given in which treatment to remove the CO_2 by pH adjustment and/or aeration were effective in controlling the corrosion of copper. Aeration of water containing little natural oxygen is, of course, discouraged.

Effects of Temperature

In their previously mentioned studies (105, 106) Tronstad and Veimo also investigated the effect of elevated temperature. Copper concentration measurements were made at different times at 46°C and 70°C and the data were corrected for the change in solubility of CO_2 with temperature. Although the initial rate of copper dissolution increased with temperature, the maximum concentration in solution did not change much, apparently due to solubility limitations and consumption of the available oxygen.

Schafer (89) obtained copper corrosion data by sectioning and measuring pipes that had been in service up to 30 years. He found that corrosion in hot or previously heated water was usually less than in the same water before heating, presumably due to removal of dissolved CO_2 and O_2 at elevated temperatures. Most corrosion appeared to occur in the first few years after installation.

Effects of Miscellaneous Parameters

Low concentrations of iron (0.05 to 0.5 ppm Fe^{2+}) have been found to inhibit the corrosion of copper in service such as seawater desalination. This concentration of iron could be dependent on the previous contact of water with steel or cast iron pipe. The effect should be considered when evaluating or designing tests of copper corrosion rates. It also raises the possibility that much copper pipe is "protected" by upstream iron pipe.

The distinction between surface and artesian waters may also be of interest in connection with reports of a natural inhibitor of localized corrosion of copper. This unidentified, probably organic inhibitor, is said to occur in most surface waters but not in underground waters (13). However, the distinction between surface and artesian waters noted by Schafer with regard to copper corrosion rates in cold (normally 10-20°C) potable waters was that:

> Surface waters usually showed a corrosion rate below 0.5 mpy (to a first approximation) irrespective of pH, hardness, chloride, and other water properties.
>
> In low pH, high carbon dioxide artesian waters containing moderate amounts of oxygen, corrosion was more rapid than in surface water and rates up to 15 mpy were observed.

A report of high concentrations of copper from 12-14 month old copper service lines in one district in England has been given (8). The chemical composition of the initial potable water was not reported. Initial morning water samples from houses in towns 12 miles apart both showed maximum copper concentrations of 5.5 ppm. These were serviced by 65 foot lengths of 0.5 inch copper tube. Four consecutive-day samples from one of these houses had copper concentrations within the narrow range of 4.8 to 5.5 ppm. Other results given in the paper do not indicate any correlation of copper concentration with length of service pipe. Based on the limited data given, the copper concentration does appear to vary inversely with pipe diameter, which could be due to the change in surface to volume ratio. Results are shown in Table 14.

TABLE 14. COPPER IN INITIAL MORNING SAMPLE OF WATER FROM COPPER PIPES (8)

Sample	Type of Pipe	Diameter (inch)	Length (feet)	Copper Concentration (ppm)
A	copper tube, welded joints	1.25	6000	0.75
B	soft copper tube, mechanical joints	0.50	65	5.5
C	copper tube, mechanical joints	1.00	1740	1.2
D	hard (straight) copper tube	0.50	15	3.2
E	soft (underground) copper tube	0.50	60	4.8 - 5.5

NOTES:

1. Samples A, B, C - all from same main
2. Samples D, E - from 2 houses in a town 12 miles away
3. All pipe said to be to British Standard Specifications, in service for 12 - 14 months.

The same report states that copper content at any house during normal day-time use did not exceed 0.5 ppm (8). Equipment was installed at one house to dispense a "slowly soluble" hexametaphosphate to the water. This caused the copper content of initial morning water samples to drop from a consistent average 5 ppm to an average of 2 ±0.2 ppm (samples from 8 different days). This relieved complaints of "blue water" and morning vomiting.

Fluoride in the concentrations added to many domestic water supplies (up to 1 ppm) has no effect on the amount and rate of corrosion in copper distribution systems (62, 63). Sulfide is discussed in the section on iron.

Localized Corrosion of Copper--

In this report, localized corrosion of copper will include pitting, impingement, and under-deposit forms of attack. Pitting is probably the most common of these and will be discussed at some length, followed by a shorter discussion of impingement. Under-deposit attack, a form of crevice corrosion, has been observed for copper, but its occurrence should be relatively rare for copper in potable water systems. However, when it does occur in hot water systems, attack can be rapid (89).

The pitting of copper is a complex phenomenon which probably has several different causes and a number of different contributing factors. The problem is apparently not widespread, but it has recieved considerable attention since failure of practically new pipe or tubing can occur in a short time. Even in a given water distribution system, the failures often occur at random locations.

The effects of copper pitting on potable water quality are difficult to assess. The actual area of material that is affected is usually small and the pits themselves are often capped with corrosion product which prevents significant loss of soluble copper to solution. On the other hand, while active pitting is occurring, the rate of attack per unit surface area is much higher than the usual rate of uniform corrosion. Also, pits may rupture unpredictably from time to time, releasing small amounts of concentrated copper ion solutions. Copper in solution can promote the corrosion of other metals, as discussed elsewhere in this report.

Causes of Pitting--

Much of the difficulty in characterizing copper pitting is in distinguishing between factors which initiate pitting and those which are necessary to sustain it. Possible synergism between these factors also complicates matters.

There are several elaborate theories on the growth and structure of corrosion pits on copper (17, 26, 64, 80). These models do not appear to be particularly helpful in determining either the probability of pit initiation or the tendency of a given water to support pitting.

Campbell has distinguished between two forms of copper pitting (13). The first form, called "soft-water pitting," occurs only in certain soft waters containing manganese. It usually is restricted to the hottest parts of hot water systems and is associated with formation of a scale of manganese oxides which forms an unfavorable galvanic couple with any exposed copper surface. The second type, hard-water pitting, occurs in hard or moderately hard waters and is nearly always restricted to cold water pipes. Campbell states that this type of pitting is usually associated with the presence of either a carbon film or a "glassy" copper oxide scale formed when certain abnormal conditions prevail during the annealing process. These films are cathodic to the copper surface and either one can provide a galvanic driving force which induces pitting.

The second form of pitting is prevalent in England and has promoted considerable study (17, 81). Its occurrence is apparently prevented by a naturally occurring organic inhibitor which is present in many surface waters, but not in well waters. The inhibitor has not been isolated, but it is removed by activated charcoal and has been further characterized (13).

The presence of carbon films on copper pipes manufactured in the United States is said to be rare (43, 111). At least one possible case of pitting due to a carbon film has been reported (111). Evidence has been presented for pitting due to a "glassy" copper oxide scale in U.S. Service lines by Cruse and Pomeroy (20). These authors examined over 65 pipe specimens. Although carbon films were present, the correlation with tendency to pit was less for carbon content than for the presence of glassy cuprite scales. They conclude that well water containing dissolved oxygen, relatively high mineralization, and a pH below 7.5 is conducive to the rapid pitting of copper, but that such pitting occurs only where the copper surface is sensitized by relatively heavy glassy cuprite scales, carbon residues, or perhaps other deposits or scratches.

From a survey conducted in the United States, pitting of copper tubing has almost invariably been associated with cold, hard well waters, according to Coher and Lynam of the Copper Development Association (15). They state that a typical aggressive well water contains greater than 5 ppm dissolved CO_2, dissolved oxygen up to 10 to 12 ppm (which may come from storage and handling of the well water), and chlorides and sulfates. Their statistical survey also shows that pitting failures are almost evenly distributed between soft or annealed temper and hard drawn tube. They state that this refutes hypotheses based on pitting due to surface conditions of the tube. They also state that pitting can be prevented by treatment of the water to neutralize the dissolved carbon dioxide. A case history discussing the effects of O_2 and CO_2 is presented in Section 6.

Impingement Attack and Flow Rate Effects--

Copper is more susceptible than most engineering metals to a flow-velocity dependent type of corrosion generally termed impingement attack (42). This is usually a localized attack caused by excessive liquid flow velocities and aggravated by the presence of entrained solids or gas bubbles. The resulting metal pits are undercut on the downstream ends and are frequently horseshoe shaped.

The rate of attack depends to some extent on water composition. The rate increases with increasing oxygen content and chloride content. Impingement attack rate increases with decreasing pH. Prevention can be obtained by limiting flow velocity to 5 feet per second for most municipal waters and to considerably lower values if entrained solids or air bubbles are present (43).

Copper Alloys--

A wide range of copper alloys are used in potable water systems, particularly as valve parts or other components where their mechanical properties are desirable. These alloys are divided here into the general classes of brasses, bronzes, and copper-nickel alloys and each is discussed below. The discussion of copper corrosion can be used as a basis for the behavior of the alloys, but their corrosion resistance also depends on alloy composition. One major difference compared to pure copper is that selective leaching is a predominant mode of corrosion for some common copper alloys. This is a corrosion process whereby one constituent of an alloy is removed from the metal, leaving an altered residual structure. A common form of selective leaching is dezincification of brasses. The fundamental mechanisms of selective leaching are a subject of disagreement (44). One view is that the entire alloy dissolves and then one of its components is replated from solution. Another group contends that one component is selectively dissolved from the alloy leaving the porous residue of the more noble species. Others believe that both modes of corrosion occur.

Corrosion of Brasses--

The common brasses are alloys of copper with 10 to 50 percent zinc. A number of other elements may be added either singly or in combination. These elements are listed in Table 15, but not all of these are commonly added to commercial Cu-Zn alloys (108). Hundreds of modifications of the brasses are known. Zinc dissolves in copper up to 39 percent to give a single phase alloy, α brass. Another single phase alloy, β brass, is formed with 47-50 percent zinc. At intermediate zinc levels, the alloy contains both phases, α and β brass (11).

TABLE 15. RANGES OF COMPOSITION OF Cu-Zn ALLOYS (108)

	Percentage
Lead	0.1 to 12
Aluminum	0.1 to 3.0
Tin	0.5 to 6.0
Nickel	0.5 to 10 (sometimes up to 30)
Iron	0.1 to 2.0
Silicon	0.1 to 2.0
Manganese	0.05 to 5 (sometimes up to 25)
Phosphorus	0.01 to 0.10
Arsenic	0.01 to 1.0
Antimony	0.01 to 0.1
Gold	0.5 to 1.0
Bismuth	0.1 to 3.0
Vanadium	0.1 to 0.5
Tungsten	0.1 to 2
Chromium	0.05 to 0.5

Many brasses corrode in major part by dezincification. Two general types of dezincification are recognized, the layer type and plug type. Layer type attack occurs fairly uniformly along the surface while plug type attack is localized and penetration occurs perpendicularly into the metal. Brasses generally resist impingement attack better than pure copper (107). They are susceptible to stress corrosion cracking under certain conditions but this is not discussed here. Pitting can occur under some conditions but this is generally similar to copper and less important than dezincification. Generally, the brasses show very good resistance to most types of unpolluted waters, with corrosion rates averaging 0.1 to 1 mpy, in the absence of dezincification (108). Soft waters containing high CO_2 can cause higher rates, often accompanied by dezincification.

Dezincification is also generally favored by the following conditions (62, 79, 107):

- elevated temperatures,
- stagnant solutions, especially if acid,
- porous inorganic scale formation and crevices,

- residual stresses and local deformation, and
- chlorides and copper ion buildup.

The composition of the alloy is also important. Brasses containing 85 percent copper or more (red brasses) are generally resistant to dezincification. Additions of iron or manganese to brass tend to accelerate dezincification while additions of low concentrations of arsenic, antimony, phosphorus, bismuth, and tin have been used to reduce the rate of dezincification (62). Thus, for example, Muntz metal, common yellow brass, and noninhibited aluminum brass are considerably less resistant than arsenical Admiralty brass and arsenical aluminum brass (108). Layer-type dezincification tends to occur more frequently for high zinc brasses and acidic environments. Plug type attack seems to occur more often for low zinc brasses and neutral, alkaline, or slightly acidic conditions (32). Many exceptions to these general statements can occur, however.

The influence of dezincification on water quality was demonstrated by extensive early work of Clark (14). Unfortunately, the pipe material was only designated as "brass"; it was probably yellow brass (67-33 Cu-Zn) or similar high zinc alloy. Water from several northeastern water supplies was allowed to stand in, or was passed through lengths of new pipe. In general, about the same amount of copper was dissolved from either pure copper or brass pipe. Brass pipes, however, yield much more dissolved zinc than copper (14).

Detailed observations of the corrosion of yellow brass and Muntz metal pipes in domestic hot water systems using several municipal water supplies (78). After 20 to 25 years service, the rate of uniform corrosion was low, but most samples exhibited local corrosion. Corrosion proceeded in two distinct phases, dezincification and then final corrosion of the copper formed. The lag between the two phases varied considerably. In α/β brass pipes the β phase was always attacked first. Initiation of the localized corrosion was often associated with signs of residual stresses. Evidence was found that the copper deposits were residual rather than formed by the redeposition mechanism (78).

A recent study of valve stem brass corrosion in hot and cold potable water at eleven cities in the United States has been reported (18). Commonly used silicon red brasses (Cu-Zn-Si two phase alloys) were tested for one year, primarily in cities where corrosion had been noted. Dezincification of the uninhibited materials was severe and widespread. Addition of 0.03 to 0.06 percent arsenic prevented dezincification in both hot and cold water while similar phosphorous levels were effective only in cold water. No correlation could be found between water composition and dezincification for the cities studied. Temperature was a most important external variable, greatest attack occurring in hot water services (18).

Corrosion of Bronzes--

Originally, the term bronze was used for alloys of tin in copper (here called tin bronzes), but it is now generally applied to casting alloys based on copper whether or not tin is present. The tin bronzes are essentially solid solutions of tin in copper. The most common wrought forms contain 1-10 percent tin. Alloys with more than 8-10 percent tin are usually used in the cast form. There are a number of modifications of Cu-Sn alloys, most of which deal with variations in the concentrations of tin, zinc, lead, and phosphorous. About twelve other elements have been added singly or in combination, although seldom more than four are added at one time. A summary of compositions is given in Table 16 (108). Additions of iron, antimony and bismuth are said to be dangerous and are tolerated only up to 0.2 to 0.5 percent (11). Aluminum bronzes generally contain up to 9-10 percent aluminum as the important minor constituents in copper and sometimes small additions of manganese and copper. Silicon bronzes contain up to 4.5 percent silicon and minor additions of manganese, zinc, iron, or tin (62).

TABLE 16. RANGES OF VARIOUS COMPOSITIONS OF Cu-Sn ALLOYS (108)

	Percentage	
Copper	60.0 - 99.5	
Tin	0.5 - 35.0	
Zinc	0.5 - 15.0	(sometimes up to 30)
Lead	0.5 - 15.0	
Phosphorus	0.01 - 3.0	
Cadmium	0.5 - 1.0	
Nickel	0.10 - 15.0	
Iron	0.05 - 4.0	
Silicon	0.05 - 2.0	
Aluminum	0.5 - 2.5	
Arsenic	0.5 - 2.0	
Antimony	0.1 - 8.5	
Cobalt	5.0	not common
Platinum	10.0	not common
Tungsten	10.0	not common
Manganese	3.0	not common
Bismuth	0.5	not common

There is relatively little information on the corrosion behavior of the bronzes in domestic fresh waters. Corrosion considerations for the tin bronzes are generally similar to those for copper. The attack by water depends on oxygen, carbon dioxide, and dissolved salt content and the formation of protective layers. Selective leaching of tin (destannification) has been noted, but apparently occurs only under relatively extreme conditions such as in superheated steam or for pump impellers handling hot feed water (11). Alloys containing more than 5 percent tin are especially resistant to corrosion by impingement (107).

Aluminum bronzes are said to show generally superior corrosion resistance to other common copper alloys in sea water service (11). However, selective leaching of aluminum can occur under ill-defined conditions. These alloys are normally more resistant than copper and most brasses to erosion corrosion or impingement attack. Several aluminum bronze samples exposed for 12 years to stagnant New York City Reservoir water showed rather extensive pitting with some cracking (37). Several tin bronzes exposed under the same conditions showed essentially no pitting and average corrosion rates less than 0.02 mpy. Extensive attack was noted for K manganese "bronze" (65% Cu 22% Zn 2% Fe 6% Al 4% Mn). Additional results for a number of copper alloys are given in References 36 and 37.

Silicon bronzes are used where high strength is required along with corrosion resistance comparable to copper. In many environments the corrosion rates are about the same as pure copper (62). The silicon bronzes are more resistant to acid attack and corrosion resistance increases with silicon content.

Other Copper Alloys--

Alloys based on copper and containing 5 to 40 percent nickel have generally excellent corrosion resistance in sea water, brackish water, and fresh water. These copper-nickel alloys are often used in heat exchangers for sea water or for other brackish water applications where ordinary copper alloys do not perform well. The 70/30 Cu/Ni and 90/10 Cu/Ni alloys are most often found, with small additions of iron and manganese to improve corrosion resistance under flow conditions (11, 62). Limited reference has been found to the use of these materials in domestic potable water systems. Extended testing of 90/10 and 70/30 cupronickels with 1 percent iron in New York City Reservoir water showed superior resistance to pitting corrosion. Weight loss general corrosion rates were reported 0.2 to 0.4 mpy (36) and 0.01 mpy (37). Several nickel-copper alloy (Monels) specimens showed significant pitting under the same conditions. Selective leading of nickel (denickelification) form copper-nickel alloys can occur under special circumstances, but it is relatively rare (11).

CORROSION OF LEAD IN THE WATER WORKS INDUSTRY

Because of its favorable structural characteristics, lead has been used for transporting and distributing potable waters in the U.S. since the nineteenth century. From an engineering standpoint, lead is highly resistant to corrosion and attack by natural waters. It is also, however, an active accumulative toxicant, and the distribution of corrosive water through lead pipe can constitute a serious health hazard to the consumer. The current standard for lead, as established by the EPA, is 0.05 mg/ℓ in drinking water. There are indications, however, that this standard may not provide a sufficient margin of safety for the fetus and children under three years old. The National Academy of Sciences has concluded that a lead standard greater than 0.025 mg/ℓ cannot assure a no-observed-adverse-health-effect status (50). Lead corrosion and high lead levels in potable waters have been identified in several municipalities throughout the nation from systematic surveys and sampling procedures. However, because of the expense and difficulty in implementing a comprehensive sampling program, possible lead contamination in many other communities has yet to be identified.

The most common use of lead in the water works industry is for lead service and residential pipes and for lead-based solders. Lead is also widely used as a pipe lining for zinc galvanized iron pipe to enhance durability and extend the useful life of the pipe. Lead is used for "goosenecks" in smaller piping systems to prevent undue stress on water mains, but some utilities have discontinued this use in favor of less potentially harmful materials.

Lead and lead-lined galvanized pipe have a useful life expentancy of 35 to 50 years and longer. Consequently, many of the lead service lines installed are still in operation. Many of the older municipalities in the nation have a large number of lead service lines ranging in length from 30 to 100 feet, which connects the street main to the household plumbing.

The most commonly used lead-based solders are composed of 50 percent tin and 50 percent lead or 60 percent tin and 40 percent lead. During installation, this solder may flow inside the pipes at the joint, thereby providing a lead-based surface area exposed to potentially corrosive water.

A secondary use of lead in the water works industry is for lead gaskets used as flanges for joining large valves and pipes in water treatment plants and on water mains. However, exposed lead surface areas are relatively small and water contact time is usually short minimizing the potential for additional lead contamination.

Lead is also used for the production of brass and bronze. Brass is a copper-zinc alloy which contains up to 12 percent lead, and bronze is a copper-tin alloy which contains up to 15 percent lead. These materials are corrosion resistant and are not suspected of contributing significantly to lead contamination of potable waters. However, test results indicate some dezincification in some brass, and it is reasonable to conclude that lead may also be leached into the water system.

Little information has been compiled which identifies or qualifies the extent to which lead is used in the water works industry throughout the U.S. The most comprehensive survey was completed by Donaldson and the results were published in 1924 (24). His study included a survey of more than 500 water distribution systems in 41 states and concluded that approximately 48 percent contained lead lines. In the Boston metropolitan area, Worth estimated that approximately 60 percent of the residences were serviced with lead pipes (114). In a report published by EPA, it is estimated that approximately 1300 of 3300 service lines in Bennington, Vermont consist of lead (83). Bennington, Vermont is considered a typical New England community. Patterson is currently conducting a survey of the quality of potable water at 1000 consumer's taps across the nation, and because the lead content of tap water is usually proportional to the amount of lead in the plumbing system, this study may indicate the extent of lead use in potable waters (76, 92). Specific areas within a given city may contain greater percentages of lead service lines, and these areas may be identified by a historical review of plumbing codes and pipe sales correlated to the time of development of the area.

In some municipalities where lead corrosion and potable water contamination have been documented, such as in Boston, Massachusetts, local or state plumbing codes restrict the use of lead for soldering only. However, several widely adopted codes, including the Uniform Plumbing Code and the Building Officials Conference of America (BOCA) Code, currently allow the use of lead pipe for transporting and distributing potable waters. Additionally, some municipalities assume responsibility only for water mains and reserve the selection and installation of water service lines to the home owners or builders. Consequently, lead service lines are still being installed and lead solders are widely used.

The primary water quality parameters related to lead corrosion and corrosion rates are hardness, alkalinity, pH, total dissolved solids, dissolved oxygen, and carbon dioxide. At least one investigator also attempted to identify the effects of chlorine content on the corrosive nature of potable waters. Although much research has been presented relating these parameters to lead corrosion, other investigators maintain that the actual contributing factors are not "hardness" but hardness in association with the corresponding anionic component; not alkalinity but the "inorganic carbonate concentration"; and not TDS but possibly specific components or the effect of the ionic strength of the water (34). Much of the literature reported here is written in terms of the traditional parameters (hardness, alkalinity, TDS), however, the factors noted above should be kept in mind when reviewing the corrosion literature. Nevertheless, research should continue to determine the specific mechanisms that contribute to the reported correlations and contradictions. Physical characteristics of the water system such as water velocity also influence lead corrosion. The length of time the lead pipe or material has been in service will also affect the corrosion rate. New lead pipe or materials are more susceptible to corrosion than older or used materials.

The occurrence of lead corrosion in potable water supplies is most prevalent among utilities which distribute corrosive or "aggressive" waters through lead pipes. In general, soft waters containing dissolved oxygen, carbon dioxide, and organic acids are corrosive to lead. However, many investigations completed to date have addressed the corrosive effects of a combination of parameters rather than a single parameter because of the interactive effects. Thus, in reviewing the literature, the reader should reflect on (1) how the parameter being correlated to corrosion reflects the actual corrosive mechanism, and (2) whether this mechanism, as reflected by the parameter, stands alone or is influenced by other factors inherent in the investigation.

Effect of Flow Rate and Volume of Water Flushed--

Because corrosion is a rate process, lead concentrations in water exposed to lead surfaces will generally reach higher levels in standing water than in running water and, consequently, a range of concentrations can be expected from a given sampling point. In a survey of homes in Worcester, Massachusetts completed in 1975 by O'Brien, nine pairs of water samples were collected to compare lead concentration in standing water and running water (74). The results of that survey are provided in Table 17 and show that lead concentrations up to 1.90 mg/ℓ were observed in standing water while no lead concentrations were observed in running samples.

TABLE 17. RESULTS OF WORCESTER LEAD SAMPLING ANALYSIS PROGRAM (Lead Concentration, mg/ℓ) (74)

LOCATION	STANDING	RUNNING
A	0.05	0.00
B	0.08	0.00
C	0.10	0.00
D	0.06	0.00
E	0.04	0.00
F	0.00	0.00
G	0.06	0.00
H	0.17	0.00
I	1.90	0.00

Wong and Berrang investigated potential sources of lead contamination from water supply facilities using lead service pipes and lead-based soldering to join copper pipes (113). In their investigation, they developed correlations between volume of water flushed through the sytem and observed lead concentrations for water supply facilities which were used only on occasion. The results of their study are shown in Tables 18 and 19. This study showed a decrease in lead concentration with increasing volume of water flushed. It is important to note that very high lead concentrations were observed from the facilities which were not used for an extended period of time.

TABLE 18. LEAD IN WATER TAKEN FROM AN OCCASIONALLY USED TAP 24 HOURS AFTER LAST FLUSHING (113)

Volume of Water Flushed Through Tap (ℓ)	Lead Concentration (ppb)
0.005	130
0.020	130
0.060	240
0.125	410
0.310	330
0.615	60
1.220	16
2.425	17
300.000	7

TABLE 19. LEAD IN WATER TAKEN FROM A TAP NOT USED FOR ABOUT SIX MONTHS (113)

Volume of Water Flushed Through Tap (ℓ)	Lead Concentration (ppb)
0.005	300
0.025	3,000
0.055	2,300
0.105	2,100
0.210	2,500
0.415	490
0.920	190
1.920	64
6.920	15
200.000	12

Effects of Dissolved Oxygen

Slunder and Boyd (95) discuss results reported by other investigators concerning the influence of dissolved oxygen on lead corrosion. They indicate that while some authors maintain that lead is insoluble in air-free pure water, others claim that lead is noticeably soluble in pure water free from gases. Despite these discrepancies, there is general agreement that dissolved oxygen increases the corrosion of lead. Slunder and Boyd present the results of Burns' work (Figure 15) showing that lead corrosion in distilled water is directly proportional to the partial pressure of the oxygen in the atmosphere above the water. An oxygen-nitrogen mixture was used, the test waters were saturated with the mixture, and an adequate pressure was maintained over the water surface during the test.

Slunder and Boyd, also stated that lead in a carbon dioxide-free water is strongly corroded because a protective film cannot form on the lead surface. In the reported experiment, the lead surface became coated with small crystals of lead oxide and hydroxide, the water became turbid, and lead concentrations rose to as much as 100 mg/ℓ. Specifics of this investigation were not presented.

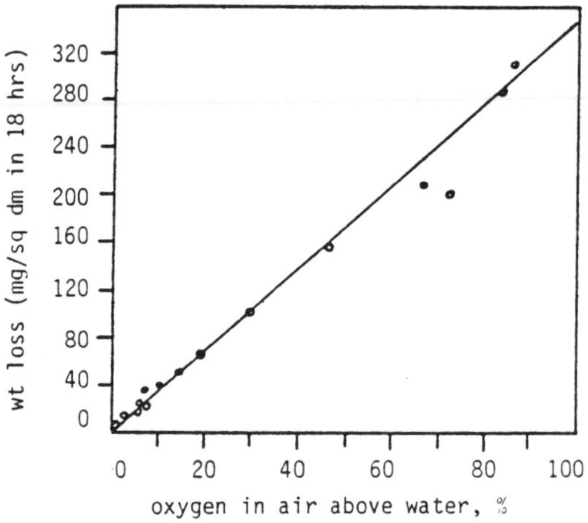

Figure 15. Effect of Oxygen on Corrosion of Lead Submerged in Distilled Water at 75°F (95).

Effect of Hardness--

M. R. Moore of the University of Glasgow attempted to develop a relationship between the rate of lead corrosion and water hardness (68). In his investigation, he synthesized hard water by the addition of calcium chloride to distilled water and measured the rate of dissolution. The results of this study are shown in Figure 16. As is shown, he determined that the dissolution rate decreased exponentially with hardness. However, the specifics of this evaluation were not presented, and it has been suggested that the increased Ca^{2+} and Cl^- contents might have altered the CO_2 absorption tendency, the pH, and hence the lead solvency (34).

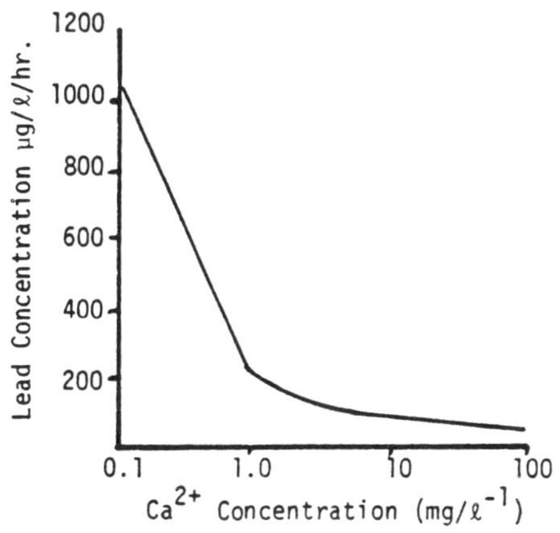

Figure 16. Effects of synthetically hard water on lead corrosion (68).

Others imply that the anions associated with the hardness component are the important factor. In a summary report, Slunder and Boyd explain that most natural waters contain some hardness components which will react with lead to form adherent films, such as calcium carbonate, on lead surfaces which will be protective and prevent further corrosion (95.) According to Slunder and Boyd, a water hardness of 125 ppm as $CaCO_3$ is sufficient to form the protective film and prevent corrosion.

To minimize possible anionic interferences, Naylor and Dague (71) used calcium nitrate and magnesium chloride to produce desired hardness levels. They determined that, in general, variations in the amount of hardness cations had only small effects on lead solubility at the experimentally maintained pH of 10.5.

In soft, aerated waters, corrosion and corrosion rates are dependent on water softness and dissolved oxygen. In general, the softer the water and the higher the dissolved oxygen concentration the greater the corrosion. Additionally, the presence of organic acids whose lead salts are soluble promotes corrosion. Water containing carbonic acids that dissolve calcium deposits will encourage corrosion by forming soluble calcium bicarbonate according to the reaction (95).

$$CaCO_3 + H_2CO_3 \rightleftarrows Ca(HCO_3)_2$$

Effects of pH--

In another experiment, Moore investigated the effects of pH on lead dissolution from a lead pipe (68). In this experiment, water was allowed to stand in a lead pipe section for one hour with pH, adjusted by HCl or NaOH addition, being measured both before and after this time period.

The results of this study are shown in Figure 17 and indicate that the rate of dissolution in distilled water increases considerably on both sides of the pH range from six to eight with a minimum of approximately 1000 μg Pb/liter/hour near pH 6.5.

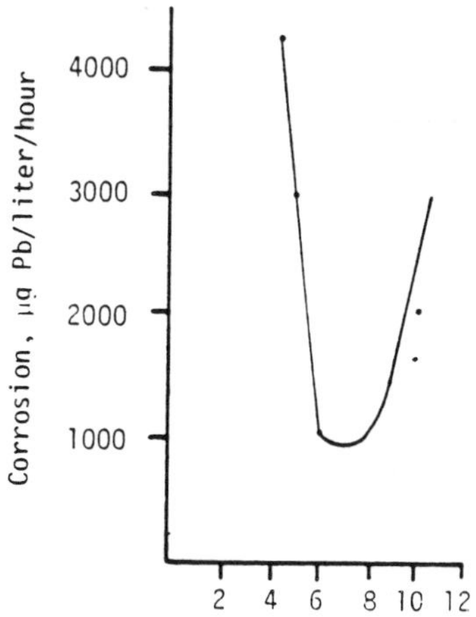

Figure 17. pH effects on lead corrosion (68).

Naylor and Dague (71) indicate that in a solution of pH 8 or less, lead ions will predominate. In the pH range of 8-11, lead precipitates as the lead oxide:

$$Pb^{2+} + 2OH^- = PbO \downarrow +H_2O$$

However at a pH > 11, this oxide will dissolve according to:

$$PbO + 2H_2O = Pb(OH)_3^- + H^+$$

Their experiments on lead control by conventional lime and lime-soda ash water treatment methods produced the lead solubility curve presented in Figure 18. Between pH 9.2 and 10.4, the lead levels were generally < 0.05 mg/ℓ although lead had been added at a rate of 2 mg/ℓ prior to pH control.

When reporting on the occurrence of lead in river systems, Hem and Durum (45) produced soluble lead-pH diagrams with respect to several concentrations of total dissolved carbon dioxide species. Their data indicated that the solubility of lead should be lower than 10 ug/ℓ above pH 8.0, regardless of the alkalinity of the water. However, at a pH near 6.5, and in water with low alkalinity (less than 30 mg/ℓ as HCO_3^-) the soluble lead concentration could range from 40 ug/ℓ up to several hundred micrograms per liter.

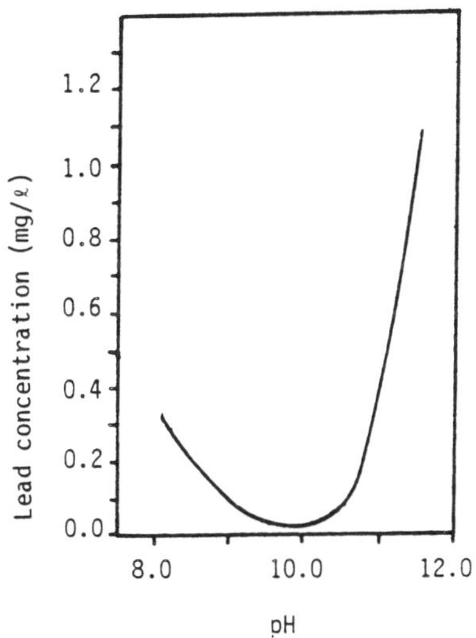

Figure 18. Effect of pH on Lead Solubility (71)

TABLE 20. RESULTS OF INVESTIGATION OF WATER QUALITY ON LEAD CORROSION (50)

Municipality	Finished Water Quality			Average Lead Concentration Observed*			Highest Lead Concentration Observed mg/ℓ
	pH	Hardness (mg/ℓ)	Alkalinity (mg/ℓ)	mg/ℓ	mg/ℓ	mg/ℓ	
Bridgeport, CT	7.1	48	18	0.010	0.011	<0.005	0.04
Marlborough, MA	6.5	14	6	0.014	0.037	0.010	0.250
Chatham, MA	6.3	20	3	0.017	0.018	0.015	0.098
New Bedford, MA	7.3	12	24	0.076	0.090	0.013	0.260
Providence, RI	10.1	40	20	≤0.005	0.006	<0.005	0.050

*SAMPLING INSTRUCTIONS PROVIDED:

After 11:00 p.m., do not use the kitchen cold water faucet until collecting the water samples the next morning. Using the following directions, in the morning, collect the water samples at that faucet before using any faucet or flushing any toilets in the house. Fill the provided containers to one inch below the top and put the caps on tightly.

SAMPLE 1. Open the cold water faucet and immediately fill bottle #1 and turn off the water, recap this bottle.

SAMPLE 2. Turn the faucet on and place your hand under the running water, and immediately upon noticing that the water turns colder, fill bottle #2. Cap bottle #2.

SAMPLE 3. Allow the water to run for three additional minutes and then fill bottle #3. Cap bottle #3.

Effects of pH and Hardness--

Several studies have been completed which correlate lead corrosion with soft acidic waters. Karalekas et al monitored lead concentrations in delivered potable waters at five municipalities and correlated the results with raw and finished water quality (50). Table 20 is a summary of the results of their study. Lead concentrations were monitored for water standing in the interior household plumbing overnight, for water standing in the service line, and for water from the main. These samples are identified in Table 20 as Samples 1, 2, and 3, respectively. Lead concentrations in finished water were below the detectable limit of 0.005 mg/ℓ for all municipalities surveyed.

As can be observed, the water systems at Bridgeport, Connecticut and Providence, Rhode Island which have higher pH values and hardness concentrations experienced the least corrosion. Although the water supply at New Bedford, Maine has a pH comparable to Bridgeport, the hardness concentration is lower and the average lead concentrations are considerably higher. In the case at Providence, Rhode Island, indications of lead corrosion are nearly eliminated by maintaining a pH of 10.1 and a hardness concentration of 40 mg/ℓ. In this study, no differentiation was made between particulate lead (detached from the pipe or an un-adhering fresh precipitate) and lead truly in solution (34).

In another study by Karalekas et al, the effects of water quality, primarily hardness, on lead corrosion were investigated for the cities of Boston, Cambridge, and Somerville, Massachusetts (49). These cities were selected for the investigation because many of the homes in these three cities are known to have lead or lead-lined water service pipes. Boston and Somerville obtain water from the same source. In this investigation, both running and standing water samples were collected. Characteristics of the finished water supplied are shown in Table 21.

TABLE 21. CHARACTERISTICS OF FINISHED WATER SUPPLIED TO CAMBRIDGE, BOSTON, AND SOMERVILLE (49)

Parameter	Cambridge	Boston & Somerville
pH	6.9-8.0	6.0-7.0
Total Dissolved Solids (mg/ℓ)	170	30
Chloride (mg/ℓ)	50	7
Hardness	56	14

Water supplied to residences in Cambridge is higher in pH and hardness concentrations than water supplied to Boston and Somerville. As expected, the percentage of samples exceeding the lead standard of 0.05 mg/ℓ was higher in Boston and Somerville than in Cambridge as shown in Figure 19. Additionally, the percentage of households having detectable levels of lead was higher in Boston and Somerville than in Cambridge as shown in Figure 20.

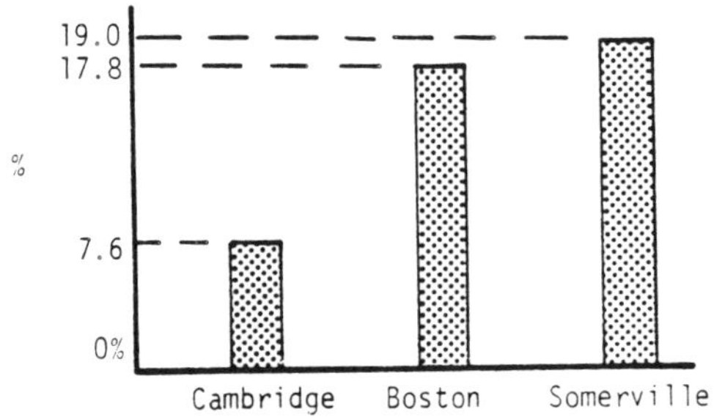

Figure 19. Percentage of samples exceeding lead standard (49).

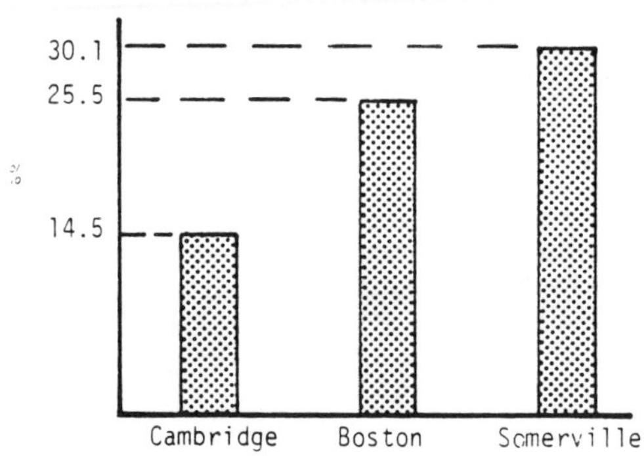

Figure 20. Percentage of households exceeding lead standard in one or more samples (49).

Effects of Alkalinity--

Moderate carbonate alkalinity concentrations have been found to be beneficial in controlling lead corrosion. The presence of this alkalinity will encourage the formation of a very insoluble lead carbonate salt film on the corroding lead surface. This film will adhere to the lead surface and form a protection from the corrosive environment as well as limit lead solubility (77).

In a sampling program to determine the extent of lead in potable water in Boston, O'Brien found that approximately 29.6 percent of the samples analyzed had lead concentrations in excess of 0.05 mg/ℓ (74). O'Brien concludes that the naturally low alkalinity of approximately 4.0 mg/ℓ as $CaCO_3$ is responsible for this high lead concentration occurrence. He also points out that the alkalinity is further depressed by an additional 2.6 mg/ℓ with the addition of hydrofluorosilicic acid to provide 1.0 mg/ℓ of fluoride ion. However, a review of the analysis of water resource characteristics for this study indicate that the water is very soft with a calcium and magnesium concentration of only 2.8 and 1.0 mg/ℓ, respectively. Additionally, the water is slightly acidic with a pH of approximately 6.3. Therefore, this source of water is highly corrosive and it is difficult to specifically identify low alkalinity as the primary cause of high lead levels in Boston's potable water.

The chemical constituency of the protective film is a function of the dissolved species, including hydroxyl, concentrations. Computer generated Eh-pH diagrams presented by Schock indicate that a lead-hydroxy-carbonate, $Pb(OH)_2(CO_3)_2$, may be the predominant solid (91). The tendency to form this hydroxyl-carbonate decreases as the dissolved Pb(II) species concentrations decrease from $10^{-6.0}$M to $10^{-6.62}$M or as the total carbonate concentration increases from $C_T = 10^{-3.7}$M to $10^{-2.7}$M. Solid $PbCO_3$ is reported to form only under low pH, high $[Pb^{++}]$ concentrations.

In a recent investigation, Thibeau et al, used Raman and infrared spectroscopy to analyze the surface film composition of potentiostatically oxidized lead samples in 0.1 M sulfate solutions (100). The results of these analyses were then compared with the composition predicted by the calculated Eh-pH diagram. Only partial agreement existed between the actual observations and the predictions based on thermodynamic equilibria. The major discrepancies were that PbO, although not predicted by the diagram, was found in neutral and basic solutions, and $PbO\text{-}PbSO_4$, the compound predicted to be stable under such conditions, was not. Thus, while Eh-pH diagrams may be an aid in interpreting free energy data, their application to the real world problem of corrosion is limited by (1) their assumption of equilibrium conditions, and (2) their disregard for the effects of other species present. For instance, the presence of ions that form soluble lead salts, such as nitrates, will interfere with the formation of protective films and result in increased corrosion (95), and the influence of sulfates or chlorides should be further investigated, especially in regard to corrosive ground waters used in areas with lead service lines, such as in the Dakotas.

While moderate levels of carbonate alkalinity have been shown to help control lead corrosion, excessive concentrations may, in fact, enhance corrosion. Schaut attempted to duplicate actual distribution conditions and study the corrosive action of various water quality parameters on lead pipe. In one test, he exposed new lead pipe to waters with various alkalinity concentrations over various contact periods. Results of this test indicated that for short contact periods, i,e., less than three hours, alkalinity concentrations have little or no effect on lead corrosion. However, for contact periods of 24 to 48 hours using cool water ranging from 35 to 40°F, he determined that lead concentrations in the water were almost doubled as alkalinity concentrations were doubled. Using water at a higher temperature, the effects of increased alkalinity concentrations were not as pronounced. Tests on old lead pipes, which were approximately 35 years old, gave slightly different results and the effects of increased alkalinity were not as obvious. However, increased corrosion rates were observed with increased alkalinity concentrations. Schaut explains that this difference is due to the formation of a basic lead carbonate which is relatively passive to the range of alkalinity used or observed in drinking water. Schaut did not report the values of alkalinity concentrations which he used (90).

In the investigations of lead removal by conventional water treatment methods Naylor and Dague (71) maintained a 200 mg/ℓ alkalinity (as $CaCO_3$) and varied the pH of their water. As expected, lead concentrations of < 0.1 mg/ℓ were found up to a pH of 8.6 in solutions initially prepared with 2 mg/ℓ lead. At higher pH, and contrary to expectations resulting from their previous work on pH effects (see Figure 18), the lead levels rose to 0.9-1.1 mg/ℓ. Because their investigations included various water treatment unit operations, Naylor and Dague felt that physical, rather than chemical, parameters were responsible for this increase in lead solubility.

Current research at the EPA, Municipal Environmental Research Laboratoreis (MERL) in Cincinnati shows similar results with respect to alkalinity concentrations (35). Studies are being conducted to evaluate the effects several water quality parameters have on lead corrosion. Dr. Marvin Gardels is currently examining changes in alkalinity and changes in pH for water treatment to minimize corrosion. Results to date indicate that corrosion may be enhanced with increasing alkalinity by carbonate addition above approximately 40 to 50 mg/ℓ as $CaCO_3$. Gardels has found that increased pH with a depressed carbonate concentration produces the least aggressive waters. From his research, it appears that an optimum combination of pH and carbonate concentration probably exists for maximum protection against corrosion. Initial results indicate, however, that the optimum pH values may be slightly above 9.0.

On the other hand, Schock presented computer generated activity ratio diagrams (constant Eh) for the total lead (II) soluble species in relation to pH, and dissolved lead (II) versus total carbonate concentrations for several pH values. The results from his studies are shown in Figures 21 and 22. He substantiated the activity ratio diagrams with laboratory data and concluded that the possibility of lead control by alkalinity-pH adjustment was not as great as previously believed, and that in the pH range of 8-9.5, there is little advantage to increasing the carbonate level above 30-40 mg/ℓ as $CaCO_3$ (91).

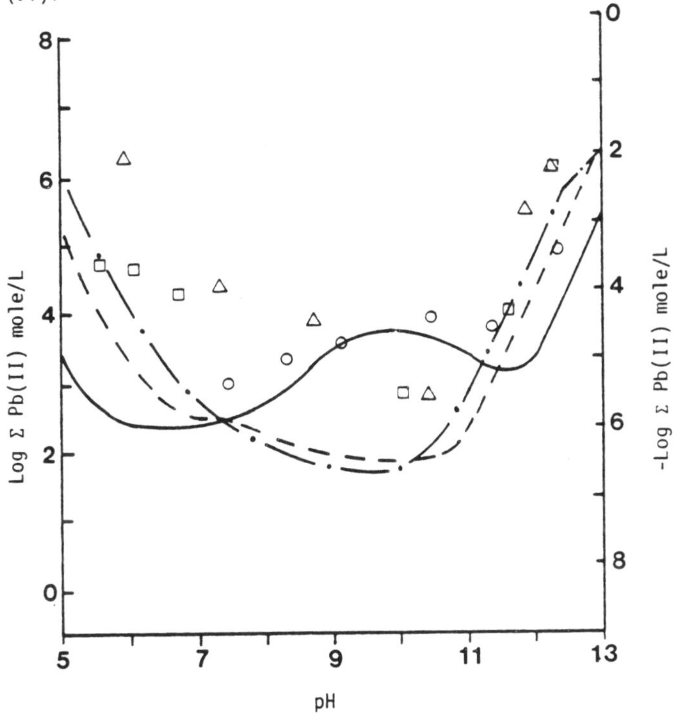

Figure 21. Revised model (I=0.1, 25°C) for log C_t = -3.6 (—— · ——); -2.7 (— —); -1.1 (————). Shown are data from precipitation experiments of Patterson et al (1977) at initial log C_t values of -3.6 (△); -2.7 (□); -1.1 (○) (91).

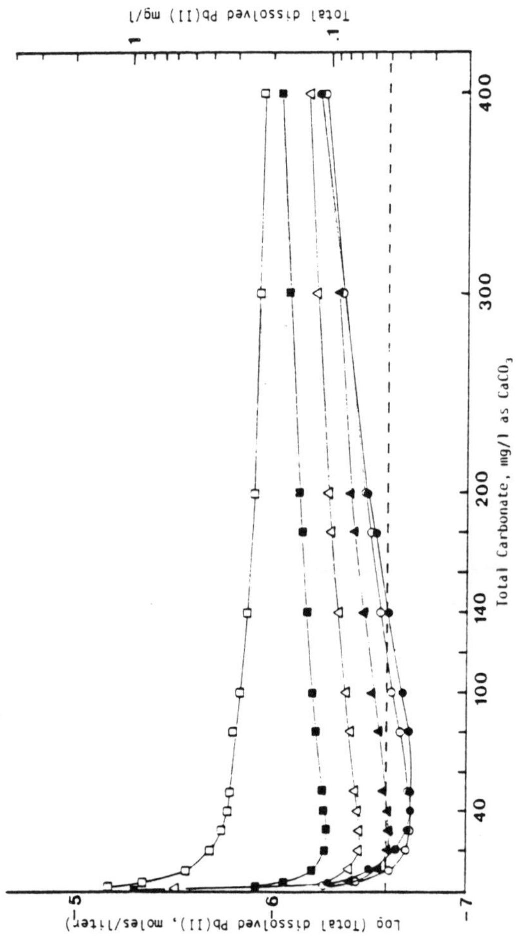

Figure 22. Plot of total dissolved Pb(II) versus C_t at fixed pH (I=0.005, 25°C). Shown are pH = 7.0 (□); 8.0 (■); 8.5 (△); 9.0 (▲); 9.5 (○); and 10.0 (●) (91).

Effects of Temperature--

Results of several investigations have been reported which correlate water temperature with lead corrosion. Using distilled water, Moore determined that the corrosion rate of lead increased exponentially with increasing temperature and developed the relationship:

$$\text{Pb corrosion rate} = 101 \exp(0.01T)$$

where T is temperature (68). Results of his experiment are shown graphically in Figure 23.

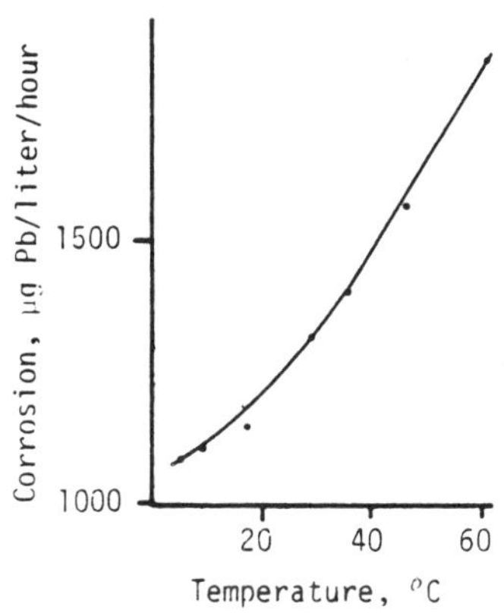

Figure 23. Temperature effects on lead corrosion (68).

Effects of Chlorination--

Schaut attempted to describe the corrosive action of chlorination on lead pipe. From his tests using municipal water with a pH range of 6.8 to 7.2 and an alkalinity of 35-50 mg/ℓ as $CaCO_3$, he concluded that in new lead pipes, the rate of chlorine dissipation is dependent primarily on temperature. Additionally, he observed that when the chlorine was exhausted with increasing temperatures, the water acquired a lead content which approximated the formation of $PbCl_2$. Under all temperature conditions the lead concentration value for new lead pipe was 0.38 ppm. Duplicating his experiment using old lead pipe, Schaut found that the lead concentration in the exposed water did not reach a value of 0.38 ppm even with up to three days contact (90). From the results of his data, Schaut concluded that chlorine contributes its lead equivalent on a percentage basis about equally in old and new lead pipe at maximum potable water temperature, at least for his expermental eight-hour contact period. Additionally, he concluded that the time it takes for water to acquire 0.1 mg/ℓ lead in new lead pipes is approximately 1/4 hour using warm water with a chlorine residual of 0.12 ppm. With cooler water the time required is increased to approximately 1/2 hour (90).

Schaut also investigated the combined effects of water temperature and chlorine concentration and concluded that the combination of chlorine and warm water is more corrosive than warm water alone. In his experiment, Schaut held alkalinity and residual chlorine concentrations constant and varied temperature. Corrosion measurements were made after an eight-hour contact period. Results from this test showed a linear rather than exponential relationship with a doubling of the corrosion rate correlated with a doubling of temperature. Again, Schaut did not provide numerical values.

Effects of Carbon Dioxide--

In a summary report by Slunder and Boyd, results of previous research on the corrosive effects of carbon dioxide content are discussed. Water containing CO_2 in the absence of oxygen has little effect on the corrosion of lead. The extent of corrosion when both CO_2 and oxygen are present is controlled primarily by the concentration of CO_2. Figure 24 is a graphical summary of results reported by several investigators and prepared by Slunder and Boyd which shows the effects of CO_2 content on the corrosion of lead (95). Unfortunately, the specifics of how the CO_2 was added to the water or how the pH was maintained are not presented. As can be seen, when less than 2 mg/ℓ of carbon dioxide is present, corrosion proceeds linearly at an appreciable rate, but at a CO_2 concentration of 60 mg/ℓ, the corrosion rate is much lower. During the tests used to develop this data, a white deposit, probably a basic lead carbonate, was formed with the water having the higher CO_2 concentration (95).

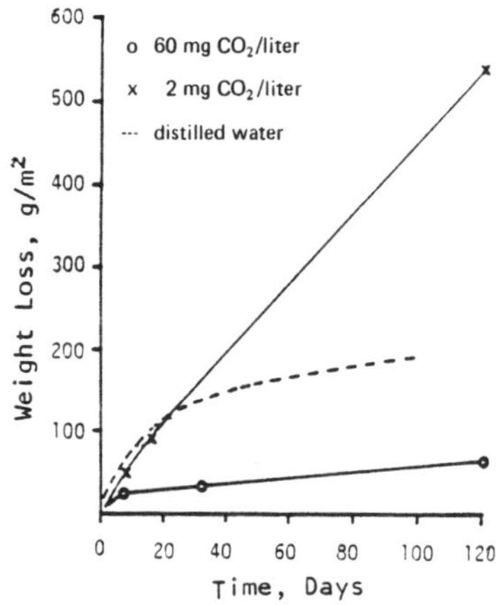

Figure 24. Effects of carbon dioxide content on corrosion of lead (95).

Lead Release from Solder Joints--

Although the primary source of lead in potable waters is thought to occur from lead service lines and lead-lined galvanized pipe used in household plumbing, several studies have been completed to quantify the rate of lead corrosion and water contamination from lead-based solders. Lyon and Lenihan measured the magnitude of lead released from solder joints of copper pipes and found the results to be much higher than expected (65). Their laboratory experiment consisted of a running water test and a static water test using deionized water. In the running water test, water was circulated through a loop constructed of copper tubing with lead-based capillary joints. From the results of the tests, they calculated a mean lead release of 322 μg/fitting/16 hours for the running water test and a mean lead release of 216 μg/fitting/16 hours for the static test. After four to five weeks, a mean lead release of approximately 20 μg/fitting/16 hours was observed from the static tests. These results were favorably compared to measurements taken from capillary joints obtained from a five-year old building which showed a mean lead release of 22 μg/fitting/16 hour period. Lyon and Lenihan concluded that an initial release ranging from 200 to 300 μg/fitting/16 hours can be expected after the first four to five weeks of operation, but that this release will decrease to approximately 20 μg/fitting/16 hours and will be maintained for a long period. It should be noted that these release rates were developed using deionized water. Lyon and Lenihan also noted that the magnitude of the pick-up experienced in the copper tubing was unexpectedly high considering the relatively small surface area of solder exposed. From this observation, it was concluded, through further experiments, that the mechanism for the corrosion process results from galvanic action.

Wong and Berrang also attempted to determine the corrosion or lead pick-up rate of lead-based solders used for joining copper pipes (113). In their experiment, they simulated household copper tubing using 50 feet of one-half inch diameter copper tubing soldered together with 20 soldered joints using 50/50, 60/40, and 95/5 (tin/lead) solder, as well as silver solder. The results of this test for various volumes of water flushed are shown in Table 22. From their experiments, they concluded that an average dissolution rate of 0.4 μg/joint/hour can be expected after one year of service. This value was favorably compared to results obtained from measurements taken from an existing system in a one-year old house. In other experiments, they determined that old lead service pipe will experience a dissolution rate of 30-240 μg/hour and that new lead service pipes will experience a dissolution rate of 480 μg/hour.

TABLE 22. LEAD CONCENTRATIONS (PPB) IN WATER STAGNANT FOR ONE HOUR IN A NEW SIMULATED HOUSEHOLD COPPER PLUMBING SYSTEM (50 Feet Copper Tubings Joined by 20 Soldered Joints) (113)

Solder	Volume of Water Flushed Through the System (ℓ) 80	1,200	12,000	25,000	150,000
50/50	1200	150	96	34	9
60/40	1100	130	49	25	7
95/5	3	2	--	1	--
Silver	2	2	--	1	--
Copper Only	1	2	--	1	--

In his work on heavy metals release from residential plumbling, Rossum (84) determined that the typical lead "sweat" fitting provided a clearance of 0.002 to 0.005 inches between the outside of the pipe wall and the inside of the fitting. Thus a ½ inch nominal size type L water tube would have an average of 2.2 square milimeters exposed solder area. Furthermore, the solder alloy was anodic to copper by 0.3 to 0.4 volts in tap water. Rossum reported that lead was released into tap water, regardless of the water quality, from new household plumbing but the length of release varied. Where a calcium carbonate film was able to deposit on the pipe, the current established by the solder-copper galvanic cell was reduced to the extent that within a few weeks the lead release was undetectable. When the film deposition did not occur, the water continued to pick up lead for longer than a year. Rossum also reported that the corrosion inhibition by formation of the film is more effective in flowing than in standing water situations. He also noted that lead pick up may occur from lead impurities used in zinc galvanizing or from brass faucets (typically composed of 6% lead) that may display an exterior of chrome but are seldom plated on the interior walls. The possibility of lead-related health disorders caused by the use of lead solder is documented and problems of excessive lead concentrations occurring in portions of Carroll County, Maryland are presented as a case history in section 6.4.

CORROSION OF ALUMINUM IN THE WATER WORKS INDUSTRY

The use of aluminum is relatively new to the water works industry so its application is presently limited. However, because its corrosive behavior is generally good, aluminum is currently being considered for more extensive use. Typical applications of aluminum in the water works industry include wier gates, storage tanks, reservoir roofs and supports, hot water systems, and water pipelines (9).

Alloys used to manufacture aluminum materials for handling fresh waters include copper, magnesium, silicon, iron, manganese, chromium, zinc, and titanium. Their composition is shown in Table 23. These alloys are sometimes clad with a sacrificial alloy to provide cathodic protection to the core metal. Corrosion induced penetration of an aluminum alloy cladding layer, anodic to the core, will spread laterally after reaching the core. However, as the area of the core is exposed, the resistance of the electrolytic path is increased and penetration of the core may proceed (6).

The corrosion behavior of aluminum is generally good owing to protection afforded by oxide or hydrated oxide films formed on the metal surface. However, corrosion of aluminum in fresh waters can be severe depending on the composition of the water and on the conditions of service.

Several investigators have reported the results of their studies of the effects of fresh waters on the corrosion of aluminum. In general, as aluminum corrodes, most of the surface is usually unaffected while the attack takes the form of small scattered pits. General corrosion or gradual uniform thinning does not occur (38). It has also been observed that, initially, corrosion occurs rapidly through the formation of a large number of pits, but slows considerably in a short period. Porter and Hadden have studied the corrosion of aluminum and have characterized this primary type of corrosion as "nodular pitting" (79). They observed that a mound of insoluble aluminum hydroxide forms on the surface of the exposed aluminum while an acidic liquid builds up underneath, causing the initiation of pits. The rate of corrosion is initially rapid but stabilizes after about two weeks as a cathodic scale forms on the surface of the pit (79).

TABLE 23. COMPOSITIONS OF TYPICAL ALLOYS USED IN FRESH WATERS (6)

Alloy	Cu %	Mg %	Si %	Fe %	Mn %	Cr %	Zn %	Ti %
1C		*	0.23	0.25		*	*	0.01
N3	0.04	*	0.18	0.32	1.18	*	*	0.01
N4	0.10	2.07	0.27	0.27	0.36	*	*	0.01
N5	0.04	3.60	0.17	0.40	0.30	*	*	0.01
N6	0.04	4.94	0.19	0.15	0.31	*	*	0.01
N8	0.03	4.25	0.14	0.24	0.70	0.12	*	0.01
H30	0.04	0.72	0.87	0.27	0.70	*	*	0.01
H19	0.03	0.65	1.04	0.27	0.02	*	*	0.01
H20	0.03	1.01	0.57	0.39	0.12	0.21	0.04	0.05

*Not normally added but may be present at an impurity level.

The predominance of nodular pitting was later confirmed by Davies, and Rowe and Walker (21, 85). Davies investigated the effects of various water quality parameters on exposed aluminum specimens. During his experiments, he observed that bubbles formed on test specimens within 2 to 4 hours after being submerged. Next, white corrosion products developed around the bubble and enveloped it with pit formation beginning within 18 to 24 hours following immersion (21).

Rowe and Walker, investigating the effects of various water quality characteristics on corrosion of aluminum, made similar observations. They also concluded that pitting corrosion was the predominant form of corrosion of aluminum. They observed that the gas bubbles collected on the test specimen soon after being submerged in the test solutions were indicative of the pit sites. At these sites, corrosion mounds began to develop and grew in size with time as gas bubbles continued to rise form the center of the mound. After a longer period, the growth rate of the mounds began to decline. Mechanically removing the corrosion product mound revealed the formation of a pit (85).

Two approaches have been utilized in aluminum corrosion investigations. Porter and Hadden used a qualitative assessment by visual examination and a quantitative assessment by measuring loss of weight, density of pitting, and depth of pitting with all pits on every specimen being measured (79). This basic methodology, with slight modification, was used by others in more recent studies. Rowe and Walker, however, studied the effects of mineral impurities in water on the corrosion of aluminum using an electrical conductance method by passing a known quantity of current through a specimen and measuring the voltage drop. This method does not produce an "absolute" corrosion rate measurement as pits may perforate the specimen. However, a loss of metal by pitting is reflected in the measurement and the method is useful for comparative studies (85).

The effects of various water quality characteristics on the corrosion and pitting of aluminum have been investigated extensively. Characteristics which have been identified as influencing corrosion include pH and total hardness as well as the presence of chlorides, dissolved oxygen, and metal ions. Conditions of service which influence corrosion of aluminum primarily include water velocity, temperature, and time of contact.

Effects of Velocity

Most investigators agree that the corrosion of aluminum occurs more readily in slow moving or stagnant waters than in fast moving waters. In corrosion tests completed by Wright and Godard, it was shown that, in general, as velocity increased corrosion decreased and, from actual field observations, no pitting occurred on aluminum which was exposed to a water velocity of 7 feet/second (115). Other laboratory tests by Godard showed similar results (38). Aluminum specimens immersed in a stagnant water pitted normally while specimens exposed to the same water but at a water velocity of 8 feet/minute showed no pitting. The results of tests by Wright and Godard using Kingston, Ontario tap water are shown in Table 24 (115).

TABLE 24. WATER VELOCITY EFFECTS ON PITTING OF ALUMINUM (115).

	Control Panels in Still Water		Test Panels in Moving Water	
Water Velocity (fpm)	Avg. # of Pits Per Panel	Avg. Max. Pit Depths (μ)	Avg. # of Pits Per Panel	Avg. Max. Pit Depths (μ)
1	95	206	244	148
2	360	156	145	107
3	126	176	26	79
4	174	227	58	90
5	119	236	26	50
6	142	156	15	35
7	347	133	50	29
8	59	155	0	0
10	85	188	0	0

Excessively high velocities may, however, enhance corrosion. From field studies conducted by Godard, it was observed that at water velocities of approximately 20 feet/second, turbulence occurred, especially at fittings, resulting in pitting (38).

Effects of Temperature

Godard also investigated the effects of water temperature on the incident and growth of pits on aluminum (38). The results of his investigation are shown in Figure 25. From Figure 25 it is shown that as the temperature rises the probability of pitting increases and the pitting rate decreases. In other experiments, Godard measured the current flow from machined pit specimens exposed to various water temperatures (38). The results of that experiment are shown in Figure 26. Godard found that the current flow reached a maximum at around 40°C and dropped off quickly as the temperature increased. At 70°C corrosion, as indicated from the current flow, was below that observed at room temperature. In other experiments, Godard found that the current flow decreased linearly over the entire temperature range and no maximum was observed. From his experiments, he concluded that at above 40°C, the service life of aluminum equipment would increase with increasing temperature (38). Consequently, aluminum is well suited for domestic hot water system applications.

Godard also determined that in the pitting of aluminum, the rate of penetration follows a rapidly decreasing rate curve that approximates a cube root function. From examination of laboratory pitting data, he concluded that the maximum pit depths (d) were proportional to the cube root of time (t) and he described the rate of penetration by the expression

$$d = Kt^{1/3}$$

where K is a function of the alloy and water characteristics. Actual time is measured from the initiation of the pit. This expression has been verified using actual field observations (38).

Water Quality Effects

As early as 1920, Seligman and Williams investigated the effects of various combinations of chlorides, sulfates, carbonates, and bicarbonates on the corrosive behavior of aluminum. From the results of their investigation, they concluded that pitting of aluminum is due to the simultaneous presence of chloride and bicarbonate in water provided there is free access of oxygen to the system (79).

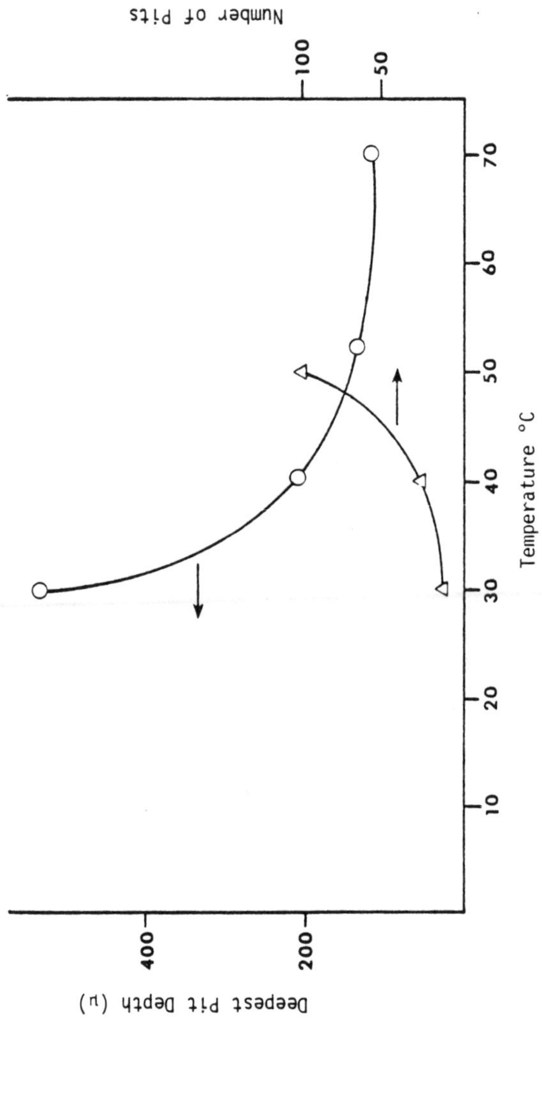

Figure 25. Temperature effects on pitting of aluminum (38).

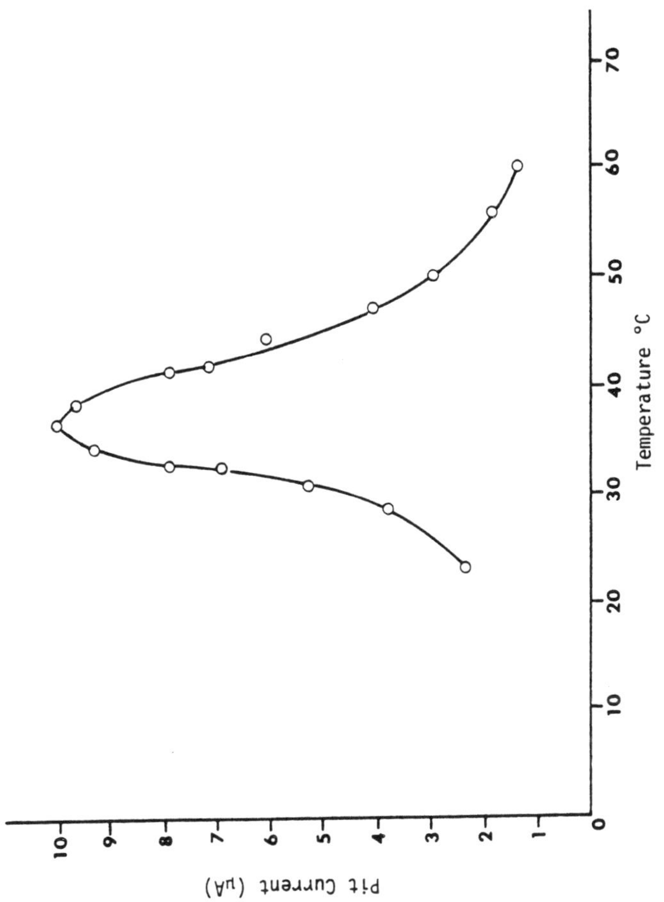

Figure 26. Temperature effects on pit current (38).

In later experiments, Porter and Hadden isolated several water quality characteristics to determine their singular effect on corrosion of aluminum. These characteristics included copper ion, dissolved oxygen, and hardness. In general, they determined that in the absence of copper ions, dissolved oxygen, and hardness nodular type pitting is prevented. They concluded that the characteristics which are necessary for the initiation of pitting include temporary hardness, chlorides, copper, and dissolved oxygen. It was also found that the water composition is more influenced on the corrosion of aluminum than is the composition of the aluminum specimen (79).

Porter and Hadden also investigated the effects of these parameters on the maintenance of pits. These tests were preformed by transferring specimens to other controlled aqueous environments after pitting was initiated. It was found that dissolved oxygen was essential for maintenance of pitting as pitting ceased in de-aerated waters. The removal or absence of copper ion, however, did not prevent the maintenance of pitting, but the rate of pitting was slowed.

In an effort to establish typical pitting curves, Godard compared the constituents of seventeen fresh waters with the resulting corrosion (38). He concluded that no simple correlation exists and, because of the wide variation in the composition of waters, it would be difficult to establish the aggressiveness of waters to aluminum from tables alone. He did conclude from his data, however, that hard waters are generally more aggressive to aluminum than soft waters. The partial composition and pitting data for fresh waters compiled by Godard are shown in Table 25 (38).

TABLE 25. PARTIAL COMPOSITIONS AND PITTING DATA FOR SEVENTEEN FRESH WATERS (In Increasing Order of Pitting Corrosivity) (38)

Order	Test No.	Weeks to 40 Mils.	Location	pH	Hardness p.p.m.	Copper p.p.m.
1	8	953	Shawinigan South, Que.	7.1	73	0.04
2	7	453	Shawinigan, Que.	7.4	18	0.04
3	12	207	Crofton, B.C.	6.7	27	0.024
4	6	205	Hamilton Bay, Ont.	7.1	205	0.003
5	6		Kingston, Ont.	7.9	160	0.005
6	5	175	Credit Valley, Ont.	7.1	0	0.028
7	4	147	Columbia River, B.C.	7.5	72	0.017
8	16	83	Canyon Meadows, Alta.	7.9	169	0.005
9	14	46	Regal Golf Course, Calgary, Alta.	8.1	331	0.005
10	1	25	N. Sask. River,Drayton Vly., Alta.	8.1	267	0.007
11	13	23	Peterborough, Ont.	7.5	86	0.11
12	15	17	R.G.May Golf Course,Calgary, Alta.	7.9	218	0.012
13	22	8	Billingham Beck, England	8.7	443	0.002
14	24	6	Jasper, Alta.	8.2	195	0.011
15	2	6	Lethbridge, Alta.	7.9	228	0.007
16	9	4.4	South Saskatchewan River, Sask.	7.6	206	0.017
17	10	2.6	Mossbank, Sask.	7.8	555	0.04
						0.005

Davies investigated the effects of sodium chloride, calcium carbonate, and dissolved copper on the pitting of aluminum under controlled conditions using water which he composed in the laboratory. All testing was performed under static conditions (21).

In general, Davies observed that pitting occurs more readily in waters containing calcium bicarbonate, chloride, dissolved oxygen, and copper salts. To further characterize the effects of these parameters on the corrosion of aluminum, Davies investigated both the singular effects and the combined effects of two and three constituents.

For the one constituent test, Davies prepared solutions of 10, 30, and 50 ppm of chloride ion in the absence of other ions, and solutions of 10, 80, and 150 ppm calcium ion, as calcium bicarbonate, in the absence of other ions. In each solution prepared, Davies exposed aluminum specimens and observed the pitting or corrosion characteristics. For waters containing chloride ions only, a negligible attack was observed. Even after six months of exposure, the appearance of the test speicmen had not sufficiently changed.

In tests with water containing calcium bicarbonate only, little corrosion was visible. However, specimens showed a slight tarnish which became more pronounced with increasing calcium ion concentration.

For experiments containing two constituents, Davies prepared test solutions by combining chloride and copper ions, calcium and copper ions, and chloride and calcium ions. He observed a slight weight loss in specimens in tests with waters containing both chloride and copper ions. Also, he observed the formation of a few nodular type shallow pits which increased in number and depth with increasing copper content.

Results of tests using solutions containing both calcium bicarbonate and copper ions showed a negligible weight loss in test specimens. The aluminum surface was essentially unchanged in appearance except for a slight dulling.

When the chloride and calcium bicarbonate ions were present in the absence of copper, Davies observed a slight weight loss in the test specimens. Additionally, only slight changes in the appearance of the specimens were observed.

In experiments where all three constituents were present, a very pronounced corrosion effect in the form of nodular pitting was observed. The results of Davies experiments are shown graphically in Figures 27 through 29. In waters where the chloride content was equal to or in excess of the calcium ion content, there was both a general attack as well as a localized attack. The general attack was in the form of a brown stain which was noticeable after two weeks and became more pronounced with time.

Davies compared his results using the laboratory solutions with tests using tap water to determine the effects of the presence of copper ions. The results of these tests are shown in Figure 30. No weight loss was observed with tap water which did not contain copper ions. However, when copper ions

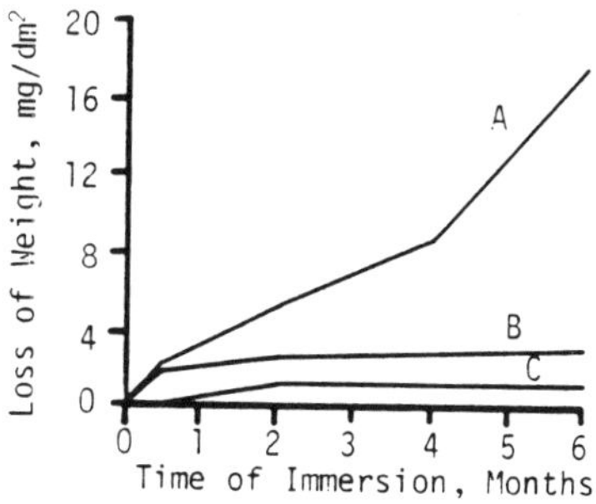

A. Cl^- = 30 ppm, Ca^{++} = 10 ppm, Cu^{++} = 0.2 ppm
B. Cl^- = 30 ppm, Cu^{++} = 0.2 ppm
C. Ca^{++} = 10 ppm, Cu^{++} = 0.2 ppm

Figure 27. Weight loss of aluminum in various water qualities (21).

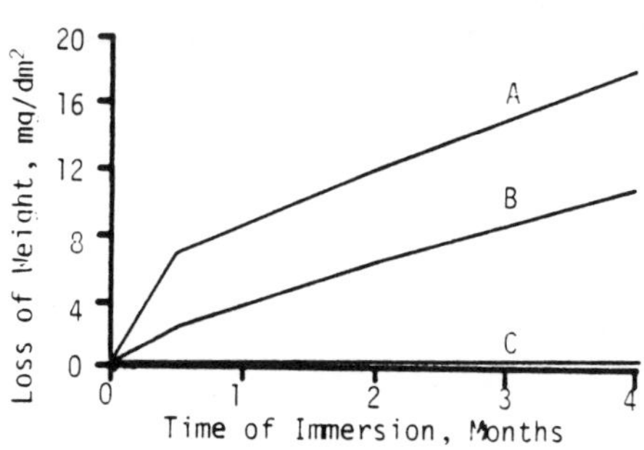

A. Cl^- = 50 ppm, Ca^{++} = 80 ppm, Cu^{++} = 0.2 ppm
B. Cl^- = 50 ppm, Ca^{++} = 80 ppm, Cu^{++} = 0.06 ppm
C. Cl^- = 50 ppm, Ca^{++} = 80 ppm

Figure 28. Effects of copper on weight loss of aluminum (21).

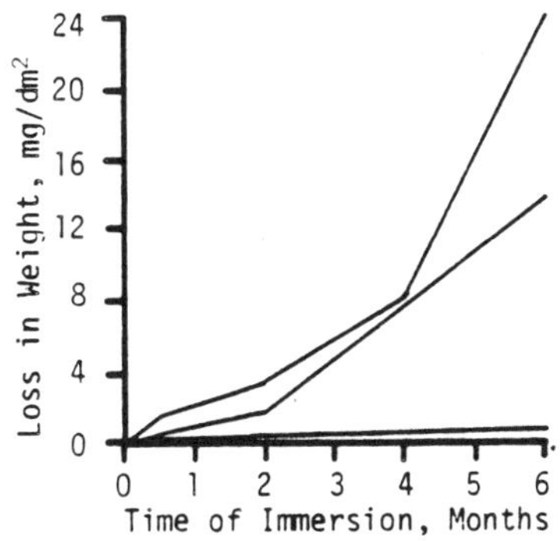

A. Cl^- = 10 ppm, Ca^{++} = 10 ppm, Cu^{++} = 0.2 ppm
B. Cl^- = 10 ppm, Ca^{++} = 10 ppm, Cu^{++} = 0.06 ppm
C. Cl^- = 10 ppm, Ca^{++} = 10 ppm

Figure 29. Effect of low calcium content on weight loss of aluminum (21).

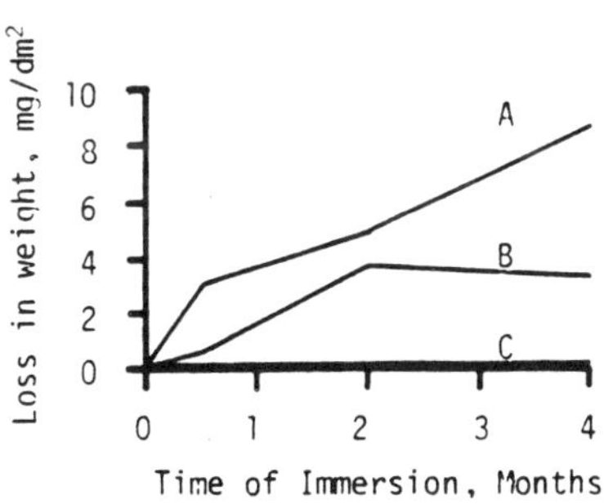

A. Tap water + Cu^{++} @ 0.2 ppm
B. Tap water + Cu^{++} @ 0.06 ppm
C. Tap water only

Figure 30. Weight loss of aluminum in tap water (21).

were added to the tap water, similar corrosion results occurred as with the laboratory test solution.

Because of the significant influence of the presence of copper ions on the corrosion of the test specimens, Davies further studied the effects of chloride and calcium ions by varying the chloride-calcium ion ratio and holding the copper ion constant at 0.2 ppm. The results of these experiments after submerging the specimens for two weeks are shown in Figure 31. The results indicate that with a two-week exposure period, a maximum weight loss is observed with a calcium ion concentrate of approximately 50 ppm. Additionally, weight losses increased with increasing chloride ion concentrations. For solutions containing less than 10 ppm calcium ion concentrations, pits were very small and not of the nodular type. At calcium ion concentrations above 150 ppm, almost no pitting was observed, but the specimens were covered with a whitish deposit. It is important to note that the above observations were made from experimental results obtained after exposure of specimens for a two-week period only.

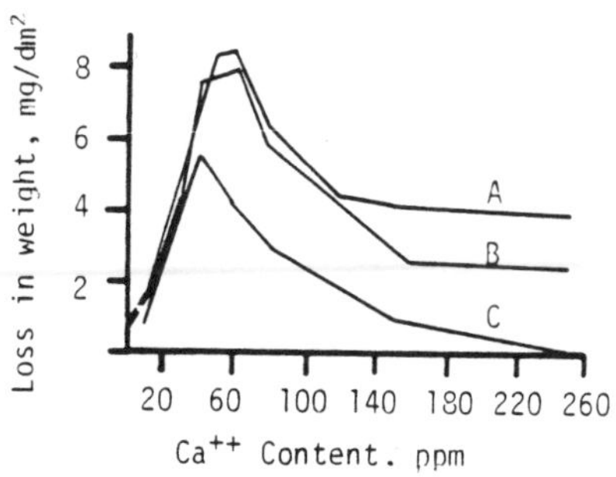

A. Cl^- = 50 ppm, Cu^{++} = 0.2 ppm
B. Cl^- = 30 ppm, Cu^{++} = 0.2 ppm
C. Cl^- = 10 ppm, Cu^{++} = 0.2 ppm

Figure 31. Effect of Ca^{++}/Cl^- ratio on weight loss (21).

Bell also investigated the effects of calcium carbonate on corrosion of aluminum in waters containing chloride and copper. The results of his tests did not provide any evidence of a maximum in the severity of corrosion at any particular calcium or calcium carbonate concentration as previously reported by Davies (5). Therefore, Bell conducted additional tests to identify the apparent discrepancy.

In his experiments, Bell exposed aluminum test specimens in water containing various concentrations of calcium carbonate ranging from 50 to 600 ppm. The chloride ion content was held at 50 ppm and the dissolved copper ion content was held at 0.2 ppm. The pH ranged from 6.5 to 7.4, with the higher values being recorded at the conclusion of the tests. The results of his experiments for various lengths of exposure are shown in Figure 32, and the maximum pit depths observed are shown in Table 26.

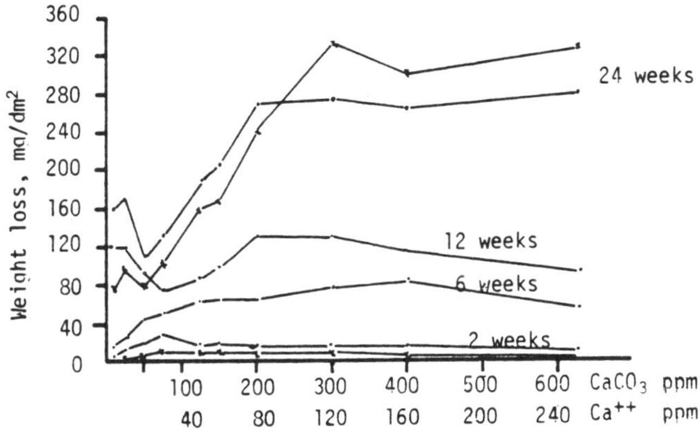

Figure 32. Effect of calcium carbonate on weight loss of aluminum specimens (solutions also contained 50 ppm chloride and 0.2 ppm copper) (5).

TABLE 26. EFFECT OF $CaCO_3$ ON MAXIMUM PIT DEPTHS (mm)
(Solutions also Contained 50 ppm Cl and 0.2 ppm Cu) (5)

Time		2 Weeks		6 Weeks	12 Weeks	24 Weeks	
Material		A	B	B	B	A	B
$CaCO_3$ p.p.m.	Ca p.p.m.						
10	(4)	< 0.02	0.08	0.15	0.13	0.22	0.27
25	(10)	0.11	< 0.02	0.19	0.18	0.22	0.28
50	(20)	0.18	< 0.02	0.28	0.26	0.48	0.25
75	(30)	0.24	< 0.02	0.32	0.41	0.50	0.37
125	(50)	0.24	0.24	0.37	0.44	0.64	0.55
150	(60)	0.17	0.22	0.42	0.48	E	0.51
200	(80)	0.19	0.39	0.32	0.47	E	0.71
300	(120)	E	0.30	0.55	0.61	0.28	0.52
400	(160)	0.12	0.22	0.42	0.42	E	0.67
625	(250)	0.11	E	0.31	E	0.48	0.61

E = attack on the edges only

The results of Bell's studies indicate that a maximum weight loss is dependent on the time of exposure for various calcium carbonate concentrations. For example, for specimens immersed for two weeks, the maximum weight loss occurred in the solution containing 75 ppm calcium carbonate. For those specimens immersed for 12 to 24 weeks, the maximum weight loss was least for waters containing 75 ppm calcium carbonate.

From his tests, Bell observed that when the calcium carbonate content was less than the chloride ion content, corrosion proceeded slowly with a slight general type of attack in the first two to six weeks. After six weeks, a film formed on the surface of the specimen and the corrosion rate increased sharply with the formation of numerous tiny mounds of corrosion product each with a pin point underneath. These pits increased in size with time of exposure. In tests where the calcium carbonate and chloride ions were approximately equal in concentrations, the rate of weight loss decreased with time as a result of film formation and very few pits were formed.

In waters containing more than 125 ppm calcium carbonate, nodular type pitting developed. It was also observed that with increasing calcium carbonate concentrations, the corrosion product mounds increased in size but decreased in number. Additionally, it was observed that these pits tended to form on the edges of the specimen with increased calcium carbonate content.

For waters containing 125 to 150 ppm calcium carbonate, Bell observed that a thick white amorphous film containing very little calcium carbonate was deposited on the specimen. At calcium carbonate concentrations of 625 ppm, no white film was produced and deposits between the pits consisted only of calcium carbonate (5).

From his experiments, Bell concluded that the relative loss of weight of aluminum specimens is dependent on the period of immersion, and for periods greater than 12 weeks, corrosion is least in waters with approximately equal concentrations of chloride and calcium carbonate. He also states that corrosion of aluminum is also dependent on sulfate content and pH and that measuring chloride, calcium carbonate, and copper concentrations is not sufficient to adequately characterize the corrosion phenomena of aluminum (5).

In their experiments, Rowe and Walker investigated the effects of chloride, sulfate, bicarbonate, calcium, and copper on the corrosion of aluminum. They observed that the corrosion rate was low in aerated distilled water. Additionally, no substantial increase in corrosion rate was observed when either chloride, sulfate, bicarbonate, or calcium were added to the distilled water at concentrations as high as 300 ppm, or copper ion up to 2 ppm. The addition of a combination of any two of these constituents also did not produce a substantial increase in corrosion rate. However, the combination of chloride, bicarbonate, and copper ions in the presence of air did produce a significant corrosion rate increase (85).

With these initial results, the additional testing by Rowe and Walker focused on the combined effects of chloride, bicarbonate, and copper. They concluded that a near maximum contribution to corrosion of aluminum occurs at an ion concentration of approximately 300 ppm for chloride and bicarbonate and approximately 2 ppm for copper. The results of these tests are shown in Figure 33 (85).

It should also be noted that while copper has been cited as the most aggressive metal to aluminum, other metal ions such as tin and nickel (79) and mercury (38) have been found to have detrimental effects on the corrosion of aluminum.

Davies conducted corrosion tests on anodized aluminum specimens with oxide films ranging in thickness from 0.05 to 1.35 μ (21). These specimens were immersed in a water containing 40 ppm chloride ion, 40 ppm calcium ion, and 0.2 ppm copper ion for a period of two weeks. This water is known to be aggressive to aluminum and gives rise to pit formation. Some specimens were sealed while others were not.

Results of tests on the sealed specimens indicated that corrosion rate is decreased even for specimens anodized for only a few seconds. One specimen with an oxide thickness of 0.4 μ, which was formed after 50 seconds of anodizing, was found to be immune to corrosion. However, Davies states that if tested longer, pitting would probably occur. The same results were not observed for unsealed specimens and pitting occurred. Davies concluded that for anodizing to be effective, the specimen must be sealed (21).

Booth et al conducted studies to obtain data in an effort to predict the service life of aluminum, primarily pipelines exposed to fresh waters (6). They concluded that, in general, severe pitting of aluminum pipelines, short of perforation, will not significantly affect their service life. From pipe sections which were severely pitted, the materials tensile strength was only slightly affected. The pipe bursting strength was affected, but not to the point that would constitute a failure. Failure would occur first by perforation. Booth et al also reported that for small to moderate diameter aluminum water pipes, the hydraulic efficiency will be reduced approximately one percent per annum over the first ten years of service life (6).

ASBESTOS-CEMENT PIPE PERFORMANCE IN THE WATER WORKS INDUSTRY

Asbestos-cement pipe was first manufactured in Europe in 1913 and was introduced in the U.S. in 1929 (98). Approximately one-third of all water distribution pipe currently being sold in the U.S. is manufactured of asbestos-cement. Since its introduction approximately 200,000 miles of asbestos-cement pipe has been placed into service for transporting potable waters (41).

Asbestos-cement pipe is composed of 15 to 20 percent asbestos fiber, 48 to 51 percent cement, and 32 to 34 percent silica. The cement portion is either Portland cement, Portland blast furnace slag, cement, or Portland pozzolena cement (98).

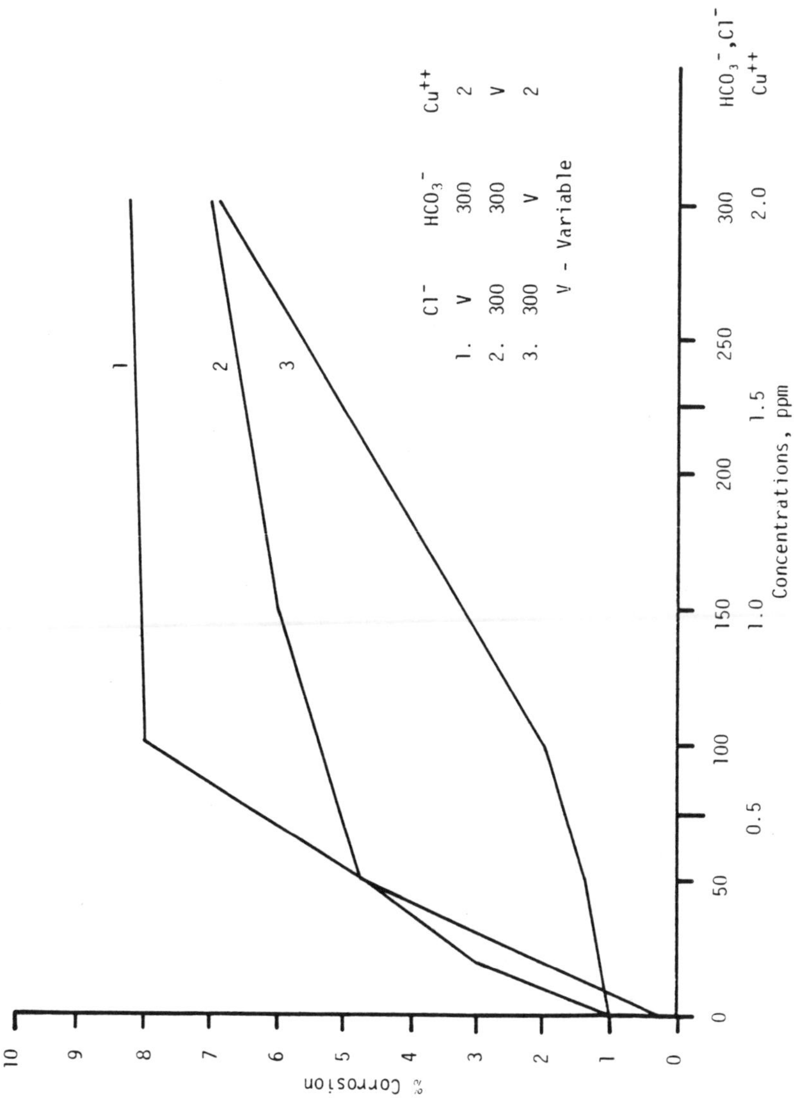

Figure 33. Corrosion of aluminum after 22 hours of exposure in solution containing chloride, bicarbonate, and copper (85).

Asbestos is a generic term representing a number of fibrous silicate minerals. These minerals vary in their metallic content, fiber diameter, flexibility, tensile strength, and surface properties. Six asbestos minerals have been identified for their commercial importance and are chrysotile, amosite, crocidolite, anthophyllite, amphiboles, and actinolete. Chrysotile referes to the serpentine [$Mg_6Si_4O_{10}(OH)_8$] variety of asbestos and comprises 80 percent or more of the asbestos used in asbestos-cement pipe (41). Chrysotile accounts for approximately 95 percent of the world's asbestos production and is mined and quarried primarily in Quebec and Vermont (88). It is identified physically as a tubular or hollow fibrous material. The remaining type of asbestos used for manufacture of potable water pipe is the crocidolite variety. This mineral is a fibrous blue or bluish green silicate of iron and sodium.

Asbestos-cement pipe is available in sizes ranging from 4 to 36 inches in diameter. Type II asbestos-cement pipe is autoclaved while Type I is not. The interior base of the pipe is polished during production and, therefore, is very smooth and requires no interior coating. During installation, asbestos-cement pipe lengths are joined by special couplings which are also made of asbestos-cement. These couplings are machined to fit over the machined ends of the pipe using two flexible O-rings to ensure a watertight seal. Asbestos-cement pipe can be easily drilled and tapped to provide additional service when necessary. Advantages of asbestos-cement pipe include an immunity to electrolysis due to its non-conductor status, and a low hydraulic resistance due to its smooth interior.

Several investigations have been initiated to determine if asbestos minerals are released from asbestos-cement pipe into potable water. In general, these investigations have attempted to correlate various water quality conditions and pipe ages with asbestos fiber releases or occurrences in potable waters after passing through a specified length of asbestos-cement pipe. The results of these investigations will be discussed subsequently. First, however, it is important to note the difficulty in conducting these investigations and obtaining meaningful results. Therefore, a brief discussion of the problems encountered in attempting to qualify or quantify the potential release of asbestos fibers from asbestos-cement pipe into drinking waters follows to provide an appreciation or understanding of the reported results and their limitations.

Asbestos is ubiquitous and asbestos fibers of one variety or another are present in soils throughout the U.S. The most frequent occurrences of near surface asbestos fibers in soils are found in the western and Atlantic seaboard states (51). These asbestos fibers are also present in our nation's water supply sources as they are leached from the soils by runoff and recharge water. Natural wind erosion and earth disturbances will also act to transport asbestos fibers into water ways. Therefore, it can be aniticipated that appreciable amounts of asbestos fiber may exist in potable water as it enters a distribution system.

To determine if asbestos fibers are released from asbestos-cement pipe in potable water distribution systems, it is necessary to quantify incremental changes in asbestos concentrations or fiber counts as the water enters and passes through the pipe. Observed incremental changes in asbestos concentrations or counts, however, do not necessarily indicate a release from the asbestos-cement pipe. Increases in fibers or fiber concentrations may result by contamination from the surrounding serpentine soil which remains in the pipe following construction or repairs.

In one study, conducted by the Vermont Department of Health, Sargent reported that asbestos fibers can appear in potable water distribution systems which do not use asbestos-cement pipe (88). The objective of their study was to compare various sources and to determine if asbestos fibers were picked up in distribution systems. To eliminate the effect of the ubiquitous nature of asbestos and its possible presence in source water, the investigators took samples of both the source and the distribution system and compared the results of the analysis. The results showed that in 17 out of 23 systems initially sampled, the number of asbestos fibers increased from source to distribution, while in six systems it actually decreased.

Incremental increases may also result from drilling and tapping operations when the interior surface of the pipe is disturbed. Although the release of asbestos fibers via drilling and tapping is directly associated with the use of asbestos-cement pipe, measured incremental increases observed during field or laboratory studies should not be construed as normal release of asbestos fibers from the smooth interior surface. Therefore, during such investigations, any drilling and tapping or other pipe disturbances must be identified for corrections.

The analytical procedures used to determine asbestos fiber release in asbestos-cement pipes is presented in Section 5. However, it should be noted that current techniques are based on microscopic quantification may be specific to a certain type of fiber and may not report fiber size. While the effect of ingested asbestos fibers on health has not been determined, it is assumed that the type and size may be important fiber parameters (51). Also, these techniques are often imprecise and generally valid to within an order of magnitude.

Causes of Asbestos Fiber Release

Several investigations have been conducted to ascertain that asbestos-cement pipe does undergo deterioration resulting in the release of asbestos-cement fibers in potable water systems. Other more recent investigations have been initiated in an attempt to identify and quantify the various characteristics which affect asbestos-cement pipe performance. The results of most studies reported to date indicate that structural deterioration is usually negligible even with apparently high asbestos fiber counts, although a measurable decrease in pipe thickness may occur. Primary characteristics identified for examination include water quality, detention or pipe-water exposure time, pipe age, and installation practices.

Hallenbeck et al investigated the effects of pipe age on the release of asbestos fibers (41). For this study, paired samples were taken from 15 public water systems in Northeast Illinois. The transmission electron microscope analysis technique was used to detect and count chrysotile fibers. Paired samples were collected for comparison and were representative of before and after passing through asbestos-cement pipe. The field data collected and the results of this study are shown in Table 27.

As can be observed, a wide variety of water quality characteristics were investigated. Consequently, the authors performed statistical tests on the before and after sample pairs. Although some increases occurred, the authors concluded that no statistical significant release of chrysotile fibers was observed. In some analyses, it was found that the fiber counts increased. From fiber length measurements, it was determined that this increase was probably due to breakage in fiber as the fibers were generally shorter in the after exposure samples.

Tracy also investigated the effects of pipe aging on the release of asbestos fibers from asbestos-cement pipe (104). In this investigation, water quality samples for pH, hardness, and alkalinity were collected from various locations in the distribution systems and changes were observed. Asbestos-cement pipe sections of various ages were selected for this study from three communities in Vermont which were Brattleboro, South Shaftsbury, and South Burlington. Water quality observations were continued over nearly a four-year period to identify any effects of pipe aging. Water quality sampling and analytical results from this study are shown in Table 28.

In Table 28 it is shown that significant increases in pH, alkalinity, and hardness were observed from samples collected from the Brattelboro and South Shaftsbury distribution systems which were approximately five years old. Samples collected from these same facilities after they had been in service for approximately nine years showed less significant changes. Additionally, Tracy observed from the results of samples taken from the nine-year-old South Burlington system that only slight changes in water quality occurred in portions of the distribution system where circulation was good as compared to dead-end sections where circulation is minimal. From the results of this study, Tracy concluded the asbestos-cement pipe may stabilize with age and become more resistant to water quality characteristics (104).

Buelow et al investigated the behavior of asbestos-cement pipe under various water quality conditions (9). The specific objective of their study was to determine if asbestos-cement pipe would be attacked and asbestos fibers released under the various conditions. Their approach was to select ten water supply systems throughout the U.S. which utilized asbestos-cement pipe and which had various water quality characteristics with respect to pH, calcium hardness, and alkalinity. Pipe sections from most of the systems were visually inspected and the samples were analyzed using the electron microscope technique. The results of their study were reported as a correlation between a water quality aggressive index, calculated from the value of the water quality parameters listed above, and the incremental increase in asbestos fibers observed from samples selected.

The aggressive index used for the Buelow et al study follows that procedure identified by AWWA Standard C400-77 which establishes criteria for

TABLE 27. FIELD DATA STUDY - DATA COLLECTED AND RESULTS (41)

WATER SYSTEM	PIPE AGE (Years)	PIPE LENGTH (Feet)	WATER QUALITY		MASS OF CHRYSOTILE FIBERS OBSERVED 10^{-5}g/Grid Square	
			pH	Aggressive Index	Before	After
Groundwater Systems:						
Westmont	0.5	200	7.1	11.2	0.15	0.92
Lisle	1.0	3,000	7.7	12.5	0.13	0.44
Hoffman Estates	18.0	3,700	8.2	12.7	0.28	0
Rolling Meadows	20.0	4,400	7.5	12.2	1.06	0
York Center	27.0	1,100	7.7	12.8	0.03	0.27
Lake Michigan Systems:						
Bannockburn	1.0	1,300	8.4	12.4	0.53	3.78
Bradley Road	14.0	1,000	8.4	12.4	5.97	0.41
Zion-Benton	18.0	10,000	8.1	12.1	0.41	1.25
Waukegan	19.0	600	DATA NOT AVAILABLE		17.21	4.71
Zion	26.0	1,150	8.1	12.1	0.28	0.34
Midlothian	27.0	1,500	8.2	12.2	1.27	1.05
Blue Island	30.0	3,248	8.2	12.2	29.90	6.63
Brook Field	35.0	40,000	8.2	12.2	0.92	0.28
Glenview	37.0	1,500	7.6	11.5	0.39	0.82
Highland Park	40.0	275	7.9	11.8	0.24	0.10

TABLE 28. EFFECTS OF PIPE AGING (104)

BRATTLEBORO
(Installed in 1941)

	Center of Village			N. 1.5 mi on Cement-Asbestos Pipe			N. 2 mi at Dead End		
	9/45	8/46	4/49	9/45	8/46	4/49	4/45	8/46	4/49
pH	7.2	7.2	7.5	9.2	9.1	8.2	9.9	9.6	8.6
Hardness	30.0	31.0	46.0	48.0	48.0	40.0	57.0	51.0	42.0
Total Alkalinity	30.0	24.0	24.0	41.0	36.0	27.0	55.0	50.0	30.0

SOUTH SHAFTSBURY
(Installed in 1940)

	Reservoir			N. 0.6 mi on Cement-Asbestos Pipe			N.W. 1.2 mi at Dead End		
	10/45	9/46	'49 ave.	10/45	9/46	5/49	10/45	9/46	5/49
pH	6.6	7.3	7.2	6.8	7.8	8.3	9.6	8.4	9.5
Hardness	24.0	29.0	32.0	26.0	29.0	24.0	45.0	50.0	40.0
Total Alkalinity	12.0	20.0		15.0	22.0	12.0	35.0	41.0	26.0

SOUTH BURLINGTON
(Installed in 1936; extended in 1948)

	City Line		E. 1.1 mi on Cement-Asbestos Pipe		E. 1.9 mi at Dead End		E. 3.1 mi at New Dead End After 1948
	11/45	8/49	11/45	8/49	11/45	8/49	8/49
pH	7.0	7.4	7.2	7.4	8.2	7.5	8.2
Hardness	54.0	46.0	54.0	50.0	54.0	74.0	72.0
Total Alkalinity	37.0	41.0	41.0	42.0	44.0	45.0	48.0

determining the quality of water that can be transported through asbestos-cement pipe without any adverse structural effects. Although this parameter is often presented in asbestos-cement studies, it is not always accurate in predicting a tendency to release fibers or to allow $Ca(OH)_2$ leaching (34). The aggressive index (AI) is calculated as:

$$\text{Aggressive Index} = \text{pH} + \log\,[\text{AH}]$$

where,

pH = index of acidity or alkalinity in standard pH units
A = total alkalinity in mg/ℓ as $CaCO_3$
H = calcium hardness in mg/ℓ as $CaCO_3$

Values greater than 12.0 identify non-aggressive water; values between 10.0 and 11.9 identify moderately aggressive water; and values less than 10.0 identify highly aggressive waters.

Three of the systems investigated had a water quality aggressive index in excess of 12.0 and are, therefore, considered non-aggressive. Samples collected from these systems were, in general, free of asbestos fibers. Only two samples collected from the three systems which had passed through asbestos-cement pipe had asbestos fiber counts which were statistically significant. The highest value reported was 0.3 million fibers per liter (MFL). In this analysis, a fiber count of 0.2 MFL was also indentified in the water source or at the treatment facility.

Two of the water systems investigated had a water quality aggressive index between 10.0 and 11.9 and are considered moderately aggressive. The first system reported had an aggressive index of 11.56 and the second had an aggressive index of 10.48. Only two samples collected from the first system had fiber counts which were statistically significant. Both values were 0.2 MFL. A third sample taken from the well pump had an asbestos fiber count of 0.1 MFL.

In the second system which had a moderately aggressive water (aggressive index = 10.43), changes in water quality with respect to pH, calcium hardness, and alkalinity were also monitored at two sampling locations. It was observed that pH and calcium concentrations increased as the water passed through the asbestos-cement pipe. This increase indicates that calcium hydroxide or other calcium products in the cement binder were being dissolved resulting in an increase in pH and calcium concentrations in the water, and demonstrates that water aggressive to asbestos-cement pipe will continue to increase in pH and calcium with time of exposure as the water seeks its calcium saturation level (9). In this system significant asbestos fiber counts ranging up to 4.6 MFL were observed. However, because of the large fluctuations in the number of fibers found in various samples, the authors explained the high fiber counts as originating from pipe tapping in the sample collection area.

Five of the ten systems investigated had a water quality aggressive index less than 10.0 and are considered highly aggressive to asbestos-cement pipes. For these five systems surveyed, the aggressive index ranged from 5.34 to 9.51. From the results of this investigation, several important observations were made. In general, water samples taken from the system showed that pH and the aggressive index increased as the aggressive water passed through the asbestos-cement pipe indicating that the asbestos-cement pipe serves a source of pH adjustment. With only one exception, high fiber counts were measured in these water systems having highly aggressive waters as was anticipated. In these tests pipe sections were removed for inspection and pipe deterioration and loosened fibers were apparent where high fiber counts were observed. In one test where asbestos-cement pipe was exposed to a water having an aggressive index of 8.74, the pipe inspection showed that the cement binder had been dissolved to a depth of 1/8 inch.

In another test by Buelow et al, asbestos-cement pipe was exposed to a water having an aggressive index of 6.0 to 7.5 and a pH ranging from 4.5 to 6.0. Although a high asbestos fiber count was expected, very few were actually observed. Additionally, a visual inspection showed little deterioration, but instead the presence of an iron rust-like coating. It is suspected that this iron rust-like coating actually provides a protective coating against pipe deterioration from aggressive water. Susequent laboratory testing confirmed this speculation (9). A summary of the results of the field test completed by Buelow et al is shown in Table 29.

The Environmental Protection Agency Drinking Water Research Division also conducted laboratory studies to investigate the performance of asbestos-cement pipe under various water quality conditions (9). In the initial testing, full lengths of four-inch and six-inch diameter pipes were used in an effort to simulate actual conditions and minimize problems associated with laboratory scale down. However, during the testing, water quality conditions were difficult to maintain as a drift in pH and alkalinity concentrations were observed owing to the exposure of the water supply source to carbon dioxide in the atmosphere. Despite the problems encountered, some interesting qualitative results were observed. For example, it was observed that iron, dissolved in the water from some of the experimental equipment, precipitated and provided a protective coating on the asbestos-cement pipe and halted calcium leaching. From this initial experimental test it was also verified that drilling and tapping of asbestos-cement pipe will generally result in increased fiber counts in water and this increase can be significant (9).

Because of the difficulties in controlling water quality conditions in this initial experimental test, a laboratory scale coupon test experiment was performed. The objective of this study was to investigate the effects of controllable water quality conditions on asbestos-cement pipe deterioration. This study included the use of chemical additives as a corrosion control strategy. A summary of the water quality conditions used in the experiments and general observations made are shown in Table 30.

A comparison between Tests 1 and 2 indicated that the addition of zinc orthophosphate to a concentration of 0.3 to 0.5 mg/ℓ provided protection for the asbestos-cement pipe. It was observed that zinc was gradually depleted but the phosphate was not.

Experimental Tests 3 and 4 were companion tests to further study the potential of zinc orthophosphate for protection at a lower pH and lower aggressive index. The results indicated that the use of zinc orthophosphate at a lower pH or aggressive index was not as effective for preventing

TABLE 29. SUMMARY OF FIELD DATA COLLECTED BY BUELOW ET AL (9)

System	Initial Aggressive Index	pH	Alkalinity mg/ℓ as $CaCO_3$	Calcium Hardness mg/ℓ as $CaCO_3$	Consistently Quantifiable Fibers	Pipe Wall Deteriorated as Determined by Inspection	Significant Observations
1	5.34	5.2	1.0	1.4	Yes	Yes	Water pH and A.I. increased as water passed through A/C pipe; A/C pipe served as source for pH adjustment.
2	5.67	4.8	3.0	2.5	Yes	Yes	High fiber counts were observed in water samples; observation on pipe section removed confirmed pipe deterioration.
3	7.46	6.0	4.0	7.5	No	No	Asbestos fibers were generally absent from water samples; observations of pipe section suggested that an iron rust-like coating provided protection from attack of this highly aggressive water.
4	8.74	7.1	89.0	0.5	Yes	Yes	High fiber counts were observed in water samples; observation on pipe section removed confirmed pipe deterioration.
5	9.51	7.2	14.0	14.5	Yes	Yes	Water pH increased with exposure time to A/C pipe.

Continued

TABLE 29 (Continued)

System	Initial Aggressive Index	pH	Alkalinity mg/ℓ as $CaCO_3$	Calcium Hardness mg/ℓ as $CaCO_3$	Consistently Quantifiable Fibers	Pipe Wall Deteriorated as Determined by Inspection	Significant Observations
6	10.48	8.3	20.0	7.5	Yes	N.I.*	Large fluctuations in water sample fiber counts indicated that pipe tapping may be responsible for the presence of some asbestos fibers.
7	11.56	7.5	88.0	82.0	No	N.I.	Water samples collected were generally free of asbestos fibers as is expected from this moderately aggressive water.
8	12.54	7.8	220.0	250.0	No	N.I.	Water samples collected were free of asbestos fibers.
9	12.74	9.4	50.0	44.0	No	N.I.	Water samples collected were free of asbestos fibers.

*N.I. = Not Inspected

TABLE 30. WATER QUALITY CONDITIONS AND GENERAL OBSERVATIONS FOR SMALL SCALE EXPERIMENTS (9)

Experiment No.	pH	Calcium mg/ℓ as $CaCO_3$	Total Alkalinity mg/ℓ as $CaCO_3$	Aggressive Index	Corrosion Control Method	General Observations
1	8.2	6	20	10.28	None	Alkalinity and calcium concentrations increased significantly during experiment; coupon was softened.
2	8.2	6	20	10.28	Zinc Orthophosphate	Alkalinity and calcium concentrations increased slightly during test; coupon retained hard surface; light gray coating on the pipe surface was observed.
3	7.0	10	20	9.30	None	Alkalinity and calcium concentrations increased; coupon was softened.
4	7.0	10	20	9.30	Zinc Orthophosphate	Alkalinity and calcium concentrations increased; coupon was softened.
5	8.2	6	20	10.28	Zinc Chloride	Alkalinity and calcium concentrations increased slightly; coupon retained hard survace.
5	7.5	145	125	11.76	None	(Unsaturated with respect to $CaCO_3$); coupon was softened.
7	7.9	145	125	12.16	Slightly Positive Langlier Index	Coupon retained hard and clean surface.
3	9.0	25	40	12.00	$CaCO_3$ Saturation	Coupon was slightly softened.

asbestos-cement pipe deterioration. It does, however, appear to offer some protection.

Experiment 5 was performed to determine if zinc alone, not phosphate, was repsonsible for providing protection. Comparison of the results between Experiments 2 and 5 verified that previous observation.

Experiments 6 and 7 were performed to demonstrate the performance of $CaCO_3$ as a protection mechanism under conditions of saturation and unsaturation. For these experiments, pH was used as the controlling variable for $CaCO_3$ saturation. From Experiment 6, it was shown that the asbestos-cement pipe was attacked by a water which was unsaturated or unstable with respect to $CaCO_3$, although the aggressive index was high. Alternatively, Experiment 7 showed that a water which was saturated with respect to $CaCO_3$ did not attack the asbestos-cement pipe.

Experiment 8 was a test of the aggressiveness of water at the point of saturation. This condition is between the conditions tested in Experiment 6 and 7. Results of this test, as expected, showed a slight softenting of the coupon.

Subsequent investigations have developed an asbestos-cement pipe protection model to alleviate problems of improper predictions based on the A.I. by considering the overall water chemistry, and not just the $CaCO_3$ saturation (34).

Organic Release from Asbestos-Cement Pipe

The appearance of significant concentrations of tetrachloroethylene in potable water has recently been associated with the use of lined asbestos-cement pipe. In an investigation performed by the Environmental Protection Agency, pipe sections of lined and unlined asbestos-cement pipe were immersed in a beaker of water and water samples were analyzed at the start, one hour, six hours, and 24 hours later. In these experiments no detectable level of tetrachloroethylene was observed in samples taken from the unlined pipe beaker. However, in the experiments using the lined asbestos-cement, the following results were observed (55):

TETRACHLOROETHYLENE CONCENTRATION (μg/ℓ)

Exposure Time	Test 1	Test 2
0 hour	Not Detectable	Not Detectable
1 hour	8	14
6 hours	25	25
24 hours	41	20

Water quality samples have been collected from the field where lined asbestos-cement pipe sections have been installed. Tetrachloroethylene concentration as high as 2508 μg/ℓ were observed from samples collected at Brenton Point Park in Newport, Rhode Island, in October 1977 (55). Samples collected from a new lined asbestos-cement service line in Newport showed a level of 56.7 μg/ℓ (1). Results showing levels in excess of 30 μg/ℓ have recently been reported in Vermont (55).

In an effort to identify the source of tetrachloroethylene, the Environmental Protection Agency has investigated the techniques used in fabrication and installation of asbestos-cement pipe. Tetrachloroethylene is used to clean the internal surface of asbestos-cement pipe prior to application of the liner. Therefore, it is concluded that the quantity or concentration of tetrachloroethylene which is released to the water is at least paritally dependent on the durability and integrity of the lining (55). It should be noted that this process has been stopped, and no pipes manufactured with the process are being sold.

CONCRETE PIPE

Concrete pipe was first used for transporting potable waters in 1910, but widespread use of concrete pipe did not occur until after 1930. Concrete pipe is composed of Portland cement, sand and gravel aggregates, water, and reinforcing steel. Three types of concrete water pipe are available and are classified in accordance with the method of reinforcement. These three types are steel cylinder, not prestressed; steel cylinder, prestressed; and noncylinder, not prestressed.

Concrete pipe for transporting potable waters can be either prefabricated at a central plant or manufactured on site. Concrete pipe can be constructed in any size, but pipe diameters generally range form 12 to 96 inches. Concrete pipe sizes up to 180 inches in diameter have been produced for water systems.

Concrete pipes are usually coated or lined internally with a specified mixture of mortar or concrete. If the pipe will be exposed to aggressive water, an internal coating of cutback asphalt is sometimes spray applied. Concrete pipe sections are joined with a modified bell and spigot joint, and a gasket is used to ensure a watertight fit. The space between the pipe and the two joining pipes is filled with mortar (98).

Concrete pipe has been used extensively for water distribution with pipe being in service for 50 years or more in some locations. The suitability and acceptance of concrete pipe for water mains is well established, but concrete pipe can be attacked in some circumstances by aggressive waters or soil conditions (94). Additional coatings are applied in such cases.

Although it is not strictly a concrete because aggregate is not present, Portland cement coatings can be applied to protect cast iron or steel water pipe on either the water or soil side or both. The cement protects the underlying from corrosion by the aggressive environments. The coating which may be applied by centrifugal casting, trowelling, or spraying ranges in thickness from 0.25 to greater than one inch. The cement coatings are subject to the same types of attack as concrete pipe. A disadvantage of cement coatings is the sensitivity to damage by mechanical or thermal shock.

However, small cracks in cold-water pipes may be automatically plugged with a reaction product of corrosion combining with alkaline products leached from the cement.

A series of investigations during the 1950's were based on visual inspection and surface layer analysis of cement lined or concrete pipe (29, 30). The samples were removed from various water supply service lines and the following conclusions regarding their deterioration resulted:

1) Concrete pressure pipe is only slightly affected by even aggressive water over service periods of 25 years or longer.

2) As seen in the cement-to-calcium oxide ratios shown in Table 31, the removal of calcium oxide from concrete pipes is limited to a surface layer less than 0.25 inches deep.

TABLE 31. CEMENT-TO-CALCIUM OXIDE RATIO
(With Respect to Depth from Pipe Surface) (29)

City / Depth (inches)	Inside 0.075	Next 0.150	Next 0.150	Next 0.150	Next 0.150	Remaining
Portland ME (3 yrs/service)	1.77	1.54	1.53	1.51	1.54	1.56
Milton PA (9 yrs/service)	1.76	1.71	1.59	1.58	1.63	1.60
St. Petersburg FL (25 yrs/service)	2.24	1.59	1.50	1.48	1.48	1.47

3) Reduction in CaO content is not the controlling factor in determination of the service life of the pipes.

4) The limiting factor in leaching CaO from concrete pipe may be the formation of a surface deposit of magnesium silicate and calcium carbonate.

5) There appeared to be no difference in the amount of CaO leached from either fine or coarse ground cement.

Dissolution of calcium compounds by aggressive waters are the primary concern on the water side of concrete pipe, but attack by soil conditions is also important, primarily to maintain structural integrity. Some soils will react with the cement in the concrete or mortar. Alkali soils contain sulfate compounds that cause gradual deterioration of concrete made with standard Portland cement but there are formulations of sulfate-resistant cement for use in these areas (4). Acid soils may contain sufficient acid to react with concrete pipe or mortar. Cut-back asphalt, coal applied tar, or coal tar epoxy may be used to coat the exterior of the concrete pipe to protect it from the acid content of the soil (4).

PLASTIC PIPE

Commercial plastic pipe was first introduced in 1930 in Germany and later in 1940 in the United States. The first type of plastic pipe commercially available was polyvinyl chloride (PVC). Large-scale production of plastic pipe, however, did not begin until after 1948 with the production of polyethylene (PE) for application in various water uses. Plastic pipe was initially used in the water works industry for service lines and household plumbing, and most pipe was two inches in diameter or smaller. However, with continued development, a larger plastic pipe is now available and is used for water distribution mains, service lines, and in-plant piping systems.

The use of plastic pipe and fittings is steadily increasing in potable water systems as well as in other more corrosive environments. Several varieties of plastics are used in making pipe. Characteristics and physical properties of plastics can vary within a chemical group as well as from one group to another. The two major classifications of plastics are thermoplastics and thermosets, and both are used in the manufacture of pipe. However, thermoplastics are the material of choice for potable water systems. Thermoplastics soften with heating and reharden with cooling which allows them to be extruded or molded into components for piping. Thermosets are permanently shaped during the manufacture of an end product and cannot be softened or changed by reheating.

Total use of thermoplastic piping in 1978 exceeded 3 billion pounds which was approximately one-third of the footage of all piping (69). Approximately two-thirds of the thermoplastic piping manufactured in the United States is used for water supply and distribution, including community and municipal systems and for drain, waste, and vent piping (116). The principal thermoplastic materials in piping are as follows:

1) polyvinyl chloride including chlorinated polyvinyl chloride,
2) polyethylene,
3) acrylonitrile-butadiene-styrene,
4) polybutylene,
5) polypropylene,
6) cellulose acetate integrate, and
7) styrene-rubber plastics.

Other thermoplastics can also be made into piping for special applications. The fist four plastics above account for approximately 95 percent of the total plastic pipe and fittings produced (33). Polyvinyl chloride, polyethylene, and polybutylene are the plastics most often used for potable water supplies. Short descriptions of the various plastics are given below. Typical physical properites of the major thermoplastics are summarized in Table 32.

Polyvinyl Chloride (PVC)

PVC is a good example of the variations that can occur within a chemical group. The properties of the thermoplastic depend on the combinations of PVC resins with various types of stabilizers, lubricants, fillers, pigments, processing aids, and plasticizers. The PVC resin is the major portion of the materials and determines the basic characteristics of the thermoplastic but the amounts and types of additives influence such properties as rigidity, flexibility, strength, chemical resistance, and temperature resistance.

Rigid PVC or Type I PVC are the strongest PVC materials because they contain no plasticizers and the minimum of compounding materials. Type II PVC materials are made by adding modifiers or other resins and are easier to extrude or mold, have higher impact strengths, lower temperature resistance and lower hydrostatic design stresses, and are less rigid and chemically resistant. Chlorinated polyvinyl chloride (CPVC) is a Type IV PVC made by the post chlorination of PVC. CPVC is similar to Type I PVC but has a higher temperature resistance. Both Type I PVC and CPVC materials have a hydrostatic design stress of 2000 psi at 75°F. Type I is useful up to 140°F while CPVC is useful to 210°F.

The long-term strength and higher stiffness of PVC makes it the most widely used thermoplastic for both pressure and non-pressure application. PVC is used in water mains, water services, drain, waste, and vent, sewerage and drainage, well casing, and communication ducts. The higher temperature resistance of CPVC makes it applicable for hot/cold water and industrial piping.

Polyethylene

Polyethylene is a polyolefin formed by the polymerization of the ethylene. Polyethylene plastics are waxy materials that have a very high chemical resistance. The resistance of polyethylenes is such that piping structures must be joined by thermal or compression fittings rather than solvent cements or adhesives. Carbon black may be added to polyethylene to screen ultraviolet radiation.

Polyethylene compounds are classified by the density of the natural resins. Type I materials are low density, relatively soft, flexible, and have low heat resistance. Type I materials have a low hoop stress of 400 psi with water at 73°F and are seldom used for pipe. When used for pipe, Type I is used for low head piping or open-end piping; therefore, it is seldom used in potable water systems. Type II polyethylenes are medium density compounds. These materials are harder, more rigid, resistant to higher temperatures, and more resistant to stress cracking. The high density polyethylenes, Type III, have maximum hardness, rigidity, tensile strength, chemical

TABLE 32. TYPICAL PHYSICAL PROPERTIES OF MAJOR THERMOPLASTIC PIPING MATERIALS (69)

Property @ 75°F	ASTM Test No.	ABS I	ABS II	PVC I	PVC II	CPVC	PE II	PE III	PB	PP	PVDF
Specific Gravity	D-792	1.04	1.08	1.40	1.36	1.54	0.94	0.95	0.92	0.92	1.76
Tensile Strength psi (10^3)	D-638	4.5	7.0	8.0	7.0	8.0	2.4	3.2	4.2	5.0	7.0
Tensile Modulus psi (10^5)	D-638	3.0	3.4	4.1	3.6	4.2	1.2	1.3	0.55	2.0	2.2
Impact Strength, Izod ft-lbs/inch notch	D-256	6	4	1	6	1.5	>10	>10	>10	2	3.8
Coeff. of Linear Expansion in/in-F (10^5)	D-696	5.5	6.0	3.0	5.0	3.5	9.0	9.0	7.2	4.3	7.0
Thermal Conductivity Btu-in/hr-ft-F	C-177	1.35	1.35	1.1	1.3	1.0	2.9	3.2	1.5	1.2	1.5
Specific Heat Btu/lb-F	-	0.32	0.34	0.25	0.23	0.20	0.54	0.55	0.45	0.45	0.29
Approx. Operating Limit*											
F, nonpressure	-	180	180	150	130	210	130	160	210	200	300
F, pressure	-	160	160	130	110	180	120	140	180	150	280

*Exact operating limit may vary for each particular commercial plastic material (consult manufacturer). Effects of environment should also be considered.

resistance, and temperature resistance. Their hydrostatic design stress for water at 73°F is 630 psi.

Many water utilitites use polyethylene for cold water distribution and service lines. The pipe most often used is two inches or less. The toughness, low flexural modules, and chemical resistance are important considerations in water service connections. It is most often used outside buildings.

Polybutylene

Polybutylene is also a polyolefin. Its use in potable water systems has been expanding considerably. Polybutylene is similar to low density polyethylene in rigidity, but its strength is greater than that of high density polyethylene. However, its significant characteristic is its ability to retain strength with increasing temperature. Polybutylene has a hydrostatic design stress of 1000 psi for water at 73°F and 500 psi for water at 180°F. Polybutylene is used for hot and cold water distribution, water distribution and service, gas distribution and services, and industrial piping. The flexibility of polybutylene makes it useful for main-to-meter water service tubing and well piping. It also protects against hot water backup into cold water systems. Polybutylene is used inside buildings for hot and cold water lines.

Acrylonitrile-Butadiene-Styrene (ABS)

ABS plastics are manufactured from the three monomers from which the class name is derived. ABS piping materials are similar to Type II PVC but vary according to the ratios of the component monomers. Acrylonitrile provides rigidity, strength, hardness, and chemical resistance. Butadiene makes the plastic tougher. Styrene contributes gloss, rigidity, and easier processing.

ABS plastic piping is relatively rigid with good impact shrength. The hydrostatic design stresses for water at 73°F range from 800 to 1600 psi. ABS plastic piping may be used up to 180°F in non-pressure applications. ABS may be used to convey potable water but its most common use is for drain, waste, and vent.

Polypropylene

Polypropylene is another polyolefin but it is not as widely used in potable water systems as polyethylene or polybutylene. It is similar to high density polyethylene, but it is more rigid and temperature resistant. Its good chemical resistance makes it more useful in environments harsher than potable water systems.

Deterioration and Release from Plastic Piping

Very little direct information exists on the corrosion or, more appropriately, deterioration of thermoplastic materials in potable water systems. One of the significant features of thermoplastics is the good chemical resistance of the compounds; this feature was responsible for many of the early applications of thermoplastics in handling highly corrosive materials. The ability of thermoplastics to withstand harsh chemical environments has received most of the attention directed toward the corrosion of these materials. Most testing has concentrated on physical properties. Consequently, little attention has been focused on thermoplastics in such relatively mild environments as potable water systems. A recent study by the National Bureau of Standards acknowledges the widespread acceptance of themoplastic piping for residential plumbing and the absence of recent reports of failures due to chemical attack or environmental stress cracking. This trend suggests that these failures have ceased to be of significant concern in the use of thermoplastics in residential and related applications (116).

There are two general types of chemical attack on plastic pipe (33). One is a solubility reaction where a chemical is removed from the plastic, contaminating the fluid flowing in the pipe. The leached chemical may be non-reacted components, reaction products, or impurities, but their leaching should not significantly alter the physical properties of the pipe. From steric considerations, the leachable components probably lie close to the pipe surface. The second type of chemical attack is where a polymer or base resin molecule is altered by chain breakage, cross linkage, oxidation, or substitution reactions. In these cases, the properties of the plastic may be irreversibly altered, and the fluid flowing in the pipe may or may not become contaminated. The chemical resistance of plastics may vary within differenct grades of the same type as a result of minor chemical or process differences. In general, a better chemical resistance exists when smaller amounts of compounding additives are used. Most plastic pipe compounds conforming to ASTM specifications use a minimum amount of compounding ingredients, although CAB plastics may use chemically susceptible monomeric plasticizers while PVC Type II uses chemically resistive impact modifiers.

Compared to metals and other construction materials, thermoplastics are generally superior in resisting corrosion. Thermoplastics are not subject to electrochemical corrosion because they are not conductors. Such electrochemical effects as galvanic corrosion do not occur with thermoplastics. As examples, soils which are corrosive to metal pipes or in which stray currents are present do not present problems for buried thermoplastic pipe. The resistance of thermoplastics alleviate the need for such measures as cathodic protection and special coatings.

Inorganics do not present significant threats to thermoplastics; most are not affected by acid and alkaline salts. Thermoplastics are resistant to polar active compounds such as acids, bases, and brines. The thermoplastics are resistant to chemical concentrations in normal household operations or potable systems. Although most plastics absorb water to a slight extent, water does not produce corrosion or other types of deterioration. Under some circumstances direct chemical attack by inorganic species such as oxygen, chlorine, other strong oxidizers, very strong acids or alkalis, and ultraviolet radiation may lead to deterioration of the plastics. Some thermoplastics such as PVC have additives such as carbon black to protect against ultraviolet rays which might otherwise degrade the long chain structure upon long duration exposure (72). However, it is unlikely that chemical attack by these types of species would be significant in potable water systems because they would have to be present in such large concentrations that hazards greater than thermoplastic deterioration would exist.

Since thermoplastics are organic materials, they are subject to deterioration by reaction with some organic compounds, primarily via a solution mechanism. The solvent cementing of plastic pipe is based on solution. The effect of organic species on thermoplastics varies with the organic compounds and plastics. For example, PVC is not affected by most esters and ketones but cellulose acetate butyrate readily dissolves in most esters and ketones. Aromatic species are the most likely class of compounds to attack thermoplastic piping. However, if organic compounds are present in sufficient concentrations to deteriorate thermoplastic they present other more significant and immediate problems from a water quality standpoint.

Environmental stress cracking is another form of degradation that may affect thermoplastics in piping systems. The process is believed to occur when a surface active agent such as an alcohol or detergent acts on surface flows in a stressed or strained plastic (69). Some degree of stress concentration, particularly at joints or fittings, might arise from 1) forced alignment of pipes and fittings, 2) building settlement, 3) lumber shrinkage, 4) thermal expansion or contraction, or 5) long-term dimensional changes (116). A chemical test for potable water pipe and fittings has been suggested by the Federal Construction Council of the Building Research Advisory Board, but this test as well as others suffers from 1) uncertainties in the representativeness of the conditions and 2) effects of exposure duration (116).

Although the data are limited, there have been several studies of thermoplastic pipe deterioration in potable water and simulated environments (72, 93, 99, 101, 102, 110). Tiedeman conducted studies to determine the possible effects of plastic pipe on the safety, quality, and palatability of water (101, 102). He conducted extraction tests to determine the aggressiveness of several water systems on various thermoplastic pipes. The results of the tests showed that no undesirable substances were extracted from the plastic pipe, with the exception of three samples that were known to contain substances which might be extractable. A typical set of results are shown in Table 33. With a pH 9.6 in test waters, 0.34 ppm lead was extracted from a plastic pipe in which a lead compound was used as a stabilizer. Lowering the pH to 1.0 by adding hydrochloric acid extracted 2.0 ppm lead. However, the results were obtained under extreme conditions of temperature, exposure duration, and area of plastic exposed per unit volume of test waters (101).

Over the course of a three-year study, it was found that the most aggressive potable water was a relatively soft water with the pH adjusted to 5 by adding carbon dioxide (102). This water extracted lead compounds from specially prepared test plastics. However, the extraction results were negative for all specimens of plastic pipe recommended for use with potable water. A Soviet study of the extraction of lead from PVC pipe materials also confirmed that lead stabilizing compounds could be leached from the PVC in potable water supplies (93). However, these results were on Soviet pipes which do not apply in this country. Changes in plastic pipe exposed to

TABLE 33. TYPICAL EXTRACTION TEST RESULTS (101)

Plastic Pipe No.	Color ppm	Turbidity ppm	Odor	Taste	Total Solids Residual	Total Solids Dissolved	Alkalinity Phenol-phthalein	Alkalinity Total	pH	Fe	Al	NO_2	NO_3	Cl	SO_4	Total Hardness	Residual Cl	DO
					ppm					ppm								
none	6	3	0	0	152	156	10	36	9.70	0.1	0.003	0.02	0.02	10.9	53.9	95	0.3	11.2
C	6	5	0	0	168	172	11	35	9.70	0.1	0.003	0.03	0.15	10.9	54.7	108	0.02	8.4
120	8	5	0	med	164	164	10	40	9.65	0.1	0.003	trace	0.08	12.4	54.2	110	0.01	9.0
160	6	7	0	m	172	176	10	39	9.65	0.25	0.014	0.01	0.04	11.5	56.2	97	0.01	9.6
170	6	6	0	med	176	172	9	36	9.60	0.1	0.003	0.04	0.05	11.8	55.2	102	0.01	9.8
none	0	5	0	0	176	176	23	62	9.90	0	0	0.01	0.02	10.9	64.3	114	0.15	10.8
C	0	5	0	0	176	192	20	59	9.70	0	0	0.0	0.04	10.9	62.9	114	0.01	6.6
110	0	5	6.7	med	184	184	21	60	9.85	0.2	0	0.1	0.02	10.2	64.3	116	0	8.6
150	0	5	0	med	188	176	20	58	9.85	trace	0	0.1	0.02	12.1	71.0	118	0.01	9.0
180	0	7	2	med	184	182	20	58	9.75	0	0	0.2	0.02	12.1	63.4	114	0	8.2
none	0	1	–	0	144	144	24	54	9.50	0.1	0	0.002	0.12	10.9	48.2	90	0.8	9.8
C	0	2	–	0	160	128	12	50	9.45	trace	0	0.004	0.20	10.9	41.3	92	0	9.2
200	2	2	–	med	168	140	8	42	9.22	0.1	0	0.004	0.28	10.3	49.4	90	0	9.0
210	0	3	17	med	168	144	6	46	9.20	0.1	0	0.005	0.16	10.7	34.1	91	0	9.0
220	0	1	–	med	160	148	24	68	9.50	trace	0	0.005	0.40	10.7	44.2	92	0.3	9.2

outdoor conditions or buried in soil at pH 2.0 and held at 35°C were slight after exposures of one year. Discoloration was the principal change in both exposures (102).

One concern is the extraction or leaching of organic species from pipe cements into water supplies. A recent study indicated that it is possible to leach such solvents as 2-butanone (MEK) and tetrahydrofuran (THF) from PVC pipe cement (110). Two sets of water samples were collected six and eight months after PVC pipe installation and usage in a laboratory. About 40 gallons of water were used daily in the laboratory. The water temperature was about 21°C. Seven water samples at different residence times in the PVC pipe were taken for analysis. Results are summarized in Table 34. A comparison of the data from the two sets of samples indicates that concentration of both MEK and THF in the second set were reduced to 1/2 of the concentration in the first set. About 2,400 gallons of water were used during the period of samples taken between Set I and Set II. This water presumably removed some of the MEK and THF from PVC pipe cement in the pipe.

TABLE 34. CONCENTRATION (PPM) OF MEK AND THF IN WATER SAMPLES AT VARIOUS RESIDENCE TIMES IN THE PVC PIPE (110)

Residence Time (h)	Samples Taken 6 Months After Pipe Installation		Samples Taken 8 Months After Pipe Installation	
	MEK	THF	MEK	THF
0	0	0	0	0
4	0.4	1.0	0.1	0.7
8	0.6	1.7	-	-
16	1.8	5.8	0.6	2.4
24	2.2	8.9	1.1	3.7
48	3.9	12	2.1	6.3
64	4.5	13	-	-
72	-	-	2.2	7.5
96	4.5	13	-	-

Another series of tests, however, found that concentrations of MEK, THF, cyclohexanone, and dimethylformamide (DMF) did not attain hazardous levels in static water or usage simulation tests (103). An analysis based on results of the tests stated that levels of the four solvents declined to less than three parts per million in less than three weeks of static exposure and that no significance in solvent leaching appears between poorly constructed solvents cement joints and well constructed solvent cement joints. Testing was performed by a private consulting engineering firm while the analysis presented was performed by representatives of the plastic resins, pipe, fittings, and solvent manufacturers. Research in this area is currently proceeding and should help to clarify the reported discrepencies concerning release extents and possible health concerns from organic solvent leaching.

REFERENCES

1. Adams, W. R., Jr., Regional Administrator, EPA, Region I, J. F. Kennedy Federal Bldg., Boston, MA 02203, Letter to Dr. J. E. Cannon, Director, Department of Health, Office of the Director, 75 Davis St., Providence, RI 02098, dated January 14, 1980.

2. Anderson, E. A., Reinhard, C. E., and W. D. Hammel, "The Corrosion of Zinc in Various Waters," J. Am. Water Works Assoc., Vol. 26, No. 1, 1934, pp. 49-60.

3. ASTM Special Technical Publication 516, Localized Corrosion - Cause of Metal Failure, American Society for Testing and Materials, 1916 Race Street, Philadelphia, PA 19103, 1972.

4. Bald, R. E., "Corrosion Resistance of Concrete Pipe," Water Wastes Engineering, Vol. 5, No. 11, 1968, pp. 50-52.

5. Bell, W. A., "Effects of Calcium Carbonate on Corrosion of Aluminum in Waters Containing Chloride and Copper," J. Appl. Chem., Vol. 12, 1962, p. 53.

6. Booth, F. F., Murray, G. A. W., and H. P. Godard, "Corrosion Behavior of Aluminum in Fresh Waters with Special Reference to Pipeline," Br. Corros. J., Vol. 1., No. 2, 1965, pp. 80-86.

7. Bopp, C. D., and S. A. Reed, "Stabilization of Product Water from Sea Water Distillation Plants," U.S. Office of Saline Water Research and Development Progress Report, No. 709, 1971.

8. Brighton, W. D., "Dissolved Copper form New Service Pipes," Water and Water Engineering, Vol. 59, July, 1955, pp. 292-293.

9. Buelow, R. W., Millette, J., McFarren, E. and J. M. Symons, "The Behavior of Asbestos-Cement Pipe under Various Water Quality Conditions," A Progress Report, Presented at the American Water Works Association, 1979 Annual Conference, San Francisco, June 27, 1979.

10. Burgmann, G., Friehe, W., and W. Schwenk, "Chemical Corrosion and Hygienic Aspects of the Use of Hot-Galvanized Threaded Pipes in Domestic Plumbing for Drinking Water," Pipes Pipelines Int., Vol. 23, No. 2, 1978, pp. 11-15.

11. Butler, A. and H. C. K. Ison, Corrosion and Its Prevention in Waters, Reinhold Publishing Corporation, New York, 1966.

12. Caldwell, D. H. and J. B. Ackerman, "Anaerobic Corrosion of Steel Pipe Due to Nitrate," Journal-AWWA, Vol. 38, January 1946, pp. 61-64.

13. Campbell, H. S., "A Natural Inhibitor of Pitting Corrosion of Copper in Tap-Waters," J. Appl. Chem., Vol. 4, 1954, pp. 633-647.

14. Clark, H. W., "The Effect of Pipes of Different Metals upon the Quality of Water Supplies," Journal-New England Water Works Association, Vol. 41, 1927, pp. 31-51.

15. Cohen, A., and W. S. Lyman, "Service Experience with Copper Plumbing Tube," Materials Protection and Performance, Vol. 11, No.2, February 1972.

16. "Cold-Water Corrosion of Copper Tubing," Task Group Report, J.A.W.W.A. Vol. 52, August 1960, pp. 1033-1040.

17. Cornwell, F. J., Wildsmith, G., and P. T. Gilbert, "Pitting Corrosion in Copper Tubes in Cold Water Service," Br. Corros. J., Vol. 8, No. 5, September 1973, pp. 202-209.

18. Costas, L. P., "Field Testing of Valve Stem Brasses for Potable Water Service," Materials Performance, Vol. 16, No. 8, August 1977, pp. 9-16.

19. Cruse, H., "Dissolved-Copper Effect on Iron Pipe," Journal-AWWA, Vol. 63, No. 2, 1971, pp. 79-81.

20. Cruse, H., and R. D. Pomeroy, "Corrosion of Copper Pipes," Journal-AWWA, Vol. 66, No. 8, August 1974, pp. 479-483.

21. Davies, D. E., "Pitting of Aluminium in Synthetic Waters," J. Applied Chemistry, Vol. 9, December 1959, pp. 651-660.

22. Davies, D. H., and G. T. Burstein, "The Effects of Bicarbonate on the Corrosion and Passivation of Iron," Corrosion-NACE, Vol. 36, No. 8, August 1980, pp. 416-422.

23. De Waard, C., and D. E. Milliams, "Carbonic Acid Corrosion of Steel," Corrosion-NACE, Vol. 31, No. 5, May 1975, pp. 177-181.

24. Donaldson, W., "The Action of Water on Service Pipes," J. Am. Water Works Assoc., Vol. 11, No. 3, 1924, p. 649.

25. Eliassen, R., Clemente, P., Romeo, A. J., and R. T. Skrinde, "Effects of pH and Velocity on Corrosion of Steel Water Pipes," Journal-AWWA, Vol. 48, August 1956, pp. 1005-1018.

26. Elzenga, C. H., and H. J. Boorsma, "Corrosion of Water Pipes in Various Types of Drinking Water," International Water Supply Association, 10th Congress, August 1974, pp. 1-7.

27. Feigenbaum, C., Gal-Or, L., and J. Yahalom, "Microstructure and Chemical Composition of Natural Scale Layers," Corrosion, Vol. 34, No. 2, February 1978, pp. 65-70.

28. Feigenbaum, C., Gal-Or, L., and J. Yahalom. "Scale Protection Criteria in Natural Waters," Corrosion (Houston), Vol. 34, No. 4, 1978, pp. 133-137.

29. Flentje, M. E., and R. J. Sweitzer, "Further Study of Solution Effects on Concrete and Cement Pipe," Journal-AWWA, Vol. 49, November 1975, pp. 1441-1451.

30. Flentje, M. E., and R. J. Sweitzer, "Solution Effects of Water on Cement and Concrete in Pipe," Journal-AWWA, Vol. 47, 1955, pp. 1173-1194.

31. Foley, R. T., "Role of the Chloride Ion in Iron Corrosion," Corrosion-NACE, Vol. 26, No. 2, February 1970, pp. 58-70.

32. Fontana, M. G., and N. D. Greene, Corrosion Engineering, McGraw-Hill Book Company, New York, © 1978.

33. Ford, K. C. ed., Plastics Piping Manual, Plastics Pipe Institute, 355 Lexington Ave., New York, N. Y., 10017, pp. 26-28.

34. Gardels, M., and M. Schock, (E.P.A. Cincinnati Laboratory), Personal Communication, via P. Lassovszky, January 12, 1981.

35. Garrels, R. M., Thompson, M. E., and R. Siever, "Control of Carbonate Solubility by Carbonate Complexes," American Journal of Science, Vol. 259, January 1961, pp. 24-45.

36. Geld, I., and C. McCaul, "Corrosion and Coatings Test Program of the NYC Board of Water Supply," Materials Protection and Performance, Vol. 11, No. 2, February 1972, pp. 41-44.

37. Geld, I., and C. McCaul, "Corrosion in Potable Water," JAWWA, Vol. 67, No. 10, October 1975, pp. 549-552.

38. Godard, H. P., "The Corrosion Behavior of Aluminum in Natural Waters," The Canadian Journal of Chemical Engineering, Vol. 38, No. 5, October 1960, pp. 167-173.

39. Goetchins, D. R., "Porcelain Enamel as a Protective Coating for Hot Water Tanks," J. Am. Ceramic Society, Vol. 25, 1942, pp. 164-168.

40. Hale, F. E., "Relation of Copper and Brass Pipe to Health," Water Works Eng., Vol. 95, 1942, pp. 84-86, 156-159, 187-189, 207-208, 240-243, 264-265.

41. Hallenbeck, W. H., et al, "Is Chrysotile Asbestos Released from Asbestos-Cement Pipe into Drinking Water?," Journal-American Water Works Association, Vol. 70, No. 2, 1978, pp. 97-102.

42. Hatch, G. B., "Copper Corrosion, Quality Aspects of Water Distribution Systems," Univ. Ill. Eng. Expt. Sta. Cir. #81, 1963, pp. 32-40.

43. Hatch, G. B., "Unusual Cases of Copper Corrosion," Journal-AWWA, Vol. 53, 1961, pp. 1417-1429.

44. Heidersbach, R. H., and E. D. Verink, Jr., "The Dezincification of Alpha and Beta Brasses," Corrosion-NACE, Vol. 28, No. 11, November 1972, pp. 397-418.

45. Hem, J. D., W. H. Durum, Solubility and Occurrence of Lead in Surface Water, JAWWA, August 1973, pp. 562-568.

46. Hilbert, F., Mizoshi, Y., Eikhor, G., and W. J. Lorenz, "Correlations Between the Kinetics of Electrolytic Dissolution of Iron, I Anodic Dissolution of Iron," J. Electrochem. Soc., Vol. 118, No. 12, 1971, pp. 1919-1926.

47. Hoover, C. P., "The Corrosive Action of Various Types of Water on Household Plumbing," Water Works and Sewerage, Vol. 83, 1936, pp. 384-387.

48. Hubbard, D. J., and C. E. A. Shanahan, "Corrosion of Zinc and Steel in Dilute Aqueous Solutions," British Corros. J., Vol. 8, No. 6, November 1973, pp. 270-274.

49. Karalekas, P. C., Jr., Craun, G. F., Hammonds, A. R., Ryan, C. R., and D. J. Worth, M.D., "Lead and Other Trace Metals in Drinking Water in the Boston Metropolitan Area," Journal-New England Water Works Association, Vol. 90, No. 2, 1976, pp. 150-172.

50. Karalekas, P. C., Jr., Ryan, C. R., Larson, C. D., and F. B. Taylor, "Alternative Methods for Controlling the Corrosion of Lead Pipe," J. New England Water Works Assoc., Vol. 92, No. 2, 1978, pp. 159-178.

51. Kuschner, M., et al, "Does the Use of Asbestos-Cement Pipe for Potable Water Systems Constitute a Health Hazard?," Journal-AWWA, Vol. 66, September 1974, pp. 4-22.

52. Lane, R. W., and C. H. Neff, "Materials Selection for Piping in Chemically Treated Water Systems," Materials Protection, Vol. 8, No. 2, February 1969, 27-30.

53. Langelier, W. F., "The Analytical Control of Anti-Corrosion Water Treatment," Journal-AWWA, Vol. 28, No. 10, 1936, pp. 1500-1521.

54. Langelier, W. F., "Chemical Equilibria in Water Treatment," Journal-AWWA, Vol. 38, No. 2, February 1946, pp. 169-178.

55. Larson, C. D., Chief, Technical Support Section, EPA Region I, J. F. Kennedy Fed. Bldg., Boston MA 02203, Letter to J. Hagopian, Principal Sanitary Engineer, Rhode Island Dept. of Health, 75 Davis St., Health Bldg., Providence, RH 02908 dated January 31, 1980.

56. Larson, C. D., Chief, Technical Support Section, EPA Region I, J. F. Kennedy Fed. Bldg., Boston MA 02203, Letter to J. Hagopian, Rhode Island Dept. of Health, 75 Davis Street, Health Bldg., Providence RH 02908 dated November 14, 1979.

57. Larson, T. E., "Corrosion by Domestic Waters," Illinois State Water Survey, Urbana, Bulletin 59, 1975.

58. Larson, T. E., and R. M. King, "Corrosion by Water at Low Flow Velocity," Journal-AWWA, Vol. 46, No. 1, January 1954, pp. 1-9.

59. Larson, T. E., and R. V. Skold, "Current Research on Corrosion and Tuberculation of Cast Iron," Journal-AWWA, Vol. 50, November 1958, pp. 1429-1432.

60. Larson, T. E., and R. V. Skold, "Laboratory Studies Relating Mineral Quality of Water to Corrosion of Steel and Cast Iron," Corrosion, Vol. 14, June 1958, pp. 43-46.

61. Leckie, H. P., and H. H. Uhlig, "Environmental Factors Affecting the Critical Potential for Pitting in 18-8 Stainless Steel," Journal of the Electrochemical Society, Vol. 113, No. 12, December 1966, pp. 1262-1267.

62. Leidheiser, H., Jr., The Corrosion of Copper, Tin, and Their Alloys, John Wiley & Sons, Inc., New York, © 1971.

63. Levelton, B. H., and D. G. Kilburn, "Accelerated Corrosion Tests on Copper Water Tubing," Materials Protection Journal, Vol. 5, No. 8, August 1966, pp. 37-40.

64. Lucey, V. F., "Mechanism of Pitting Corrosion of Copper in Supply Waters," Corrosion J., Vol. 2, No. 5, 1967, pp. 175-185.

65. Lyson, T. D. B., and J. M. A. Lenihan, "Corrosion in Solder Jointed Copper Tubes Resulting in Lead Contamination of Drinking Water," Br. Corros. J., Vol. 12, No. 1, 1977, pp. 41-45.

66. Hahato, B. K., Voora, S. K., and L. W. Shemilt, "Steel Pipe Corrosion Under Flow Conditions - I. An Isothermal Correlation for a Mass Transfer Model," Corrosion Science, Vol. 8, No. 3, 1968, p. 173.

67. McCauley, R. F., and M. O. Abdullah, "Carbonate Deposits for Pipe Protection," Journal-AWWA, Vol. 50, 1958, pp. 1419-1428.

68. Moore, M. R., Plumbosolvency of Waters," Nature, Vol. 243, No. 5404, May 25, 1973, pp. 222-223.

69. Mruk, S. A, "Thermoplastics Piping: A Review," Managing Corrosion Problems with Plastics, Vol. 4, National Association of Corrosion Engineers, P. O. Box 218340, Houston TX 77218, © 1979, pp. 3-14.

70. NACE Technical Unit Committee T-7 on Potable Waters, Task Group T-7B-2, Second Corrosion Study of Pipe Exposed to Domestic Waters, NACE Publication 7B170, Materials Protection & Performance, Vol. 9, No. 6, June 1970, pp. 34-37.

71. Naylor, Lewis M., Richard R. Dague, Simulation of Lead Removal by Chemistry Treatment, JAWWA, October 1975, pp. 560-565.

72. Nesbitt, W. D., "PVC Pipe in Water Distribution: Reliability and Durability," Journal-AWWA, Vol. 67, No. 10, October 1975, pp. 576-581.

73. Nielsen, K., "Contamination of Drinking Water with Cadmium and Lead from Brazed and Soldered Joints and from Other Metals in Plumbing Systems," Second International Brazing and Soldering Conference, London, Okt. 1975.

74. O'Brien, J. E., "Lead in Boston Water: Its Cause and Prevention," Journal of the New England Water Works Association, Vol. 90, No. 1, January 1976, pp. 173-180.

75. Patterson, J. W., "Corrosion Inhibitors and Coatings," Proc.-AWWA Seminar Controlling Corros. Water Syst., Paper No. 6 (4 pp.) 1978.

76. Patterson, J., Illinois Institute of Technology, Personal Communication, December 1979.

77. Patterson, J., and J. E. O'Brien, "Control of Lead Corrosion," Journal of the American Water Works Association, Vol. 71., No. 5, May 1979, pp. 264-271.

78. Polushkin, E. P., and H. L. Shuldener, "Corrosion of Yellow Brass Pipes in Domestic Hot-Water Systems - A Metallographic Study," Corrosion, Vol. 2, No. 1, March 1946, pp. 1-19.

79. Porter, F. C., and S. E. Hadden, " Corrosion of Aluminium Alloys in Supply Waters," J Applied Chemistry, Vol. 3, September 1953, pp. 385-409.

80. Pourbaix, M., "Recent Applications of Electrode Potential Measurements in the Thermodynamics and Kinetics of Corrosion of Metals," Corrosion (Houston), Vol. 25, No. 6, 1969, pp. 267-281.

81. Pourbaix, M., "Theoretical and Experimental Considerations in Corrosion Testing," Corrosion Science, Vol. 12, 1972, pp. 161-191.

82. Reedy, D. R., "Corrosion in the Water Works Industry," Materials Protection, Vol. 5, No. 9, September 1966, pp. 55-59.

83. Report on Lead Contamination of Bennington, Vermont Drinking Water. Report by USEPA Region I, Water Supply Branch, Boston MA, January 4, 1977.

84. Rossum, J. R., Pickup of Heavy Metals from Residential Plumbing, Rough Draft prepared for the California Department of Health, December 1, 1975.

85. Rowe, L. C., and M. S. Walker, "Effect of Mineral Impurities in Water on the Corrosion of Aluminum and Steel," Corrosion-National Assoc. of Corrosion Engineers, Vol. 13, December 1961, pp. 105-113.

86. Rudek, R., Blankenhorn, R. and H. Sontheimer, "Verzogerung der Eisenoxidation durch natürliche organische Wasserinhaltsstoffe und deren Auswirkung auf die Korrosion von schwarzen Stahlrohren," Vom Wasser, Vol. 53, 1971, pp. 133-146.

87. Russell, R. P., Chappell, E. L., and A. White, "The Effect of Velocity on Corrosion of Steel under Water," Ind. Eng. Chem., Vol. 19, 1927, pp. 65-68.

88. Sargent, H. E., "Asbestos in Drinking Water," Journal-New England Water Works Association, Vol. 88., No. 1, 1974, pp. 44-57.

89. Schafer, G. J., "Corrosion of Copper and Copper Alloys in New Zealand Potable Waters," New Zealand Journal of Science, Vol. 5, December 1962, pp. 475-484.

90. Schaut, G. G., "The Action of a Chlorinated Water Supply Upon Lead Pipe," American Journal of Pharmacy, Vol. 114, 1942, pp. 241-249.

91. Schock, M. R., "Response of Lead Solubility to Dissolved Carbonate in Drinking Water," Pre-Publication Copy, Physical and Chemical Contaminants Removal Branch, Drinking Water Research Diviison, U.S. Environmental Protection Agency, 26 W. St. Clair Street, Cincinnati OH 45268, October 1979

92. Scholefield, R. J., "Metal Corrosion Products in Municipal Drinking Waters," Thesis for Master of Science in Environmental Engineering at Illinois Institute of Technology, Chicago IL, August 1979.

93. Sheftel, V. O., "The Lixiviation of Lead Stabilizers from Polyvinyl Chloride Water Pipes," Hygiene and Sanitation, Vol. 29, 1964. pp. 121-122.

94. Simmonds, M. A., "Effect of Aggressive Waters on Cement and Concrete with Particular Reference to Cement-Lined Mains," The Journal of the Institution of Engineers, Australia, Vol. 26, January-February 1954, pp. 9-16.

95. Slunder, C. J., and W. K. Boyd, Summary Report on Lead - Its Corrosion Behavior to ILZRO, Battelle Memorial Institute, Columbus OH.

96. Streicher, L, "Effects of Water Quality on Various Metals," Journal-AWWA, Vol. 43, No. 3, March 1956, pp. 219-238.

97. Stumm, W., "Investigation on the Corrosive Behavior of Waters," Proceedings of the American Society of Civil Engineers, Vol. 86, No. SA-6, November 1960, pp. 27-45.

98. Symons, G. E., Ph.D, "Water Systems, Pipes and Piping, Part 1/Piping Systems Design," *Water and Wastes Engineering/Manual of Practice Number Two*, Vol. 4, No. 5, May 1976, pp. M3-M50.

99. Thermoplastic Piping for Potable Water Distribution Systems, Federal Construction Council, Building the Search Advisory Board, Division of Engineering, National Research Council, Technical Report No. 61, National Academy of Sciences, Washington, D.C., 1971.

100. Thibeau, R. J. , Brown, C. W., Goldfarb, A. Z. and R. H. Heidersbach, "Raman and Infrared Spectroscopy of Aqueous Corrosion Films on Lead in 0.1 M Sulfate Solutions," *J. Electrochem Soc.*, Vol. 127, No. 9, September 1980, pp. 1913-1918.

101. Tiedeman, W. D., "Studies on Plastic Pipe for Potable Water Supplies," *Journal-AWWA*, Vol. 46, August 1954, pp. 775-785.

102. Tiedeman, W. D., and N. A. Milone, "Effects of Plastic Pipe on Water Quality," *Journal-AWWA*, Vol. 48, 1956, pp. 1019-1023.

103. Toxicological Task Force of the Plastic Pipe and Fittings Association, Toxicological Analysis of James M. Montgomery, Consulting Engineers, Inc., Plastic Pipes Study Draft Report, Glen Ellyn, Illinois, September 1980.

104. Tracy, E. L., "Observations on Water Samples form Cement-Asbestos Pipe Systems in Vermont," *Journal-New England Water Works Association*, Vol. 64, 1950, pp. 164-178.

105. Tronstad, L. and R. Veimo, "The Action of Water on Copper Pipes," *Water and Water Eng.*, Vol. 42, May 1940, pp. 189-191.

106. Tronstad, L. and R. Veimo, "The Action of Water on Copper Pipes," *Water and Water Eng.*, Vol. 42, June 1940, pp. 225-228.

107. Uhlig, H. H., *Corrosion and Corrosion Control, an Introduction to Corrosion Science*, John Wiley & Sons, Inc., New York, 1963.

108. Uhlig, H. H, ed., *The Corrosion Handbook*, John Wiley & Sons, Inc., New York, 1948.

109. Wagner, I., "Influence of Water Quality and Water Treatment on Corrosion and Coatings in Steel and Galvanized Steel Tubes," *EUROCOR '77, 6th European Congress on Metallic Corrosion*, (Met. A., 7807-72 0184), 1977, pp. 413-419.

110. Wang, T. C., and J. L. Bricker, "2-Butanone and Tetrahydrofuran Contamination in the Water Supply," *Bull. Environm. Contam. Toxidol.*, Vol. 23, 1979. pp. 620-623.

111. Waters, D. M., "Internal and External Copper Corrosion in Domestic Water Services," Proc. - AWWA Annu. Conf. 97th, Vol. 1, 1977.

111a. Wells, S. W., "Hydrogen Sulfide Problems of Small Water Systems," Journal - AWWA, Vol. 46, Feb. 1954, pp. 160-170.

112. Whitman, G. W., Russell, R. P., and V. J. Altieri, "Effect of Hydrogen-Ion Concentration on the Submerged Corrosion of Steel," Industrial and Engineering Chemistry, Vol. 16, No. 7, July 1924, pp. 665-670.

113. Wong, C. S. and P. Berrang., "Contamination of Tap Water by Lead Pipe and Solder," Bulletin of Environmental Contamination and Technology, Vol. 15, No. 5, 1976, pp. 530-534.

114. Worth, D., "Relationship of Blood Lead Levels to Lead in Household Drinking Water," Proceedings - New England Water Works Association, Walthan, MA, December 18, 1975.

115. Wright, T. E. and H. P. Godard, "Laboratory Studies on the Pitting of Aluminum in Aggressive Waters," Corrosion, Vol. 10, June 1954, pp. 195-198.

116. Wyly, R. S-, Parker, W. J., Pierce, E. T., Rorrer, D. E., Shaver, J. R., Sherlin, G. C., and M. Tyron, "Investigation of Standards, Performance Characteristics, and Evaluation of Criteria for Thermoplastic Piping in Residential Plumbing Systems, NBS Building Science Series 111, National Bureau of Standards, Washington, D.C., May 1978.

5. Corrosion Monitoring and Detection

Detection of degradation and measurement of corrosion will be desirable for assessing the corrosivity of a given water, determining the efficacy of water treatment or inhibitor programs, and evaluating health effects of water system corrosion. The procedures involved in corrosion testing are deceptively simple in the sense that measurements can be obtained using relatively simple procedures. The detailed preparation of specimens and apparatus, however, is critical to obtaining reliable numbers. And the design of the experiment and use of the results for prediction requires consideration of many aspects of corrosion. This section describes the basic test methods applicable to corrosion in potable waters and gives references to more detailed procedures.

The following general methods are discussed in this section.

- specimen exposure for an extended duration followed by examination and weight-loss determination,
- electrochemical measurement of "instantaneous" corrosion rates, and
- chemical analysis for changes in concentration of a chemical species resulting from corrosion.

As with all corrosion tests, the value and reliability of these methods will depend on proper planning and execution of the details involved in the procedures. The applicability of a given procedure will depend on the objectives of the tests.

This discussion is intended to apply primarily to testing under field conditions (in the water treatment plant or distribution system). Testing under laboratory conditions requires careful preparation and control of the corrosive environment in addition to the other precautions. As in the rest of this report, external corrosion will not be considered.

SPECIMEN EXPOSURE TESTING

Placement of a test specimen in the corrosive environment and examination after some exposure duration is the oldest corrosion test method. While fundamentally simple, there are a number of details which must be considered. One of the most basic considerations is that the test specimen should "see" the same environment as the equipment of interest. This environment includes the chemical content of the fluid, the temperature, flow rate, galvanic coupling, periodic environment fluctuations, entrained solids or gases, etc. While the test specimens cannot be exposed to exactly the same environment as a given material in a water supply system, placement should be chosen to be representative of the application of that material. It is often necessary to consider the effect of specimen placement on the properties of the environment such as flow patterns and chemical content. Because corrosion is a function of electrochemical kinetics and surface phenomena, it is not surprising that surface preparation of specimen and careful documentation of metallurgical history are important procedural considerations. Planning and evaluation of tests should be done after careful review of factors affecting the known corrosion behavior of the materials in similar environments.

The general procedures used for corrosion testing can be delineated as follows:

- Selection of materials and specimens. Care should be taken that factors such as heat treatment and chemical composition are known and representative of the actual pipe or equipment of interest.
- Surface preparation. Actual equipment surfaces generally cannot be duplicated, but efforts to approach them with a reproducible preparation method must be made.
- Measuring and weighing. Both surface area and weight must be accurately measured with care taken to avoid fouling the surface.
- Exposure technique. Proper placement should be maintainable for the entire test period.
- Duration. Exposure time and an examination program should be carefully planned before starting the test period.
- Examination and cleaning of specimens after test. This step is important where documentation and use of proper technique is critical.
- Interpretation of results.

Details of these steps are discussed in large part by Fontana and Greene (4). Procedures are also given in standards or recommended practices by the American Society for Testing and Materials (ASTM) and the National Association of Corrosion Engineers (NACE). The main ASTM publication is the Standard Recommended Practice designated G4 on Conducting Plant Corrosion Tests which gives general guidelines and information on apparatus, test specimen preparation and placement, test duration, specimen removal and examination, and reporting (2). The ASTM Standard Recommended Practice G1 gives additional details on preparing, cleaning, and evaluating corrosion test specimens (1). Another useful guide is the NACE Standard TM-01-69 (1976 Revision) on Laboratory Corrosion Testing of Metals for the Process Industries (12). Use of this guide in potable water corrosion control testing has been described by Mullen and Ritter (11).

The size and shape of test specimens depends on several factors and cannot be rigidly set. It is generally desireable to have a high ratio of surface area to mass to obtain maximum corrosion loss. While the sample should be as large as possible, it should not exceed the weight limitations of the usual analytical balances (about 160 grams) or present problems in placement in pipes or equipment. Thin sections can be used to satisfy several of these requirements but the specimen should not be so thin as to be perforated by corrosion or to lack reasonable mechanical stability. The edges of specimens should be finished by polishing or machining to eliminate cold-worked metal. Specimens with sheared edges should not be used. Any dirt or heat-treated scale should be removed and the specimens should be freed from water breaks by suitable cleaning. Metal specimens should be abraded to at least 120 grit surface finish. The specimen should be stamped for identification, weighed to the nearest 0.1 mg on an analytical balance, and their surface area accurately determined.

A number of methods can be used for supporting specimens for exposure. The main considerations are that the corrosive media should have easy access to the specimens, the supports should not fail during the tests, the specimens should be insulated or electrically isolated unless the study of galvanic effects is intended, and the desired degree of immersion should be obtainable. Ready access to the specimens is also desireable. Apparatus for mounting specimens is described in detail and with mechanical drawings in ASTM G4-68 (2). They describe a spool rack in which specimens with a hole drilled through their center are positioned on a metal support rod which is covered with insulating plastic. Plastic tubing spacers also spooled on the center rod keep the specimens separate and supported. Insulating end disks are provided and the assembly is completed by nuts which are tightened on either end of the support rod. Other support methods are based on similar principles. They should be tailored to fit the equipment and operating conditions at hand.

Misleading results may be obtained if exposure duration and number of exposure periods are not carefully selected. It is often found that initial corrosion rates are considerably higher than those obtained after some time. However, in some cases pitting or crevice corrosion may not occur until after a certain incubation period. In general, tests run for long periods are considerably more realistic than short term tests. For uniform corrosion, a very rough guide for minimum exposure time suggested by both ASTM and NACE is given by:

$$\text{duration of test (hour)} = \frac{2000}{\text{corrosion rate (mpy)}}$$

This guideline is based on the general rule that the lower the corrosion rate, the longer the test should be run. The guide can be used with an estimated lower limit of corrosion rate or used to decide if tests should be repeated for a longer period based on existing results.

Most sources recommend using the planned-interval test originally proposed by Wachter and Treseder for setting up tests and evaluating results. This procedure allows evaluation of the effect of time on corrosion of the specimen and also on the corrosiveness of the environment. The procedure and evaluation of results are given in Table 35 along with an example of its application. This procedure is recommended by NACE TM-01-69 and also by Fontana and Greene (4).

After removal from the test environment the appearance of the test specimens and the rack should be noted. Specimens should be washed in water to remove soluble materials from the surface. Color photographs of the specimens should be made. The appearance and degree of adhesion of any coatings or films or the surface should be noted. If possible, samples of the corrosion product films should be preserved for future study. Specimens are not generally weighed until corrosion products are totally removed, since metal converted to corrosion product is structurally lost. But for potable water studies, additional information on the addition of species to the water stream might be obtained by also weighing the dried specimens at this point. Following this, the corrosion layers should be removed by a method that does not affect the base metal. The cleaning procedure is critical and will depend on the base material as well as the nature of the corrosion products. Procedures may include light mechanical cleaning (eg. rubbing with a rubber stopper), electrolytic cleaning, and chemical cleaning. Detailed procedures are given in ASTM G1-72 and in Fontana and Greene (1, 4). The possibilty of solid metal removal should be checked by applying the proposed method to fresh and to already cleaned, dried, and weighed specimens to determine any additional weight loss. After cleaning, the specimens should be dried and weighed to the same accuracy as the initial pre-test weighing. Weight-loss corrosion rates should be calculated for uniform corrosion cases. The specimens should be carefully examined visually and any modes of degradation such as pitting, crevice corrosion, dealloying, or other attacks noted. Photographs of the specimens should again be made since cleaning will often disclose more features of attack. If pitting occurs the maximum and average pit depths should be measured and also the number, size, general distribution, and shape of the pits should be noted. Distinction should be made between pits which occur under insulating spacers and those on exposed surfaces. The former is probably related to crevice corrosion. The depth

TABLE 35. PLANNED INTERVAL TEST (4)

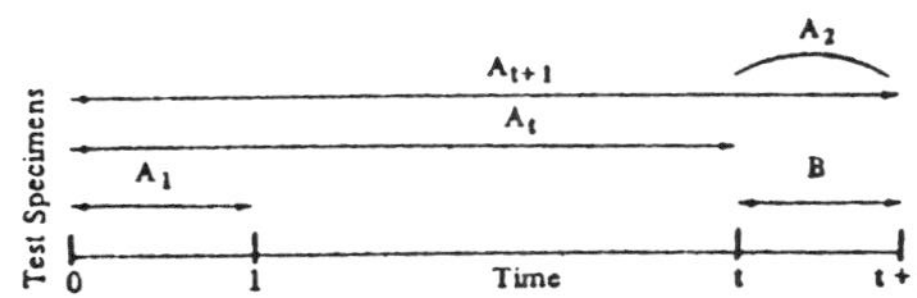

Identical specimens-all placed in the same corrosive fluid. Imposed conditions of the test kept constant for entire time t + 1. Letters, A_1, A_t, A_{t+1}, B, represent corrosion damage experienced by each test specimen. A_2 is calculated by subtracting A_t from A_{t+1}.

Occurrences During Corrosion Test		Criteria
Liquid corrosiveness	unchanged	$A_1 = B$
	decreased	$B < A_1$
	increased	$A_1 < B$
Metal corrodibility	unchanged	$A_2 = B$
	decreased	$A_2 < B$
	increased	$B < A_2$

Combinations of Situations

Liquid Corrosiveness	Metal Corrodibility	Criteria
1. unchanged	unchanged	$A_1 = A_2 = B$
2. unchanged	decreased	$A_2 < A_1 = B$
3. unchanged	increased	$A_1 = B < A_2$
4. decreased	unchanged	$A_2 = B < A_1$
5. decreased	decreased	$A_2 < B < A_1$
6. decreased	increased	$A_1 > B < A_2$
7. increased	unchanged	$A_1 < A_2 = B$
8. increased	decreased	$A_1 < B > A_2$
9. increased	increased	$A_1 < B < A_2$

Example of Planned Interval Corrosion Test

Conditions: Duplicate strips of low-carbon steel, each 3/4 by 3 inches (20 x 75 mm) immersed in 200 ml of 10% $AlCl_3$ - 90% $SbCl_3$ mixture through which dried HCl gas was slowly bubbled at atm. pressure. Temperature 90 C.

	Interval, days	Wt. Loss, mg	Penetration, mils	Apparent Corrosion Rate, mils/yr	mmpy
A_1 . .	0-1	1080	1.69	620	15.7
A_t . .	0-3	1430	2.24	270	6.9
A_{t+1} .	0-4	1460	2.29	210	5.3
B . . .	3-4	70	0.11	40	1.0
A_2 . .	calc. 3-4	30	0.05	18	0.5

$A_2 < B < A_1$
$0.05 < 0.11 < 1.69$

Therefore, liquid markedly decreased in corrosiveness during test, and formation of partially protective scale on the steel was indicated.

of any crevice corrosion should be determined. Pitting rates are often ambiguous and no extrapolations should be made. The actual pit depth and length of exposure should be reported. Selective or localized attack can be examined and recorded in greater detail by metallographic and microscopic techniques.

Metals which may be susceptible to dealloying or stress cracking should be bent after other examination and the development of any cracks should be noted. These results should be compared to similar bend tests with unexposed specimens (2).

ELECTROCHEMICAL TEST METHODS

Electrochemical methods for corrosion measurement are more complicated than specimen exposure testing with respect to both equipment and interpretation. Electrochemical methods are also relatively new and less established than the conventional exposure tests. When properly applied, however, the newer methods offer several advantages. The electrochemical methods of main interest are very rapid and can be used for near-continuous monitoring of corrosion rates under proper conditions. They are adaptable to measurement of low corrosion rates which are most difficult to measure by weight loss. Because the most often used electrochemical methods do not significantly affect the specimen, time profiles of corrosion rates can be obtained. Also, the effects of various water treatment methods can be monitored on a given specimen.

The electrochemical measurements will be most reliable when the metal of interest undergoes uniform corrosion in systems where scale formation is minimal. The care required for sample preparation and placement is as important for these methods as for the simple specimen exposure methods discussed above. One important limitation on the use of electrochemical methods to obtain rapid corrosion measurements is that the corroding behavior of a metal often depends on the length of time it has been exposed to a given environment. Electrochemical methods can provide "instantaneous" corrosion rates, but variation of corrosion rates with time must also be considered.

The basis of these methods is the electrochemical nature of the corrosion of metals in aqueous solutions. The methods are not applicable to nonmetallic materials used in water supply systems. The rate of electrochemical reactions, and thus the rate of corrosion reactions, can be expressed as an electrical current. The driving force for obtaining the reaction giving rise to the current is an electrical potential difference. In the general case, the relation between current and potential difference for electrochemical reactions is non-ohmic (i.e., nonlinear); generally an exponential or complex mixed relationship is seen. However, for small deviations in driving force from some steady state, or open circuit corrosion potential, the current and potential difference are approximately ohmic or linear. Use of such small potential deviations (i.e., small polarization) methods forms the basis for practically all commercial electrochemical corrosion rate instruments and methods proposed for field or routine use. They are often referred to as "linear polarization" or "polarization resistance" methods. It is noteworthy that the use of these small perturbation methods also causes less change in the specimen surface and makes possible multiple measurements with the same specimen. Even small perturbations are of some concern. However, tests of repeatedly polarized cast iron specimens in a potable water environment gave generally the same results as freely corroding samples (10). The possibility of differences occurring with other metals should be considered.

The derivation of current-potential relationships for linear polarization conditions has been described for various degrees of model sophistication (4). A widely used form is given by the equation:

$$\frac{\Delta E}{\Delta i} = \frac{\beta_a \beta_c}{(\log 10) i_{corr} (\beta_a + \beta_c)}$$

In this equation, ΔE is the deviation from the corrosion potential, Δi is the current density (current per unit area of electrode specimen), i_{corr} is the corrosion current density which can be related to the corrosion rate, and $\beta\alpha$ and βc are parameters (so-called Tafel slopes) associated with the electrochemical kinetics of the individual anodic and cathodic corrosion reactions. At least to a first approximation, $\beta\alpha$ and βc are constants for a given corrosion system. For model systems, β values are such that the following simple form is obtained.

$$\frac{\Delta E}{\Delta i} = \frac{RT}{F} \cdot \frac{1}{i_{corr}}$$

where RT/F consists of fundamental constants and the absolute temperature. The value of RT/F is about 0.026 volts at 25°C. These equations are applicable for ΔE values of 20 millivolts at most, and smaller ΔE values are probably preferable. The simple form often gives useful approximations; although absolute errors may be large in case of unequal Tafel slopes, relative measurements may still be useful. The quantity i_{corr} (microamperes) can be related to the corrosion rate by the following equation:

$$\text{Corrosion Rate (mpy)} = \frac{0.129\ i_{corr}\ A}{nD}$$

Here A is the atomic weight of the metal, n is the number of electrons lost by the metal atom during corrosion (eg. n = 2 for iron going to Fe^{+2}), and D is the metal density (g/cm^3). In some commerical instruments the conversion factors are set by specimen size and electronic means to give direct readouts of corrosion rates.

There are a number of means of actually carrying out the linear polarization measurement. There does not appear to be a definitive study to determine if any particular method is superior for use in potable water type systems. Schematics of two types of electrode circuits are shown in Figure 34. The three-electrode method is widely used and employs a test specimen electrode,

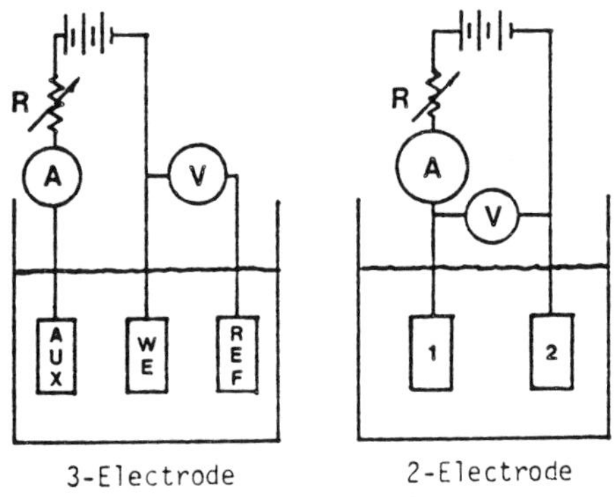

Figure 34. Schematic of 3 and 2 electrode circuitry.

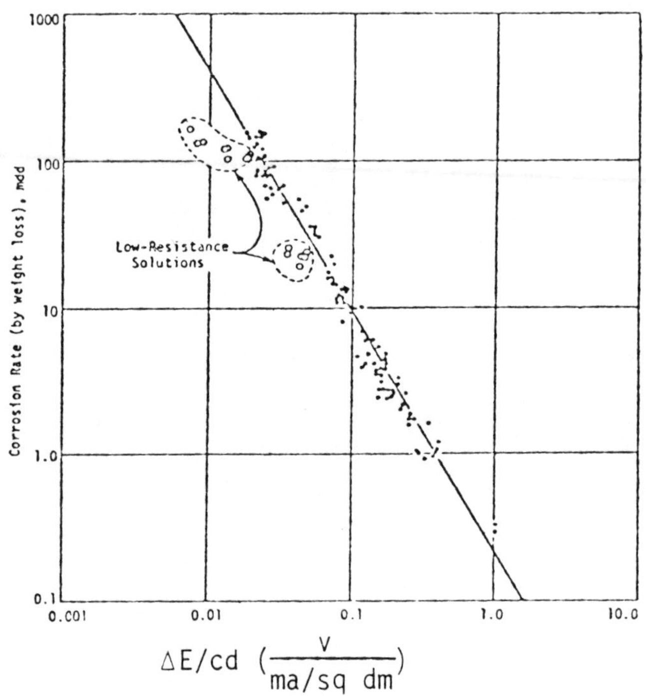

Figure 35. Empirical relationship between initial slope of polarization curve (resistance) and corrosion rate determined by weight loss. (7)

an auxiliary electrode and a reference electrode. External current flows only through the test and auxiliary electrodes while the polarization of the test electrode is measured by way of the referenced electrode. The degree of polarization may be set at some constant value, 10 millivolts is often used,or varied in the vicinity of the corrosion potential. It is also possible to use two identical electrodes for test and auxiliary and measure potential difference between these two along with the current, without using a reference electrode. Comparison of three-electrode and two-electrode methods have been given, along with a general analysis of errors in linear polarization methods, by Bandy and Jones (3). This method is intended to measure only the equivalent resistance of the corrosion reaction rate and, therefore, care should be taken that other resistances such as that of the solution or connectors are small and that abnormal scale or fouling of the test electrode does not occur.

The applicability of the linear polarization method to certain potable water systems was first demonstrated by Larson (7). The close correlation found between corrosion rates obtained by weight loss measurement and the linear polarization resistance is shown in Figure 35. These results were obtained for cast iron and steel in potable water environments of various composition. Several other evaluations of the linear polarization method in potable waters have been reported (10,9).

Polarization curves can be recorded by the potentiodynamic method in which the potential of the test specimen is electronically varied at a preset linear rate and recorded along with the resulting current. For general use the potential excursions can be limited to about ± 10 millivolts of the open circuit voltage to decrease the chance of modifying the surface by polarization. The potential scan rate should be low enough that current due to double layer charging is negligible compared to the faradaic current. The method has the advantages of providing a better measure of the current-voltage slope which is the desired measurement and a readout by which any departure from linearity of this plot can be obtained. This is obtained at the cost of increased complexity in instrumentation and operation.

The linear polarization method has promise for rapid estimates of the corrosivity of potable waters. It must be applied with care and time must be taken to ensure that metal specimens have reached a steady-state corrosion rate. With careful use, it should be possible to monitor the relative effects of various treatment methods using these procedures. A number of probes of the same material should be used and periodic inspection for localized corrosion or scaling should be done. Integration of the instantaneous rates and comparison with weight-loss specimens should be done for confidence in absolute rates derived from linear polarization methods.

Linear polarization methods are applicable to uniform corrosion or at best averaging small areas of pits with large uniform areas. Electrochemical methods for determining localized corrosion susceptibility are being evaluated under laboratory conditions, but have not been standardized. Potential applications of such methods to field testing has recently been discussed by Martin (8). Additional work is needed to develop these methods for use in potable water environments.

Another electrical, but not electrochemical, method for determining corrosion rates is based on measuring the change in electronic resistance of a metal specimen. As the relatively thin wire, tube or strip specimen element corrodes, its electrical resistance increases due to the decrease in cross sectional area. Measurements of resistance over a period of time (several days or weeks) give an estimate of the change in specimen thickness and, thus, a corrosion rate can be obtained. Results can be obtained more quickly than with coupon tests, but are not "real time" in the sense of linear polarization tests. For uniform corrosion behavior, the method is relatively direct.

Reedy has described the field use of this type of resistance probe for testing a number of metals in two natural water sources (13). A stabilization period of 10-14 days immersion was required before consistent corrosion rates were obtained. This result was probably due to the normal surface changes that occur on exposure of a fresh metal specimen to a corrosive environment. Long-term testing (50 days) confirmed that reproducible results could be obtained after the initial period.

CHEMICAL ANALYSES FOR CORROSION PRODUCTS

Corrosion can be inferred from the increase in concentration of a metal species in solution from one point in a water distribution system to another point downstream of the first. Although this analysis is not a conventional means of corrosion measurement, it provides a direct measure of the quantity of interest and, in many cases, may be the only way of determining a health hazard if construction materials are unknown or access to the distribution system is difficult. Due to the uncertainties involved, this procedure will probably serve best as an indicator of potential corrosivity of a water system and the need for application of more quantitative corrosion rate methods.

Sampling points and procedures will be dictated by the information desired. To obtain baseline samples without any contamination by piping, raw water samples should be dipped from surface sources, or for groundwater supplies collected as close to the well as possible. In all cases samples for trace metal analysis should be preserved with ultrapure nitric acid in thoroughly cleaned sample bottles. Finished water samples can be obtained after treatment at the treatment plant. Sampling in homes can be set up to provide some differentiation between sources of metal contamination. Karalekas and coworkers describe the following sampling program (5, 6). The water samples were collected at the kitchen sink the first thing in the morning before any water was used in the house.

- Interior plumbing sample; this is collected immediately upon opening the faucet and represents water that has been standing overnight in the fixture and plumbing serving the faucet,

- Service line sample; this is obtained after the sample collector feels the water temperature change from warm to cool, representing the water in the service line to the house,

- Water main sample; this is collected after allowing the water to run for several minutes and represents water from the main which has had minimum contact time with the service line and interior plumbing.

These samples provide a representation of the range of trace metal concentrations the consumer is likely to experience as well as an indication of the source of the contaminant.

Although the comparison of chemical analyses from several points seems to provide a simple procedure, there are several complications. Because a difference of two experimental numbers is required, the analytical methods must be both accurate and precise. Surface water supplies may have trace metal contents from a geochemical origin. Trace materials may also be introduced by airborne or waterborne pollutants or as impurities in chemicals added to the water during the treatment process. In analyzing any trace component in water, care must be taken during sampling, sample storage, and sample pretreatment to avoid large systematic errors due to such problems as contamination by equipment, precipitation or adsorption of the measured species, or contamination by impurities in reagents used for sample preservation or pretreatment. The dependence of material concentration on system flow history is well documented with regard to "stagnant" versus "free-flowing". More complicated behavior may also occur in distribution or household systems as corrosion products precipitate or undergo redox chemistry or adsorption. Corrosion products may spall from walls or be released from pits at irregular intervals. The sorption of relatively large quantities of lead and copper ion by hydrous ferric oxide has been studied (14). Materials such as hydrous ferrous oxide could exist either as a layer attached to iron pipes or in suspension. In view of these considerations, reliable results may often require taking a large number of specimens or continuous monitoring over a period of time.

If the materials in use in the water system are unknown or are not well defined, the chemical analysis method for corrosion products is further complicated. Some chemical analyses are expensive, and performing tests for large number of possible contaminants may be cost prohibitive for small public or private water systems. Additionally, specific corrosion products can result or appear from the use of various materials.

Analytical procedures used to quantify the existence of asbestos minerals in water are severely limited and are subject to produce erroneous results. Primary reasons for the extreme difficulty in determining asbestos fiber concentrations in water include 1) asbestos fiber concentration in potable water is generally very low, 2) chemical analytical methods are not applicable because elements present are common to all rock-forming minerals, 3) asbestos fibers cannot be concentrated or separated from other inorganic solids present in the water, and 4) fiber sizes are often below the resolution of the optical microscope (15). Also, a knowledge of field operations is necessary to determine the possibility of fiber release resulting from drilling and tapping as opposed to regular deterioration.

One technique employed to analyze asbestos fiber concentrations or counts in water utilizes the electron microscope. Using this technique, solids are removed from the water sample by filtration on a membrance filter and the entire sample is ashed to destroy the filter and any organic and oxidizable inorganic solids that may be present. The inorganic residue is then rubbed out in a dilute solution of nitrocellulose to reduce the particle sizes and transferred to a standard electron-microscope grid. This sample is examined under an electron microscope and the presence of chrysotile fibers is quantified by measuring the length and diameter of each fiber, calculating the total mass and finally relating this mass to the original amount of water sampled. This technique is specific to chrysotile fibers because of their recognizable hollow-tube structure.

The accuracy and precision of this anlysis is very poor because only an extremely small fraction of the sample can be examined and this sample generally contains only a small amount of asbestos. Analysis of samples using the electron microscope technique cannot be duplicated by better than a factor of three, and it is estimated that the measured value is accurate to within a factor of ten of the true value. Therefore, measured values should be considered indicators or indices of the relative amount of fiber present. Another limitation of this technique is that the analysis is specific to the structurally recognizable chrysotile fibers. Several other varieties of asbestos materials may be present in water such as crocidolite which is also a component of most asbestos-cement pipe (15).

Development of analytical methods for trace material determination has coincided with growing environmental concerns. A critical evaluation of the many developing methods is outside the scope of this report. Three of the more popular methods for the metals likely to be of interest are atomic absorption spectroscopy (AAS), anodic stripping voltammetry (ASV), and neutron activation analysis (NAA). Use of NAA requires access to a reactor. AAS often requires considerable sample pretreatment as well as equipment usually restricted to a permanent laboratory. Continuous monitoring of tap water for Pb, Cd, and Cu has been demonstrated using ASV and related techniques by National Sanitation Foundation personnel (9). The number of elements that can be analyzed by this method is relatively small, but they correspond well to those of interest from a corrosion and health standpoint. The instrumentation required for ASV is somewhat complicated, but portable.

REFERENCES

1. ASTM Standard Recommended Practice for Preparing, Cleaning, and Evaluating Corrosion Test Specimens, Designation G1-72, American Society for Testing and Materials, Philadelphia, Pennsylvania

2. ASTM Standard Recommended Practice for Conducting Plant Corrosion Tests, Designation G4-68, American Society for Testing and Materials, Philadelphia, Pennsylvania

3. Bandy, R. and D. A. Jones, Analysis of Errors in Measuring Corrosion Rates by Linear Polarization, Corrosion - NACE, Vol. 32, No. 4, April 1976, pp. 126-134.

4. Fontana, Mars G. and Norbert D. Greene, Corrosion Engineering, McGraw-Hill Book Company, New York, 1978.

5. Karalekas, Peter C., Gunther F. Craun, Arthur F. Hammonds, Christopher R. Ryan, and Dorothy J. Worth, M. D., "Lead and Other Trace Metals in Drinking Water in the Boston Metropolitan Area", Journal--New England Water Works Association, Vol. 90, No. 2, pp. 150-172, 1976.

6. Karalekas, Peter C., C. R. Ryan, C. D. Larson, and F. B. Taylor, Alternative Methods for Controlling the Corrosion of Lead Pipe, J. New England Water Works Assoc., Vol. 92, No. 2, 1978, pp. 159-78.

7. Larson, T. E., Corrosion by Domestic Waters, Illinois State Water Survey, Urbane, Bulletin 59, 1975.

8. Martin, R. L., Potentiodynamic Polarization Studies in the Field, Materials Performance, Vol. 18, No. 3, March 1979, pp. 41-50.

9. McCelland, Nina I., and K. H. Mancy, Water Quality Monitoring in Distribution Systems: A Progress Report, JAWWA, Vol. 64, No. 12, 1972, pp. 795-803.

10. McClanahan, Mark A., and K. H. Mancy, Comparison of Corrosion - Rate Measurements on Fresh vs. Previously Polarized Samples, JAWWA, Vol. 68, No. 8, August 1974, pp. 461-466.

11. Mullen, Edward D. and Joseph A. Ritter, Potable - Water Corrosion Control, JAWWA, Vol. 66, No. 8, August 1974, pp. 473-479.

12. NACE Standard TM-01-69 (1976 Revision) Test Method: Laboratory Corrosion Testing of Metals for the Process Industries, National Association of Corrosion Engineers, Katy, Texas, approved March, 1969 (revised 1972, 1976).

REFERENCES - SECTION 5 (continued)

13. Reedy, Donald R., Corrosive Effects of Southern California Potable Waters, Materials Protection & Performance, Vol. 12, No. 4, April 1973, pp. 43-48.

14. Swallow, Kathleen C., David N. Hume, and Francosi M. M. Morel, Sorption of Copper and Lead by Hydrous Ferric Oxide, Environmental Science & Technology, Vol. 14, No. 11, November 1980, pp. 1326-1331.

15. Wright, G. W. (Chairman), et al, Committee Report, "Does the Use of Asbestos-Cement Pipe for Potable Water Systems Constitute a Health Hazard?" JAWWA, September 1974, pp. 4-21.

6. Corrosion Prevention and Control

Corrosion control in potable water systems is most commonly attempted by establishing a protective barrier between a corrodible material surface and a corrosive water. This protective barrier can be applied mechanically as a material coating or lining prior to installation of facilities or it can be applied chemically by adjusting the water quality characteristics of the potable water to precipitate a byproduct which forms on the surface of water system materials. The most commonly used pipe coatings include coal tar enamels, epoxy paint, and cement mortar. Water tank linings generally include coal tar enamels and paints, vinyl, and epoxy. Other coatings and linings which are also used are hot and cold applied petroleum based waxes, zinc paints, and asphalt. Any applied coating or lining must be environmentally sound, including their application procedures, and must not impart objectionable aesthetic or health effects into the water.

The addition of lime to induce calcium carbonate precipitation and form a protective coating chemically is the most commonly used and accepted method of corrosion control through water quality adjustment. Chemically applied protective coatings are also formed by the addition of sodium silicate and inorganic phosphates. These chemicals are generally referred to as corrosion inhibitors.

Other corrosion control practices include adjusting the water quality characteristics to render potable waters less aggressive or less corrosive. As previously discussed, corrosive waters are generally characterized as containing high dissolved oxygen and/or carbon dioxide, low alkalinity and hardness, and low pH. Adjusting these parameters for optimum corrosion control of a specific material is sometimes practiced or attempted, but optimum conditions are often difficult to determine.

Corrosion control alternatives currently being practiced and available to the water works industry are discussed in this section. Also included in this section is a brief description of the case history of the Seattle Water Department attempts to correct or retard corrosion in its potable water supply facilities. Their attempts emphasize the magnitude of the problems and difficulties in implementing a corrosion control program. A list of the applicability of the mechanically applied linings and corrosion inhibitors discussed is presented in Table 48 (Section 7):

MECHANICALLY APPLIED PIPE LINING AND COATINGS

As early as 1860, attempts to solve corrosion problems in water pipes resulted in dipping cast iron pipe in a bituminous coating material, and cement lining was suggested by the French Academy of Sciences. By 1930, the use of a cement mortar lining for pipes used in water distribution systems was specified in eastern and southeastern areas of the United States, while many areas in the Midwest specified coal tar coatings. Professional papers in the 1930's and 40's (13, 14,21) extolled the virtures of cement lined iron and cast iron pipes for removing "red water" problems, but inconsistencies between different systems prompted comparison investigations in the 50's (3,50). These papers reported tubercule formation and corrosion resistency for various cement, bituminous dips and enamels, pitch, tar dip, and asphalt sealed cement linings on iron pipes. Water tank corrosion was controlled with paint coatings or cathodic protection which were gradually replaced with coal tar enamel. More recent technologies have produced vinyl, epoxy, chlorinated rubber, and other tank linings. The focus of current research is to determine the extent of trace organics released and subsequent public health affects from the use of bituminous based linings.

The three principal pipe linings currently used for corrosion protection are coal enamel, epoxy, and cement mortar.

Hot Applied Coal Tar Enamel

Hot applied coal tar enamel is produced from coal tar pitch, a residue in the fractional distillation of coal tar obtained from the destructive distillation of bituminous coal. Coal tar used in potable water systems is required to meet AWWA specifications for type of coal and production process. Coal tar enamel is made by dispersing coal in a mixture of coal tar pitch and coal tar gas oil. Talc is added to provide strength.

Hot applied coal tar enamel is used in steel pipes, and it is estimated that between 50 and 80 percent of all steel pipe used for water distribution is lined this way. The coating process involves spraying a primer coat, usually a chlorinated rubber based resin, on the cleaned surface. After the primer dries, the coal tar enamel is heated to a fluid consistency and poured on the pipe surface from a trough extending lengthwise through the rotating pipe. Fittings are manually lined. Thickness of the coating is 0.094 inches (2.4mm). If connections are to be welded, field touch-up around welded joints is necessary.

Hot applied coal tar enamel was used to line steel water pipelines in New York City as early as 1914. Coal tar enameled pipes, inspected after many years of service in water supply systems, have shown no coating failures when the lining was properly applied. Service life of this lining is relatively long, often in excess of 50 years. Other advantages include good erosion resistance to silt or small amounts of sand in the water and resistance to biological attachments. Disadvantages include the need to re-apply to welded areas and to use special care in handling pipes during weather extremes as cold causes brittleness and heat may initiate cracking. An increase of trace organic compounds exists in water flowing through coal tar lined pipe, but the potential effects of their release is presently unknown. The extent of this release is currently under investigation at the EPA laboratory in Cincinnati, Ohio.

Epoxy

Epoxy paint coatings have been used to line steel water pipes since the mid 1960's and are becoming increasingly popular. Two types of epoxy systems exist: single component and double component. The former consists of an epoxy resin, pigments, drying oil, and a reactive resin. Drying results from oxidation and polymerization. The latter consists of a base containing the epoxy resin, pigment, and solvent and a polyamide or amine and solvent hardener.

The lining is usually applied in the field because welding will burn off the epoxy and re-application can be avoided. The pipe is first cleaned and a phosphate treatment or rust inhibiting zinc silicate primer is applied to the surface. This step is not necessary, but it improves the epoxy adhesion and abrasive resistance. Two coats of epoxy are then applied by spraying from a "pig" pulled through the pipe.

A major advantage to using epoxy linings is the high smoothness coefficients (Hazen-Williams coefficients of 138-172 (17)) produced and the corresponding reduced pumping costs. Also, epoxy paints can be formulated from components approved by FDA for food contact surfaces. Principal disadvantages include cost and a reduced resistance to abrasion, compared to coal tar enamel, which limits service life to under 15 years.

Recent use of powdered epoxy coatings have reduced occupational health and air pollution problems associated with solvents necessary for spray applications. The powder is sprayed onto the heated surface and melts to a thin film that fuses to the pipe. Because epoxy linings have been used to line steel water pipes for fewer years than their anticipated service life, case histories are currently just being developed. In 1964, however, the installation of 20,000 feet of epoxy lined 30-inch force main by an eastern Pennsylvania water company resulted in lower construction costs and a more beneficial Hazen-Williams coefficient. As a result, the company installed an additional 28,000 feet of epoxy lined 20- and 30-inch lines and, 10 years after the 1964 installation, was planning further installations (17). It is also reported that three pipeline companies use only epoxy to line their steel pipes (19).

Cement Mortar

Although being used less to line steel pipes, cement mortar is the only coating used to line cast iron and ductile iron pipes. Cement linings are applied by introducing a slurry into a rotating pipe. Centrifugal action distributes the mix to a uniform coating. An alternative process involves using a rod with a revolving head. The rod is pushed through the pipe, and the slurry is sprayed onto the surface of the pipe. Composition of the applied slurry depends on the size of pipe being lined. Pipes under 20 centimeters in diameter use a cement : sand : water weight ratio of 2 : 3 : 1, whereas larger pipes use a ratio of 2 : 4 : 1 to control possible cracking.

Proper curing of the lined pipe is important. Steel pipes are cured by sealing the ends, spraying water on the lining, or alternating applications of steam and water. Mortar in cast iron and ductile iron pipes is coated with a thin (0.001 inch, 0.0254mm) layer of an asphalt, mineral spirit, and xylene sealant. This process provides a moisture barrier, constraining water in the slurry, promoting a proper slow cure. This sealant has the added advantage of preventing decalcification of the lining in soft waters.

Advantages to using cement mortar lining include its cost and low sensitivity to variations in the substrate quality or application procedures. It can be applied in situ on pipes whose lining has failed provided the loosely coated areas are removed. Because the corrosion abatement mechanism is a result of the calcium hydroxide released in addition to the physical barrier, uncoated metal at pipe joints is protected. Also, surface fissures will heal themselves when immersed in water.

The rigidity of the lining is a disadvantage because pipes subject to deflections may experience lining cracking and sloughing. Also, the depth of the coating reduces the cross sectional area of the pipe, and hence its carrying capacity significantly. For example, the area of a 20-inch I.D. pipe with a cement mortar lining is reduced 6 percent. A coal tar enamel lining reduces the area 2 percent, and an epoxy lining reduces the area by less than 0.2 percent. Also, decalcification in soft, calcium dissolving waters may impart an objectionable taste to the water.

The first applications of cement mortar linings for steel water pipes was in the late 1800's and some of these pipes have been in use since that time. Cement lined sheet-iron pipes were also common at that time and, if designed and installed properly, provided a 40 to 50 year service life. The first cement lined cast iron pipe was installed in America in 1922, so the articles published in the 1930's and 1940's citing the longevity of cement lined pipes were in reference to the steel lined pipes. Published literature of the 1950's (2, 46) indicated the widespread use and acceptance of cement lined cast iron pipe, but academic decisions were being deferred pending further use. Today, cement mortar is the only lining used in cast iron and ductile iron pipe.

TANK LININGS AND COATINGS

The principal types of water tank linings are coal tar enamels and paints, vinyl, and epoxy. Other coatings exist and are mentioned, but they are either no longer or not extensively used. Water tank linings should exhibit ease of application, effective corrosion control, and good erosion resistance.

Coal Tar Based Coatings

Hot applied coal tar enamel is prepared and used as discussed in the pipe lining section. This coating has a tendency to sag or ripple when applied above the waterline when the tank walls are heated. Hot applied coal tar enamel is the primary coal tar based coating used to line water tanks.

Coal tar paints are often used to reline existing water tanks. Cold applied coal tar paint is prepared by adding coal tar solvents such as xylene or naptha to coal tar enamel. It is brushed or sprayed on the surface to form a relatively thick film resistant to sags or runs. Because it can impart unpleasant tastes and odors to the water, its use is restricted to above waterline surfaces. It is less durable than hot applied coal tar enamel and has a service life of 5 to 10 years. A tasteless and odorless cold applied coal tar paint is produced by combining coal tar enamel with solvents free of phenols and other taste and odor constituents. This paint may be used below water, but should not be exposed to sunlight or ice.

Coal tar epoxy paints used to line potable water tanks are generally two component systems. One part consists of a coal tar pitch base with a polyamide resin, magnesium silicate, xylene, ethyl alcohol, a gelling agent, and a catalyst added. The second part is a liquid epoxy resin. Some types of coal tar epoxy paints sold do not conform to the Steel Structure Painting Council specifications and may produce taste and odor problems in the water. Coal tar epoxy paints are less resistant to abrasion than coal tar enamel. Exposure to sunlight causes chalking to occur, but if this exposure is eliminated, coal tar epoxy paints may provide a service life of over 20 years.

Coal tar urethane paints have been inconsistent in their service lives. Some applications have provided a 25 year service life, but other applications have failed within one year. This type of lining is not presently marketed.

Coal tar emulsion paint is a water based suspension of coal tar pitch, magnesium silicate or other mineral filler, and a rust inhibitor. Although it has good adhesive characteristics, is practically odorless, and resists sunlight degradation, it is not as watertight as organic solvent coal tar coatings. Consequently, its use below the waterline is limited.

Investigations have shown that trace organics are released into water stored in coal tar base lined tanks. The results of analytical tests on water stored in a coal tar lined tank are shown in Table 36a. Possible health affects associated with these organics in the concentrations observed are not presently known. Research presently being done on the release of these organics, including rates, solubilities in water, and identification of decomposition products will have to be combined with toxicological evaluations to determine future uses or restrictions on coal tar based linings.

Vinyl

Vinyl paints are a mixture of a vinyl chloride-vinyl acetate copolymer with a hydroxyl compound and/or a carboxy compound. These paints are applied on steel in either a three-coat or a four-coat system, with different formulations used in each system and within the five-coat system. Vinyl paints are inert in water and produce a hard, smooth surface. Soft water conditions may reduce the expected service life of 20 years. Recent vinyl failures in California have been blamed on formula changes made to meet that state's air pollution criteria. Another report stated that vinyl paints do not wear well in soft waters, but one engineering consultant estimated (1977) that vinyl paints are specified for 90 percent of their storage tank projects. Although vinyls are one of the more popular water tank liners, their intricacies of application prohibit their use in pipes.

Epoxy

Epoxy is produced and applied as previously discussed in The Pipe Lining And Coating section. Like vinyl, epoxy produces a hard, smooth surface that exhibits low water permeability and strong adhesion to steel when properly formulated and applied. Reformulation necessary to conform to the California air pollution control regulations has adversely affected their performance. While most reports on epoxy linings are for pipe applications, where flow resistance co-effecients are important concerns, tests by Bethlehem Steel have indicated that epoxy wears as well as hot applied coal tar enamel in water immersion, non-abrasive, testing (17).

Other Mechanically Applied Tank Linings

Hot and cold applied wax coatings are also used to line water tanks. These coatings are blends of petroleum waxes and oil based corrosion inhibitors. These coatings may be applied directly over old rust or paint, but commercial blast cleaning of the surface prior to application is preferred. Application either by stiff bristled brushes or spray equipment is followed by torch flaming used to smooth and thoroughly close the surface. The major disadvantage to using wax coatings is a relatively short service life of approximately five years.

TABLE 36a

ESTIMATED CONCENTRATIONS OF COMPOUNDS DETECTED IN THE WATER IN THE BAYOU CASSOTTE GROUND STORAGE WATER TANK USING GAS CHROMATOGRAPHY/MASS SPECTROMETRY (19)

Compound	Sample Date and Concentration (μg/l)				
	9/6/77		1/16/78	2/21/78	
	A	B	A	A	B
naphthalene	5.4	6.7		1.3	2.7
methyl naphthalene	0.75	1.4	0.63	<1	1.3
biphenyl	0.21	0.40		<1	<1
acenaphthene	2.8	4.6	1.3	3.1	8.0
dibenzofuran	3.1	5.0	1.1	2.3	6.3
fluorene	3.4	5.1	1.5	2.9	8.0
phenanthrene/anthracene	8.7	9.3	4.5	14	35
carbazole	0.70	1.3	0.44	3.9	11
bromoform	<10	<10	<10		
C_4 alkylchlorobenzene	<50	<50	<50	2.7	7.3
indene	<10	<10			
C_3 alkylbenzene	<10	<10			
anthraquinone	<10	<10	<10	2.3	8.3
methyl benzofuran	<10	<10		<1	1.0
quinoline	<10	<10		<1	<1
methyl styrene/indan/indene	<10	<10		1.7	2.6
methylene phenanthrene/methyl phenathrene	<10	<10	<10	1.2	4.0
pyrene	<10	<10	<10	2.2	7.0
2,5-diethyltetrahydrofuran			11		
dimethyl naphthalene				1.2	2.7
fluoranthene				2.7	9.7

A Sample obtained from a valve approximately 3 feet above the bottom of the tank

B Sample obtained from the top of the tank

Metallic sprayed zinc coating is a relatively expensive process where zinc wire is melted, atomized at a high pressure, and sprayed onto the surface. The application requires special skills and equipment, but the coating provides excellent rust inhibition and a service life of 50 years. Surface preparation by blast cleaning with one of three specified grits is mandatory.

Zinc-rich paints, containing 80-95 weight percent zinc dust, will provide a hard, abrasion and rust resisting surface on steel. The cost is high, and surface preparation involves near-white blast cleaning.

Chlorinated rubber paints may be used when the control of fumes from the application of other linings is difficult or where their use is specified as in Baltimore County. Chlorinated rubber compounds are formed by exposing natural rubber to chlorine gas. Because the resultant material is brittle, plasticizers or linseed oil are added, producing a more flexible and adherent coating. Application requires a near-white, blast cleaned surface and a zinc-rich primer. The chlorinated rubber paint is then spray applied in several coats. The intercoat adhesion is strong enough to essentially produce a single unit. This benefit provides easy surface preparation and reapplication when the orginal coating nears the end of its service life. Service lives have been estimated as 6 to 15 years.

Asphalt based linings are absorbed by water faster than coal tar based linings, and their application is generally limited to reclining existing asphalt lined tanks or sealing "green" cement mortar coating in pipes to provide a proper curing environment.

CORROSION INHIBITORS

Corrosion protection by formation of a film on the surface of a pipe may be achieved by chemical as well as mechanical means. The importance of calcium carbonate solubility in this regard was recognized in the early twentieth century by the German chemist, Tillmans, and in the U.S. by Baylis (29). Excess calcium and carbonates would form scale in pipes, reducing carrying capacities and increasing pumping requirements. $CaCO_3$ deposition in rust forming waters, however, could form an effective rust inhibitor. The need to understand the deposition-solubilization tendency of $CaCO_3$ formed the basis of the Langelier Saturation Index, a parameter formulated in 1936 but still used in corrosion control. When silicates and polyphosphates were found to inhibit corrosion, their theory and applications were investigated. Today, orthophosphates are also being studied, but corrosion control in many public water systems is still engineered from $CaCO_3$ solubility data. A summary sheet of corrosion inhibitors is presented in Table 36b.

TABLE 36b. DOSAGES AND PREFERRED WATER PARAMETERS FOR CORROSION INHIBITORS (see text for discussion and references)

	Dosages	Velocity	pH	Alkalinity	Hardness	Additional
$CaCO_3$	Varies considerablly	Not stagnant	6.8-7.3	> 40 mg/L	> 40 mg/L	$\frac{\text{halogen + sulfate}}{\text{alkalinity}} < 0.2$ (meg basis)
Sodium Silicate	No specific concentrations usually 2-8 mg/L	Must be flowing	< 8.4 A lower pH is desired	low	very low	
Phosphates	1-2 mg/L	High velocity	< 7.0 for lead pipe	low	low	

$CaCO_3$ Precipitation

Early observations that carbonate minerals in a water supply tended to inhibit corrosion in steel and cast iron pipes led to theories on calcium carbonate deposition as an anti-corrosive mechanism as well as several indicies of corrosion that are still used in contemporary water treatment. Effective $CaCO_3$ protection depends on the presence of anodic (metal oxidation) and cathodic (oxygen reduction) reaction products in the water. Reduced oxygen, as the hydroxyl ion, reacts with bicarbonate to form water and carbonate ions. The increased concentrations of carbonate ion along with the (oxidized) metal ions (Fe, Cu, Zn, Pb) subsequently exceed the solubility product of and, consquently, form metal carbonates (1) and carbonate containing solids. If this deposit satisfactorily adheres to both itself and the pipe surface, an effective corrosion barrier forms.

Of the carbonates formed, zinc carbonate forms a less porous structure than other metal carbonates. Calcium carbonate, however, forms a solid, though soft, coating that has good adherent properties. The protective ability of a $CaCO_3$ intermeshes with existing hydrous ferric oxides and ferrous carbonate (32). Thus, the most effective use of $CaCO_3$ depends on maintaining an environment where $CaCO_3$ is slowly formed, along with corrosion products, eventually producing a hard, impenetrable coating.

Corrosion waters are generally of low pH and low alkalinity, often with high CO_2 concentrations. The introduction of a carbonate to this system will affect all three parameters because of the relationships:

$$\frac{[H^+][HCO_3^-]}{[H_2CO_3]} = K_1' \quad \text{(eq. a)}$$

First dissociation constant of carbonic acid.

$$\frac{[H^+][CO_3^=]}{[HCO_3^-]} = K_2' \quad \text{(eq. b)}$$

Second dissociation constant of carbonic acid.

$$[H^+][OH^-] = K'w \quad \text{(eq. c)}$$

Solubility product of water.

$$[Alk]+[H^+] = 2[CO_3^=]+[HCO_3^-]+[OH^-] \quad \text{(eq. d)}$$

[Alk] is in equivalents/liter all other concentrations are molal.

Also, the calcium concentration may control carbonate by the solubility reaction:

$$[Ca^{++}][CO_3^=] = K_s' \quad \text{(eq. e)}$$

Empirical results have shown that optimum conditions for calcium carbonate scale formation and protection are (43):

1. Calcium carbonate oversaturation of 4-10 mg/l.
2. Calcium and alkalinity concentrations of at least 40 mg/l.
3. pH range of 6.8 to 7.3.
4. Halogen plus sulfate/alkalinity ratio of less than 0.2, on a milliequivalent basis.

By knowing the pH, alkalinity, and calcium concentration, the saturation condition of the water may be determined by using Caldwell-Lawrence diagrams (35). Additions of calcium and carbonate alkalinity needed to meet the optimum criteria are calculated from these diagrams. An alternate approach (34), if adequate alkalinity exists in the water, involves empirically determining the saturation pH of the water and adding lime to achieve that value. A third alternative is to adjust the water to achieve a zero or positive number on the Langelier Saturation Index (SI) (29), defined in section 2. Saturation index correlations with observed corrosion and red water problems have been good but not completely consistant (12). Maintenance of the SI within one unit of zero is considered satisfactory to eliminate extensive corrosion problems (16).

Methods used to increase calcium, alkalinity, and pH in corrosive waters depends on existing water constituents and raw material procurement costs, including transportation (5). One method used with desalinization waters is to blend the process stream with a naturally hard water. Because the distilled water is deaerated and of low pH, this water may need to be aerated and limed to add necessary oxygen and raise the pH.

When carbon dioxide is inexpensive, it can be added to react with either lime or limestone to produce a calcium bicarbonate rich water. A concentrated solution is prepared in a split stream and blended into the entire flow. Additional lime may be needed to raise the pH of the water to obtain a zero Saturation Index. The use of pulverized limestone, ground to 80 mesh, is generally insufficient to produce effective stabilization because of the required contact time. A pH above 6.5 is seldom achieved, but this method is attractive if iron removal is also needed. Filtration through partially calcined dolomite will raise the pH to 8.0 to 9.0, but the cost, compared to lime, makes lime more attractive except in very small systems where simplicity of operation is paramount. In areas where alkalinity is >25 mg/l as $CaCO_3$, the preferred method is lime addition. Lime readily dissolves in water and is generally cheaper than limestone on the basis of equivalents of calcium (5).

Although the use of calcium carbonate corrosion control is widespread, best results are still based on empirical data. While the optimum guidelines have been set, variables of temperature, velocity, dissolved oxygen, and other dissolved solids also affect the process. A temperature increase will decrease the calcium carbonate solubility, causing a given water to be closer to $CaCO_3$ saturation conditions. Also, reaction rates will increase. In solid-liquid reactions, a 10°C rise will generally increase the reaction speed by a factor of two (29). Saturated water heated to boiling will dissociate bicarbonate ions to the corrosive free CO_2 and hydroxyl ions (23). Water velocity affects the boundary layer and hence the transport of corrosive or protective material to the surface. In general, velocities greater than one fps are needed to deposit a protective barrier. Higher velocities provide better coatings (33). Because the strength of the $CaCO_3$ film depends on the presence of ferric oxides, the maintenance of 5 ppm dissolved oxygen is recommended (5).

Corrosion in lead pipes can be controlled by calcium carbonate deposition (38). Again, the stability of $CaCO_3$ is regulated by adding calcium and/or alkalinity or adjusting the pH. If pH adjustment alone is used, enough hardness must be present to provide the necessary calcium carbonate. Patterson and O'Brien (38) have suggested that pH adjustment without sufficient carbonate and alkalinity may be detrimental due to the preferential formation of $Pb(OH)_2$. This compound is non adherent and is potentially toxic as are all lead carbonate and hydroxide solids. Also, intermittent monitoring may not be sufficient to detect its presence. In general, soft waters, which are not amenable to pH adjustment only, occur on the eastern seaboard and in the southeastern and northwestern portions of the United States.

Dissolved solids other than calcium and carbonate species influence the rate of $CaCO_3$ formation. An increase in salinity causes an increase in the solubility of calcium carbonate ($CaCO_3$ is about 500 times more soluble in sea water than in fresh water) and explains why brackish waters are unable to form protective pipe coatings (29). Divalent ions (Mg^{++}, $SO_4^{=}$) also increase the solubility of $CaCO_3$, probably due to the increased ionic strength where fewer calcium and carbonate ions would become available for interaction (15). Because of the change in concentration of ionic species after some deposition has occurred, it is not uncommon to experience excessive deposition in and near the treatment plant with lessened protection in outlying areas. It should also be noted that increasing calcium and alkalinity too far beyond the saturation point may produce excess deposition (scaling). This is not uncommon when alkaline waters are lime treated without recarbonation. Also, where chlorination is used as a disinfectant, a pH increase means a decreased HOCl (the most effective chlorine form in water) content. Thus, lime addition for corrosion control may necessitate increased chlorine doses for effective disinfection.

Lime is available commercially as quicklime or hydrated lime. Quicklime is purchased in granular form and contains at least 90 percent CaO, magnesium oxide being the primary impurity. Hydrated lime is a powder containing approximately 68 percent CaO. Both forms of lime are dissolved into a 5 percent slurry prior to addition to the water system. Alkalinity is often applied in the form of soda ash (Na_2CO_3), a greyish-white powder containing about 98 percent sodium carbonate.

Lime addition requires the use of a slaker for quicklime while hydrated lime is often prepared in a tank with a turbine mixer. The slaker uses a gravimetric feeder to introduce the lime into a mixing chamber, where the proper amount of water is added to produce the desired slurry concentration. The ratio of water to lime is about 5 to 1, and pH controls that compensate the water to lime ratio in relation to changes in water quality or lime purity are available. Regulators protect against excessive temperatures resulting from the chemical reaction:

$$CaO + H_2O \rightarrow Ca(OH)_2.$$

A grit remover is used to remove coarse material prior to the solution being pumped to the slurry feeder. The feed rate into the water system is controlled by an automatic pH and flow control system (52).

One carbonate process that is gaining popularity is the low-carbonate method. Recently implemented in Bennington, Vermont, and soon to be used in Seattle (see section on Case History), this method raises the pH of the corrosive water to 8-8.3. Adding small amounts of alkalinity causes the formation of insoluble carbonate salts on pipe surfaces. Effective corrosion control is achieved without drastically altering the chemical makeup of the water. Results in Bennington showed an 82 percent corrosion reduction in lead and an 80 percent reduction in copper pipes (39). This method is also effective with galvanized surfaces.

Sodium Silicate

Sodium silicate, originally used to prevent metal solubilization in lead pipes, has been used for rust inhibition for over 50 years (48). Water conditioning by addition of sodium silicate is primarily applicable to galvanized piping where it is used to curtail existing red water problems and prevent corrosion, especially in hot water pipes. Sodium silicate is non-toxic and has the ability to control pre-existing conditions.

Perhaps the first consideration for using sodium silicates for corrosion control in an entire utility was to prevent the solution of lead from lead pipe in 1922. It was noted in this work that the corrosion of ferrous materials was reduced (48). Since that time, the use of sodium silicates has been attempted with some success over a wide range of applications (44)

The mechanisms of corrosion control with the addition of sodium silicate is thought to be the formation of a two-layered protective film between the material surface and the corrosive water. From results of laboratory studies, Lehrman & Shuldener determined that in dilute sodium silicate solutions, silica will exist in equilibrium between its ionic and collodial states (30). At low concentrations, such as those required for corrosion control in potable water systems, this equilibrium is attained very rapidly, if not instantaneously. The presence of silica in a zinc/iron system tends to make the zinc anodic to iron and the zinc reacts with water causing the formation of zinc hydroxide. The positively charged zinc hydroxide reacts with the negatively charged collodial silica from the water and silica is absorbed producing an amorphous silica (gel) deposit or layer capable of enmeshing compounds of iron, calcium, magnesium, and organic matter above a layer of metal corrosion products (30). In low alkalinity waters, this deposit is basically hemimorphite [$Zn_4Si_2O_7(OH)_2$ H_2O] with some ferrous or calcium silicate depending on whether the water is soft or hard (28), (11). In waters of higher alkalinity, especially at higher temperatures, Shuldener & Lehrman report that the zinc-iron potential can be reversed and the iron becomes anodic to the zinc causing the formation of ferric hydroxide (45). Although not as effective as zinc hydroxide, ferric hydroxide will also remove silica from water forming the amorphous silica layer.

Shuldener & Lehrman explain that when small amounts of silica are added to systems containing bicarbonate, especially at higher temperatures, two competing reactions can occur (45). The presence of silicate tends to make the zinc anodic to iron and the bicarbonate tends to do the reverse. The zinc corrodes to form a zinc hydroxide corrosion product which adsorbs silicate. As the silicate is removed and the concentration lowered, its effect is lowered and iron corrosion products are formed by the influence of the bicarbonate concentrations. Because the iron is not as effective at removing silica and because the silica concentration is now much lower, some rust can appear.

Shuldener & Lehrman conducted laboratory studies to investigate temperature, pH, and bicarbonate effects on the effectiveness of silicate for corrosion control (45). They simulated a galvanized pipe carrying a water at approximately 1.9 fps by rotating a pipe in water solution through which CO_2-free air was bubbled. They determined that the corrosion rate, as observed from rust formation, increased as the temperature increased from 72°F to 160°F, but that the addition of sodium silicate in the solution would act as a corrosion inhibitor and decrease the corrosion rate. From the laboratory studies on pH effects, Shuldener & Lehrman observed that in a solution containing 8.5 mg/l SiO_2 and 14 mg/l bicarbonate, decreasing the pH from 8.5 and 8.0 to 7.5 and 7.0, respectively, resulted in reducing the corrosion rate. Similar results were obtained in waters containing 8.5 mg/l SiO_2 and 54 mg/l bicarbonate when the pH was decreased from 8.5 and 7.0 to 6.5. PH adjustment was with dilute H_2SO_4 but specifics of the pH measurements in relation to the heating of the water or temperature corrections were not presented.

From their experiments it was shown that the presence of bicarbonate has an accelerating effect on the rate of corrosion. As is noted, as the pH increases in the range of 8 to 9, the concentration of bicarbonate decreases which implies that the effect of bicarbonate is closely associated with the pH of the solution (45). From their studies they concluded that bicarbonate has a very strong influence in causing reversal of potential with a resulting rust formation at 160°F (45).

Lane et al, conducted laboratory tests to investigate the effects of pH at various silica levels on the corrosion of galvanized steel at a domestic hot water temperature of 140°F. Their tests were conducted using a water with a pH range of 7.2 to 9.0, a hardness concentration of 100 mg/l (as $CaCO_3$) an alkalinity concentration of 35-55 mg/l and a velocity of 1.5 fps. Silica levels were varied from 10.0 mg/l to 20.0 mg/l. From results of their experiments, they determined that initially, the higher concentration of 20 mg/l silica was more effective than the lower concentration of 10 mg/l silica for controlling corrosion. However, after a sufficient period of time (possibly 60 days), Lane et al concluded that the effects on corrosion control is essentially the same at both levels of treatment indicating that there is no advantage to using the higher application rates (28). In both cases, however, corrosion control was most effective at the higher pH (8.5-9.0) condition. From the results of their tests, Lane et al state that the optimum silicate treatment appears to be a complement of pH and silica dosage, i.e., lower pH requires higher SiO_2 dosage and vice-versa, and is apparently influenced by water quality factors as calcium, magnesium, alkalinity, chloride, sulfate, and pH (28). Their experiments did not attempt to evaluate the effects of these water quality factors.

Sodium silicate dosages are independent of naturally occurring silica in the water and there are no specific concentrations recommended for the various conditions of water quality (51). As a general rule, however, an average concentration of 2 to 8 mg/l and possibly up to 12 mg/l SiO_2 is sufficient to maintain corrosion control in a system once a protective film is established. This inhibitor has been found to be particularly useful in waters with very low hardness, alkalinity, and pH < 8.4, and is more effective under higher velocity flow conditions. The application of sodium silicate requires the use of solution feeders, small positive displacement pumps that deliver a specific volume of chemical solution for each piston stroke or impeller rotation. The two general types of solution feeders are diaphragm and plunger metering pumps, although some rotary pumps may be used as positive displacement pumps as well. Both the diaphragm and plunger pumps may be controlled manually by adjusting the stroke length or rate of reciprocation, or they may use an automatic control unit that regulates stroking in proportion to water flow (52).

Inorganic Phosphates

Polyphosphates have been used to control calcium carbonate scale build-up in water treatment plants and irrigation systems since the 1930's. Presently, both polyphosphates and orthophosphates may be used for scale inhibition or corrosion control. The effectiveness of these chemicals is a function of flow velocity, phosphate concentration, temperature, pH, and calcium and carbonate levels.

Both ortho- and polyphosphates (especially tri-, pyro-, and higher polyphosphate anions) are known to form complexes with a number of metal ions, including calcium, iron, and lead (49). The presence of larger amounts of polyphosphates has been blamed for assisting iron and lead uptake from pipe surfaces in laboratory situations by upsetting local metal-ion equilibrium through complexation. When concentrations generally less than 10 mg/l of polyphosphates are introduced into a flowing water, however, they have a tendency to form a thin film on the metal pipe surface that protects the metal from further corrosion. This film will also adhere to calcite crystals, preventing further growth. In this way, a water supersaturated with calcium and carbonate can be prevented from scaling. The minimum concentration of polyphosphates that will prevent crystal formation for a particular water is called the threshold level. Subthreshold concentrations will produce distorted calcite crystals (33).

The threshold concentrations of polyphosphates, usually added as sodium hexametaphosphate, is pH dependent in calcium and carbonate containing waters due to the pH dependency of $CaCO_3$ deposition. Successful red water control has been achieved in waters that had unsuccessfully responded to lime teatment. In Little Rock, Arkansas, Uniontown, Pennsylvania, and Newport, Rhode Island, polyphosphate addition to lime treated waters with initially high pH ($\geq$9.0) resulted in corrosion abatement at polyphosphate concentrations of 1.0 to 2.0 ppm. However, the red water problems in Nitro, West Virginia, persisted in the lime treated water even at 10, 5, and 1 ppm polyphosphate concentrations.

Phosphates may be used to control corrosion in lead pipes, but because corrosion control in lead pipes is primarily interested in preventing lead being carried into the water as opposed to the pipe deterioration, the phosphate concentration must be low enough not to form soluble lead complexes (21). Corrosion protection by addition of sodium hexametaphosphate is best a pH $\leq$ 7.0 (9). Above this pH, and in the absence of phosphates, a lead-carbonate film will deposit on the surface and protect the pipe. If polyphosphates are present, the formation of this film is retarded and more lead may enter the water than would occur if either the addition of polyphosphates at low pH or the existence of a high pH alone were maintained. Effective corrosion control in lead pipes at pH below 7.0 has been achieved with sodium hexametaphosphate concentrations at or below 2 mg/l (21). These experiments also showed that lead levels at pH 8.9 were equivalent with or without polyphosphate addition and lower than those achieved at pH 7.0 or below with polyphosphate addition at low alkalinity.

Other pipe surfaces respond differently to polyphosphate corrosion control (9). The staining of plumbing fixtures due to the dissolution of copper or the components in brass has been lowered but not stopped by polyphosphate addition.

Aluminum protection has been obtained in laboratory tests, especially at low pH values. Zinc corrosion has also been controlled by 2 ppm polyphosphate addition to the water. At least two separate cases have been reported where the protection of galvanizing was excellent. Specifics on these tests, however, were not given (9).

One report of copper corrosion control by hexametaphosphate addition (6) reported that early morning copper concentrations were reduced from an average 4.8 to 5.5 ppm to an average of 2 ± 0.2 ppm copper. The pipe was 12 to 14 month old house lines in a district in England. Initial composition of the water was not reported. Hexametaphosphate addition relieved complaints of blue water and emetic reactions to use of the water that had stood in the pipes overnight.

A process involving orthophosphate addition has been developed (36, 27). The process involves the formation of a zinc orthophosphate, $Zn_3(PO_4)_2$, in the pipeline and its subsequent precipitation as an insoluble zinc orthophosphate film. This reaction involves adding zinc sulfate, sulfamic acid, and monosodium orthophosphate to the water:

$$3ZnSO_4 + 2HNH_2SO_3 + 2NaH_2PO_4 \rightarrow Zn_3(PO_4)_2 + 2NaNH_2SO_3 + 3H_2SO_4$$

Because zinc solubility, even in the presence of $PO_4^{\equiv}$, is pH dependent, an initial application of zinc is empirically maintained at 2 to 3 ppm.

After the coating is formed, in about 3 weeks, the zinc is reduced to 1 ppm. In highly alkaline waters, between 0.5 and 1 ppm sodium hexametaphosphate may have to be added to prevent $CaCO_3$ precipitation. This is because $CaCO_3$ weakens the $Zn_3(PO_4)_2$ film, decreasing its anti-corrosive efficiency. The results of one test using this procedure is presented in Tables 37 and 38.

TABLE 37. ANALYSIS OF A NEW JERSEY WATER

hardness, ppm	96
sulfate ($SO_4^=$), ppm	44
pH	7.0
M.O. alkalinity, ppm	52
TDS, ppm	150
Langelier Index	1.1

TABLE 38. 63-DAY CORROSION TEST ON CAST IRON USING TREATED NEW JERSEY WATER

Coupon	Exposure (days)	Corrosion Rate (mpy)	% Corrosion Reduction
Control	31	10.20	
Treated	31	8.24	19
Control	63	10.77	
Treated	63	5.24	51

Source = (27).

Other recent tests (26, 38) on the ability of zinc orthophosphate to prevent corrosion in lead pipes has presented negative results. Karalekas et al added zinc orthophosphate to Boston water (low alkalinity and hardness, pH < 7.0) for a six-month period and could not reduce the lead levels to the 0.05 mg/l standard. Patterson and O'Brien immersed lead coupons in stagnant water and determined that lead corrosion could increase from the use of zinc orthophosphate. The pH conditions during their short term test varied from 7.24 to 9.13. The long term test pH was held between 6.52 and 6.82 and showed a 60% increase in soluble lead content in the treated water. Murry (36) reported that zinc orthophosphate is insoluble in water so his process was the in situ formation of zinc orthophosphate during flowing conditions. However a Ksp of 1×10^{-32} has been reported for zinc orthophosphate (4) and others (18) have reported that orthophosphate can substantially reduce lead levels in waters of pH 7-8.2 with low alkalinities and carbonate levels. Because of these inconsistant findings, more research is needed on this corrosion control mechanism.

Other organic and inorganic phosphate-phosphonate corrosion inhibitors are used for industrial scale control but are not applicable to potable water systems. Also, the possibility of eutrophication in lakes and streams that ultimately receive the water has discouraged at least one city from using inorganic phosphates for corrosion control. The critical level for algal growth has been established as somewhere near 0.01 mg/l (42), and it was believed that the wastewater treatment plants might not be able to reduce the phosphorous content to this extent.

In summation, several important factors must be addressed prior to the use of inorganic phosphates for corrosion control in any specific water system. The effectiveness of polyphosphates, usually added as sodium hexametaphosphate, is greater with increasingly turbulent velocities, and flow velocities of 2-5 fps and above are not continually experienced in all parts of any distribution system (51). Under stagnant or nearly stagnant conditions, such as service lines, polyphosphates will not be effective. In waters saturated with calcium carbonate, polyphosphates will prevent $CaCO_3$ deposition, and larger amounts of polyphosphates may assist in iron and lead concentrations through complexation. Polyphosphates have been shown to alleviate "red water" in many, but not all, situations where the iron content was originally in the water and where the iron was a result of pipe corrosion. Polyphosphates can also prevent lead release into the water by forming a thin film of complexation compounds across the lead surface. Effective polyphosphate concentrations reported are usually $\leq$ 2 ppm, but an initial concentration of up to 10 mg/l for up to three weeks has been used to produce an initial film. Orthophosphates have produced both positive and negative results for lead control and more clarifying research is needed.

Polyphosphate addition is by use of solution feeders as discussed in the silicate section. Concentrated solutions are mildly corrosive, so stainless steel is often used for the dissolving tanks, baskets, and the liquid end of the chosen pump. The positive displacement pump(s) are generally selected from accuracy, durability, capacity, corrosion resistance, and pressure capability requirements, and most manufacturers have designs to accommodate the specific need.

Miscellaneous Methods

Other corrosion control methods are applicable to more specific situations. In instances where high sulfate concentrations are producing corrosive sulfides, which tend to accelerate the anodic dissolution of iron (43), the sulfate reducing bacteria responsible for the sulfides may be eliminated by maintaining proper chlorine residuals in the water system. This problem should not occur under normal conditions where disinfection procedures and natural aerobic conditions occur.

Dissolved oxygen removal will curtail corrosion. Deaeration may be accomplished by applying a pressure vacuum, heating and degasification, or deactivation. Deaeration is too expensive to be considered a viable corrosion control for municipal water systems.

Cathodic protection, previously a common corrosion inhibiting method in water tanks, occurs when electrons are supplied to the metal surface being protected, decreasing that metal's tendency to oxidize. Electrons may be supplied by a direct current or by a sacrificial anode. The latter consists of a metal higher in the electromotive series than the protected metal and is, therefore, preferentially corroded. Examples of a cathodic protection are magnesium rods in hot water heaters and the zinc used to galvanize steel.

ECONOMICS

The economics of corrosion abatement is often simplified to a comparison of annual costs incurred by implementing additional chemical treatment to the water supply and/or the increased cost of pipe lining versus the annual costs for replacing distribution pipes (46). Increased pumping costs due to reduction in the hydraulic efficiency of corroded piping and increased costs due to shortened pipe or appliance life are sometimes, but not always, included in cost analysis. However, health and aesthetic effects and possible increased cost due to excessive deposition (scale) are often ignored.

Benefit/Cost Analysis

In Benefit/Cost analysis, a price is determined for each aspect of corrosion control and the sum of the annual costs for imlementation of the abatement method is compared to the sum of the annual "lack of costs" experienced by the utility or by the user and the utility. Costs for implementation of the corrosion control are presented in the next sections and reflect construction, materials, chemicals, and labor cost as well as intangibles such as public acceptance and relative economy of the area. Benefits are harder to quantify, especially in terms of pricing the "lack of costs" to be incurred by the consumer.

User benefits may be classified as directly economic, aesthetic, or in terms of health. The directly economic savings results from extended service lives of piping, fixtures, and appliances, as well as lessened damage from water leaks, and are generally formulated by first determining the extent of corrosion-based annual costs and subtracting annual costs for similar repairs experienced in similar areas practicing corrosion control. Methods used to determine the extent of problems caused by corrosive waters include a review of customer complaint records, the use of questionnaires, corrosion rate testing, examination of plumbing pipes, and water quality monitoring in residences. From data collected, a service life for waterheaters, pipes, faucets, shower heads, appliances, etc. is determined, and annual cost for maintenance and repair or replacement is figured. Unfortunately, the prices determined are applicable only to the individual distribution system samples supplying the data as results from similar areas using the same water source have produced drastically different costs (8). Also, attempts to correlate water quality parameters with corrosive effects in terms of economics have been poor (37, 8). One study (37) tried to correlate the corrosive potential with the Saturation Index, the Ryznar Index, and the chloride plus sulfate concentration and ended up finding the greatest correlation (r = 0.33) between TDS and houses that had to replace interior piping. No positive correlation was found between TDS and replacement of faucets, shower heads, or toilet flushing mechanisms. Whereas that study assessed a penalty cost of 18¢/m^3 for users of Colorado River water in Southern California another study assessed a 4.5¢/m^3 value for the same water source in the same area (8). (These prices included the cost of soaps and detergents and bottled water costs as well as corrosion-induced costs) Both studies were conducted by questionnaires, but problems with this type of survey were demonstrated by the latter study. For example, substantial differences in service life estimates were found between plumber and repair shop responses as compared to consumer responses. It was also noted that increased water hardness caused greater use of home softening units that may increase the corrosiveness of waters by increasing conductivity and decreasing the Saturation Index. Results of testing designed to substantiate this claim were, however, inconclusive as seen from the results of copper testing in five residences:

	influent	after softening but prior to contact with plumbing	after contact with plumbing
Copper piping	30	440	40
" "	20	80	110
" "	130	260	530
Galvanized iron piping	10	5	<5
" " "	80	50	320

TABLE 39

Copper Concentrations (ug/l) in Domestic Waters

(10)

An example of the computed savings per residential unit resulting from the recommended corrosion control measures in Seattle (see section 6 - Case Histories) is presented in Table 40. The annual plumbing deterioration cost savings is estimated at $1,700 or $7.50 per year per residential unit. Table 40 indicates that the savings for a specific unit depends on its location, age, and type of plumbing. The savings to older homes, whose piping is presently substantially deteriorated would be reduced, and the control measures should minimize the rate of plumbing system corrosion in the future. (25)

TABLE 40
(1978 Prices)
Source (25)

Comparison of Estimated Annual Costs and Savings of Corrosion Related Deterioration with and Without Corrosion Control Treatment

Residential Category	Plumbing Type	Number of Units	South Area Cedar River: Estimated Annual Cost per Unit Without Treatment ($)	South Area Cedar River: Expected Annual Cost Savings per Unit Provided by Treatment ($)	South Area Cedar River: Expected Annual Cost per Unit With Treatment ($)	Number of Units	North Area Tolt River: Estimated Annual Cost per Unit Without Treatment ($)	North Area Tolt River: Expected Annual Cost Savings per Unit Provided by Treatment ($)	North Area Tolt River: Expected Annual Cost per Unit With Treatment ($)
Single Family and Duplex									
All units	Galvanized steel	62,051	39.35	10.40	28.95	52,551	48.50	7.20	41.30
Units built 1971-75	Galvanized steel	166	7.60	6.35	1.25	176	17.10	7.55	9.55
All units	Copper	14,183	3.95	3.90	0.05	15,151	5.10	4.60	0.50
Units built 1971-75	Copper	664	7.05	6.95	0.10	880	7.10	6.70	0.40
Multi-Family									
All units	Galvanized steel	40,700	25.30	8.75	16.55	12,988	32.90	6.45	26.45
Units built 1971-75	Galvanized steel	410	4.70	3.90	0.80	152	12.45	6.45	6.00
All units	Copper	16,812	2.90	2.85	0.05	8,720	3.20	3.15	0.50
Units built 1971-75	Copper	1,720	4.90	4.85	0.05	760	4.80	4.60	0.20

Health and aesthetic problems resulting from corrosive waters are definable but hard to quantify economically. Aesthetically, iron concentrations may present turbidity and color (red water) problems; the drinking water limit for zinc is based on taste tests; dissolved copper may cause a blue green stain on porcelain where soap residues accumulate; and manganese may cause brown-black staining. In terms of health, lead and cadmium may be toxic in low levels, and other heavy metals may be leached from pipes and solders. Furthermore, causes of the correlation between increased cardiovascular disease and soft waters have not been positively identified but may result from either the role of the bulk mineral content (Ca, Mg, Na) of hard waters or the metal constituents in soft waters resulting from its corrosive nature (24). While cost figures have not been placed on the aesthetic problems, it is estimated that the difference in cardiovascular mortality, between corrosive and non-corrosive waters is 52 deaths/100,000 population annually, which translates to 47,000 avoidable deaths attributable to a lack of corrosion control, and an income loss to the U.S. of $3 billion annually (24).

Trends and Costs of Mechanically Applied Linings and Coatings

Cast iron and ductile iron pipe, composing over 76% of the pipe used in U.S. water supply distribution systems serving over 2,500 people, is exclusively lined with cement mortar. Steel pipe, at 6%, is increasingly being lined with epoxy. Although epoxy is more expensive than coal tar, the benefit of a high smoothness coefficient, a lack of having to control coal tar enamel application fumes, and possible adverse health concerns arising from the use of coal tar based linings, have caused this increased epoxy usage. There are two manufacturers of hot applied coal tar enamel for potable water system lining in the United States with total 1976 coal tar enamel sales of $1,500,000. One company estimated that half of the steel pipes installed by the water supply industry were coal tar enamel lined.

The extent of use of various linings and coatings for water tanks is not well defined. The extremes of material use estimates reported by various vendors and manufacturers is shown in Table 41. As can be seen, a wide range of discrepancy exists.

TABLE 41.

EXTREMES OF VARIOUS ESTIMATES OF USE OF TANK LININGS AND COATINGS

COATING	EXTREMES OF ESTIMATIONS (10 FIRMS REPORTING) OF % OF NEW WATER TANKS LINED WITH A GIVEN COATING NATION-WIDE
Vinyl	47.5 – 90
Epoxy	6 – 50
Coal Tar	0 – 90*
Other	0 – 17

* 0-19% reported on all estimates except one firm indicated 90% (19)

Besides manufacturer's preferences, regional preferences exist in relation to water tank linings as seen in Table 42.

TABLE 42.

REGIONAL PREFERENCE FOR WATER TANK LININGS

(compiled from (19))

CITY/STATE	PREFERRED LINING/SPECIFICATION FOR STEEL TANKS
Seattle	~100% Coal Tar Enamel
Atlanta	Coal Tar Enamel
Houston	Vinyls and Epoxys Used
Virginia	Coatings containing coal tar, vinyl or bituminous material not approved.
Baltimore	Chlorinated Rubber
California	No longer uses coal tar enamel due to possible health affects

The cost of applying linings to water tanks varies widely and is a function of the application method, surface preparation, cost of labor, and intangibles such as regional competition and local materials preference. Table 43 presents the extremes of cost estimates for the linings discussed in the text.

Costs of Corrosion Control by Chemical Applications

After the chemical constituents of a corrosive water have been identified, and potential control mechanisms are listed, the cost of treatment for each method must be determined. Because local construction, materials, chemical, and labor costs, along with intangibles such as the local economy and water quality demands, influence cost estimating, this section should be regarded as a general overview. Where possible, weight units are used because prices quoted are apt to reflect temporal and regional values that may or may not differ from current prices. Brand names used are those reported from the articles cited and do not constitute an endorsement of the product.

Sodium silicate has been effective in reducing plumbing repair costs when used to treat water used in groups of buildings or housing developments (44). This additive is more cost effective when used in very soft, low alkaline waters, especially when compared to the use of lime, sodium hydroxide, and/or soda ash. To obtain the 8 ppm silica concentration regarded as the minimum level needed for corrosion control, a 28 ppm sodium silicate solution, equivalent to 232 pounds of solution per million gallons of water must be used (48). The cost of equipment needed is that of a solution feeder. Prices vary depending on the size, type (diaphragm or plunger metering pumps), and manufacturer; and selection is usually based on accuracy, durability, capacity, corrosion resistance, and pressure capablity.

The cost of using polyphosphates is seen in a 1966 study by the city of Richmond, Virginia. They compared the use of lime, Calgon brand metaphosphate with lime, and TG-10(a Calgon phosphate composition) with lime to stabilize their corrosive waters (7). Empirical tests coupled with manufacturer's data indicated that corrosion would be controlled using either 2.0 ppm Calgon or 1.25 ppm TG-10, each with a reduced lime consumption. An additional capital cost ($3000) would be required for the polyphosphate addition and the 1966 costs for the use of TG-10 plus lime, Calgon plus lime, and lime would have been $3.08, $2.35, and $1.00 per MG respectively. The report recommended the use of TG-10 based on lower pipe line maintenance costs, reduced valve and meter repair costs, and lower pumping costs from increased pipe diameters expected to result from reduced scale formation.

The additional equipment required for the TG-10 and Calgon systems included a ½ HP stainless steel Simplex pump (0.0-0.5 gal/min), two 400 gallon stainless steel tanks, two 5-cubic foot stainless steel dissolving buckets, two agitators (¼ HP motor, 3-three blade propellers), one Simplex pump assembly, two electric timers, and one switch. Materials, based on a 40 MGD flow, were 2000 lbs lime; 400 lbs lime plus 717 lbs Calgon; or 400 lbs lime plus 440 lbs TG-10 to be used daily.

TABLE 43

COST ESTIMATES (19)

COATING	SURFACE PREPARATION $/Sq. Ft.	MATERIALS $/Sq. Ft.	APPLICATION $/Sq. Ft.	INITIAL COST* $/Sq. Ft.	SERVICE LIFE YEARS	COST EFFECTIVENESS $/Sq. Ft./Yr.
Coal Tar Enamel (Hot)	.40- .50	.10-.15	.30- .60	.94-2.00	20-50	.019-.100
3 Coat Vinyl	.47-1.00	.11-.22	.16- .50	.95-1.11	8-23	.048-.139
Metallized Zinc	.80-1.65	.50	1.15-1.75	2.45-3.90	40-50	.049-.098
4 Coat Vinyl	.24-1.00	.09-.20	.09- .21	.55-1.21	8-23	.053-.144
Chlorinated Rubber	.30- .75	.05-.17	.15- .50	.50-1.14	6-15	.055-.190
Coal Tar Paint (Cold Applied	.30- .75	.13-.25	.30	.73-1.30	12-15	.055-.083
Coal Tar Epoxy	.30- .75	.15-.20	.25- .30	.75- .95	10-14	.068-.075
2-Component Epoxy	.30- .80	.15-.50	.16- .70	.90-1.95	10-15	.060-.115
Asphalt (Petropoxy)	.50-1.00	.24	.30- .50	1.24-1.34	15	.083-.089
Wax	.12	.04	.29	.35- .45	4- 5	.088-.090
1-Component Epoxy	.24- .80	.06-.25	.16- .50	.47-1.50	10	.086-.115
Asphalt (Inertol #49)	.30- .75	.10-.11	.30	.70- .91	6	.117-.152
Phenolic	.30- .75	.11-.13	.16- .50	.77-1.11	4- 6	.185-.193
Zinc Rich Paint	.80-1.00	.25	.05- .15	1.10-1.40	5	.220

*Initial costs are as reported and may not include figures reported in surface preparation, material, or application columns.

Estimates were obtained from published literature, paint supplies, engineering consultants, and painting contractor costs were calculated from the 33rd edition of Building Construction Cost Data, 1975, Robert S. Means Company, and the Estimating Guide of the Painting and Decorating Contractors of America, 10th edition, 1977-78.

Relative costs of calcium carbonate stabilization are presented in a 1971 report on the pacification of product water from a distillation plant (5). Costs of adding 40 ppm (as $CaCO_3$) of calcium bicarbonate to the water are presented for the use of lime (CaO) and CO_2, limestone and CO_2, calcined dolomite and CO_2, and limestone and H_2SO_4 for both 10 MGD and 50 MGD flows. The capital costs of the storage bins, dust filter, slaker unit, mixers, slurry feed pump, recarbonator, aerator, instruments, etc. were $110,000 and $357,000 for the 10 MGD and 50 MGD flows and salary related costs were $27,000 and $68,000 per year, respectively. These costs were essentially the same for all processes investigated. Chemical prices cited were a reflection of shipping distances, especially in relation to the transportation charge of powered limestone, and the method of contracting for liquid CO_2 that may vary the CO_2 price by a factor of two. The 1971 prices given were:

Lime	$30/ton
Limestone	15/ton
Calcined Dolomite	150/ton
Liquid CO_2	50/ton
H_2SO_4 (93%)	23/ton

Chemical costs in 1971 dollars per 1000 gallons are presented in Table 44.

TABLE 44

CORROSION CONTROL BY CALCIUM CARBONATE STABILIZATION

PROCESS	COST $/1000 GAL	CAPITAL AND LABOR 10 MGD	CAPITAL AND LABOR 50 MGD	CHEMICALS	TOTAL COST 10 MDG	TOTAL COST 50 MGD
Lime and CO_2		.89	.45	1.12	2.01	1.57
Limestone and CO_2		.89	.45	.82	1.71	1.27
Dolomite and CO_2		.89	.45	.95	1.84	1.40
Limestone and H_2SO_4		.89	.45	.98	1.87	1.43

from: (5) (1971)

Although these costs reflect 1971 prices, it is significant to note that in instances where limestone is half as expensive as lime and one tenth as expensive as calcined dolomite, and 93% H_2SO_4 is half as expensive as liquid CO_2, all on a weight basis, limestone and CO_2 would be the preferred process (5). Furthermore, a considerable savings would be realized if split stream treatment were used on part of the flow followed by blending with the remainder of the water.

Because of their applicability to calcium carbonate stabilization, cost curves (20) for lime feed systems and recarbonation via liquid CO_2 systems are included in this section. Construction costs for the lime feed system (Figure 36) are based on hydrated lime use up to 50 lb/hr and quicklime use at higher rates. The hydrated lime arrives in 100 lb bags, is introduced by feeder to a dissolving tank, and gravity fed to the point of application. The quicklime is stored in hoppers with a 30 day storage capacity (3 days if recalcination is used) located over the slaker. The slurry is gravity fed to the point of application. Operation and maintenace cost (Figure 37) do not include the price of lime and are based on \$0.03/kw hour and \$10/hour labor costs. Construction costs for liquid CO_2 system (Figure 38) include a 10-day storage tank, CO_2 vaporizer, a solution-type CO_2 vaporizer, a solution-type CO_2 feeder, injector pump, main header, diffuser pipes, and an automatic control system. Operation and maintenance (Figure 39) are based on \$0.03/kw-hr. and \$10/hr. labor costs. Costs on these curves are based on October, 1978 prices.

Estimates of the national cost of piping damage resulting from the distribution of corrosive waters range from \$210 million annually (1975 dollars) (47) to \$375 million annually (1976 dollars) (24). Estimates for the cost of implementing lime-addition corrosion control to stabilize these corrosive waters are from \$20 million (47) to \$27 million (24) annually. This simplified economic analysis does not account for the intangible benefits of elimination of red water, possible decreases in CVD mortality, or lessened pumping costs; yet, juding from the economics of pipe replacement alone, well-run utilities with corrosive waters can ill-afford not to implement control methods.

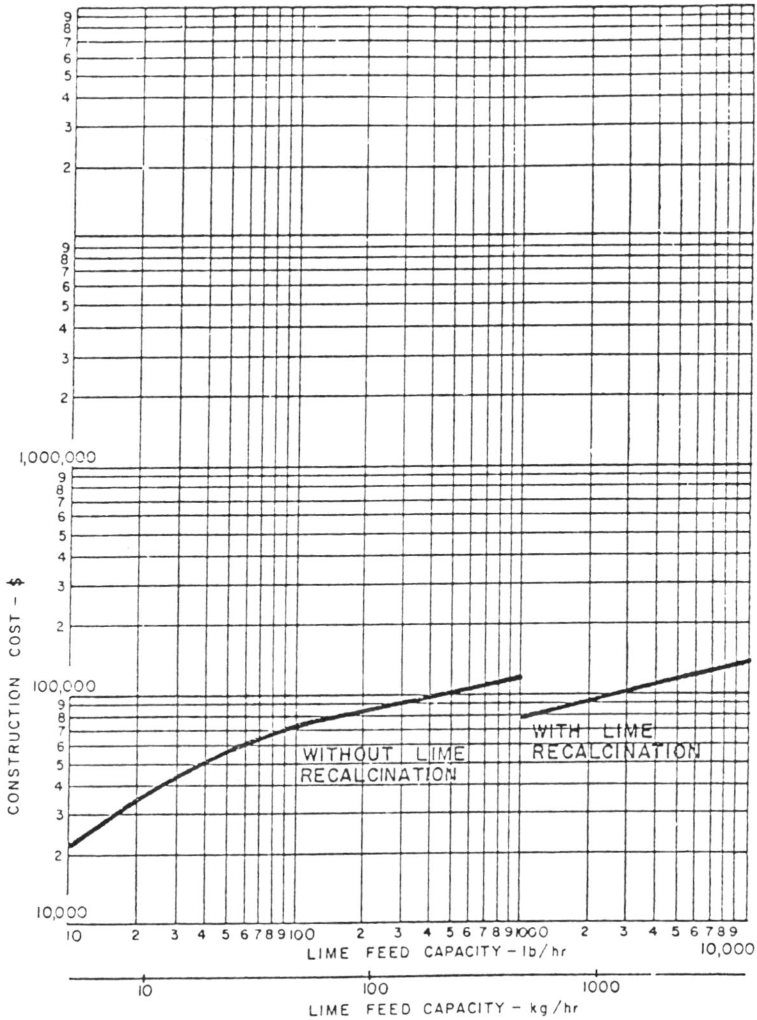

Figure 36. Construction cost for lime feed systems (20).

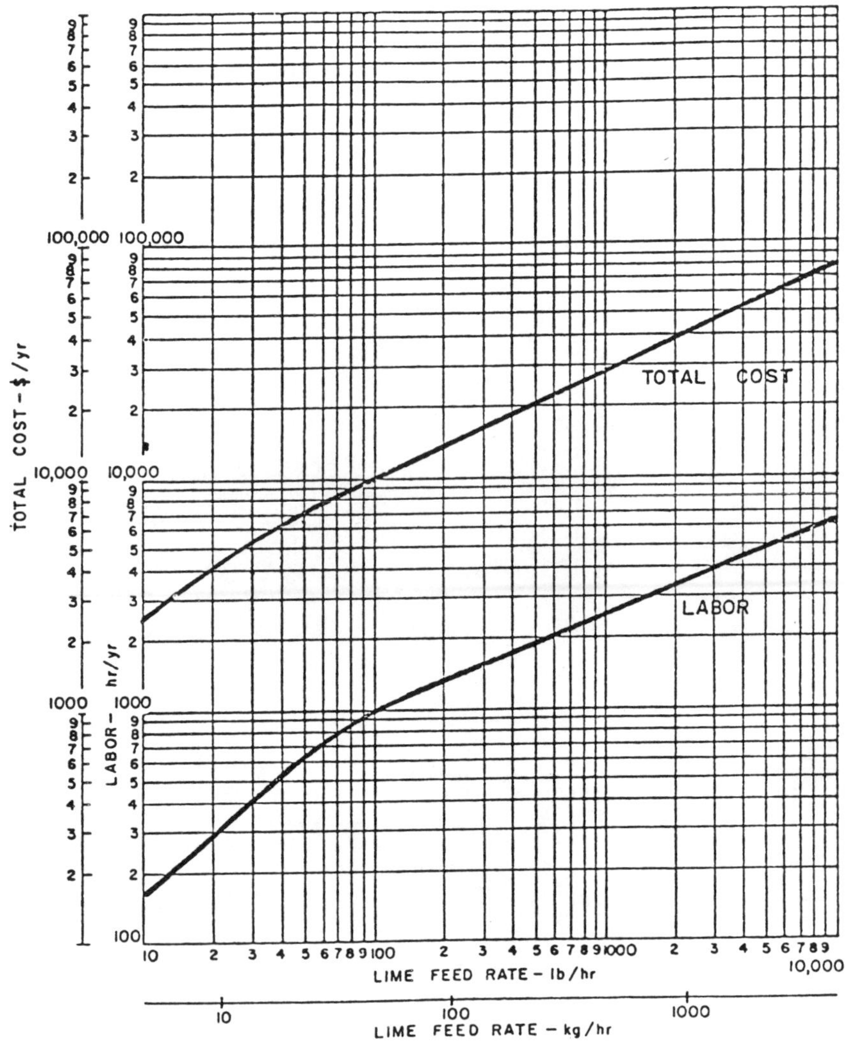

Figure 37. Operation and maintenance requirements for lime feed systems - labor and total cost. (20)

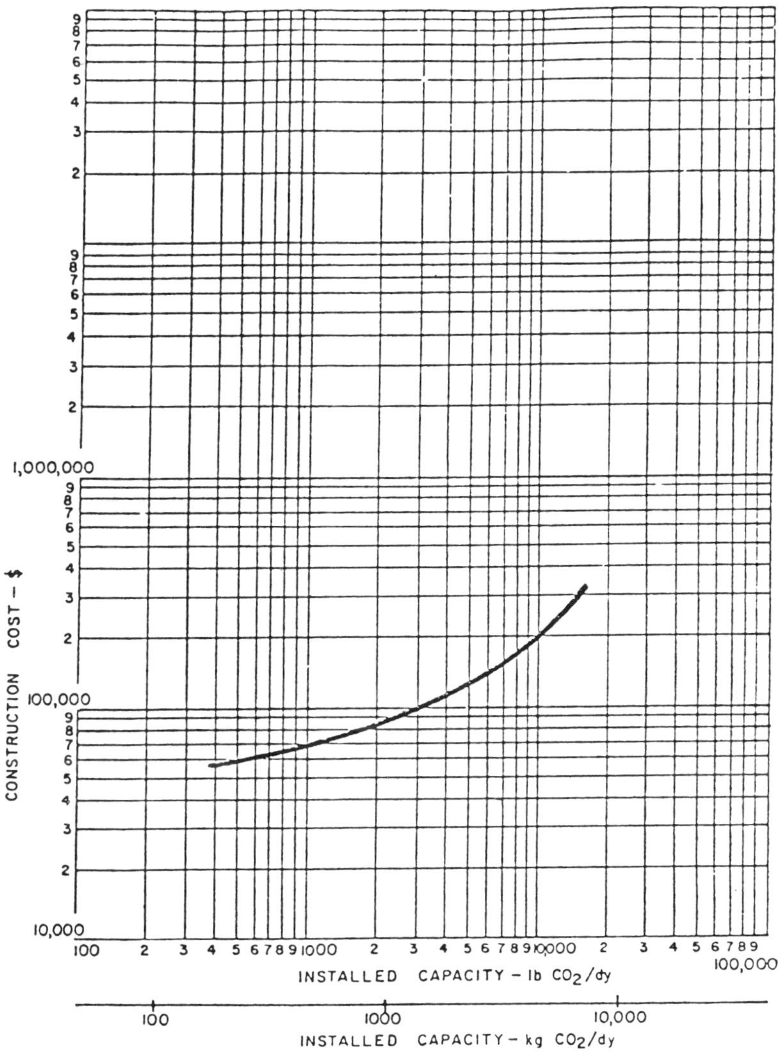

Figure 38. Construction cost for recarbonation - liquid CO_2 source. (20)

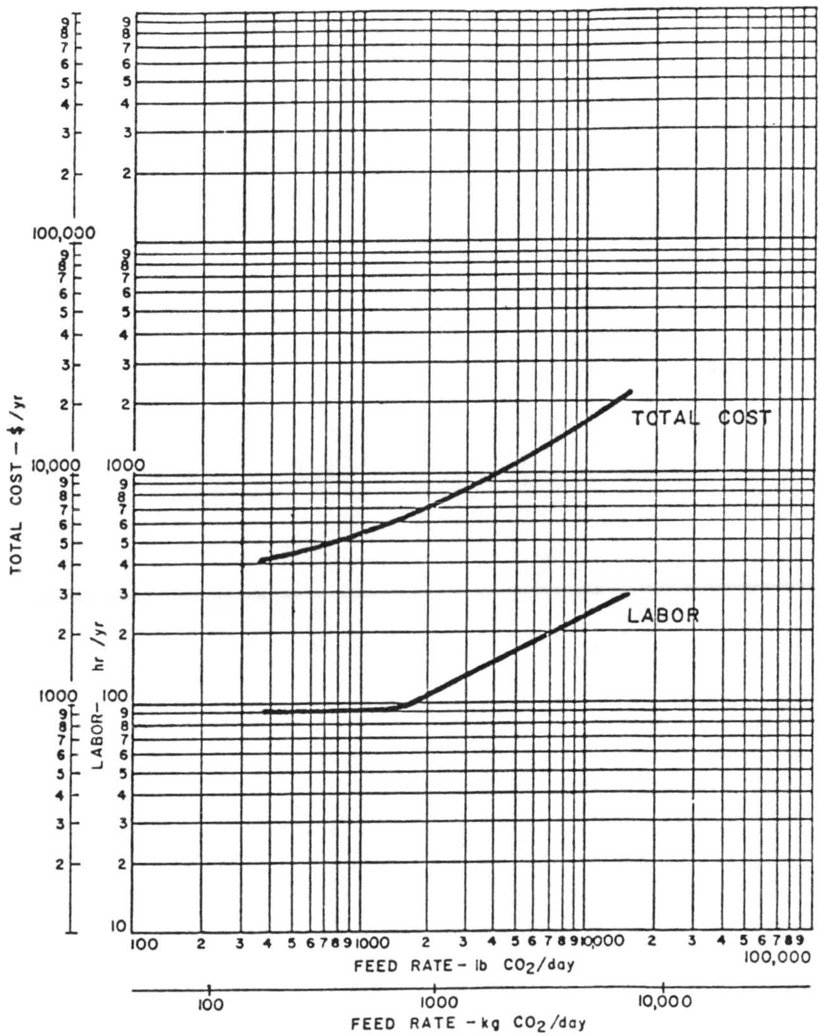

Figure 39. Operation and maintenance requirements for recarbonation-liquid CO_2 as CO_2 source - labor and total cost. (20)

CASE HISTORIES

Seattle

The Seattle Water Department has initiated a corrosion control program. Although the system is not yet operational, a review of the engineering involved in defining the problem and determining its solution demonstrates that corrosion control in the water works industry is an intricate and multidimensional process.

A slight increase in corrosion complaints occurred after Seattle started to use the Tolt River waters in addition to the existing Cedar River supply in 1964, but widespread complaints became common after 1970 when chlorination doses were increased, ammoniation was stopped to increase the free chlorine residual, and fluoridation with hydrofluosilicic acid began. Both water supplies are surface waters from rain and snow runoff in mountains east of the city. They are very soft, low TDS, low pH, highly oxygenated waters with some organics, principally tannins and lignins. Addition of free chlorine and fluorides caused this water to become excessively aggressive. Aesthetically based consumer complaints were followed by a water metals survey. Although metal increases in the distribution system mains were very low, lead, iron, zinc, and copper were being appreciably leached in buildings and residences. Lead, iron, and copper limits often exceeded federal limits when water was left standing in internal piping for as little as five hours. The predominant pipes in these structures were galvanized iron or copper. Lead increases were believed to be leached from soldering joints.

Because of adverse health and aesthetic reasons, as well as the expense water customers had to assume in maintaining residential plumbing, a study was initiated to determine the best applicable corrosion abatement process. Alternatives included:

1. Alternative disinfectant chemicals to replace gaseous chlorine.

2. Alternate fluoridation chemicals to replace hydrofluosilicic acid.

3. Blending ground water with the surface water to provide a less corrosive product.

4. Addition of corrosion inhibitors.

Alternate disinfection chemicals included calcium hypochlorite, sodium hypochlorite, and ozone. The first two would provide a less corrosive water but cost was two to six times greater than chlorine. Alternative fluoridation chemicals, sodium silicofluoride and sodium fluoride, did not provide effective corrosion control and were more expensive than the hydrofluosilic acid.

Blending ground water was deemed unfeasible because the only available ground water supply could be ecomonically used with only one of the two water systems. Additional costs of well construction, pipelines, and pumping stations coupled with the uncertainty of this means to affect a solution to the corrosion problem removed this alternative from consideration.

Corrosion inhibitors examined included lime, sodium carbonate, sodium silicate, sodium bicarbonate, caustic soda, and various phosphate compounds. The following are the results of the six alternatives selected for extensive testing.

1. Low-carbonate system. Lime and sodium carbonate were used to raise the pH to 9 and alkalinity to 10-30 mg/l. No calcium carbonate scale formed on pipe surfaces, excellent corrosion protection was provided for copper pipes and fair protection for galvanized and black steel pipes. However, the high pH attained raised questions about trihalomethane formation as the surface waters contained organic precursors and THM formation is known to be accelerated above pH 8.5. Also, the high pH was unstable in contact with air, decreasing by at least one pH unit within a day. Further testing with $NaHCO_3$ eliminated both of these problems, yet provided the same amount of corrosion protection.

2. Lime and zinc-orthophosphate. Although this method effectively eliminated corrosion, it was felt that the potential stimulant to algal growth from the phosphate and subsequent taste, odor, plugging of fine filters, and additional chlorine required did not warrant this as a desirable alternative.

3. Sodium silicate. Silica, added to a concentration of 10 mg/l, gave corrosion protection, but evidence of increased pitting rates coupled with an increased need to remove any silica added to the water by industries that were using the existing water without pretreatment made this alternative unattractive.

4.-6. These methods involved the carbonate method of corrosion control based on $CaCO_3$ precipitation. Each of these methods gave excellent protection for copper pipes and good protection for galvanized and steel pipes.

4. High pH, balanced lime and alkalinity. Adding 15-20 mg/l lime and 15-20 mg/l sodium bicarbonate produced a water with pH 8.8. Cost and THM formation possibilities made this alternative unattractive.

5. Moderate pH, balanced lime and alkalinity. By adding 35-40 mg/l lime and 55-60 mg/l CO_2 a water of about 75 mg/l calcium and alkalinity (as $CaCO_3$) and pH 8.3 was achieved. Cost of this method is excessive and the water would be classified as "moderately hard," imposing a much greater impact on industries that pretreat the water for hardness removal.

6. High alkalinity, moderate pH. By adding 10-20 mg/l lime and 65-130 mg/l $NaHCO_3$, a very soft water with pH 8-8.5 would result. Although attractive from a corrosion control and water quality aspect, the cost was excessive.

METHOD	ESTIMATED ANNUAL COST ($)
Low carbonate	510,000
Lime and zinc orthophosphate	430,000
Sodium silica	630,000
High pH, balanced lime and alkalinity	1,100,000
Moderate pH, balanced lime and alkalinity	1,700,000
High alkalinity, moderate pH	1,400,000

The low carbonate method using sodium bicarbonate was chosen because of its low cost and minimal affect on existing water and environmental quality. Additional sodium silicate will be added to the Tolt River system, bringing the silica level up to that of the Cedar River (7-9 mg/l) and providing additional corrosion protection. The mechanism of low carbonate control is a result of reactions between pipe material and carbonate, forming insoluble metal oxides that coat the pipe surface. Cost of the process and additional chlorine needed due to the increased pH is expected to be $1.13 per thousand gallons. Assuming a 100 gpcd water use rate, this amounts to $1.40/residence/year. The computed savings to residences from expected lower piping repair and service needs is about $7/year less than the present $35/year estimate.

In association with this process, city policy changes have been recommended. These include encouraging the use of less corrodible pipes, such as cement lined steel and ductile iron, for distribution systems. The use of asbestos cement and galvanized pipes is discouraged for municipal, industrial, or commercial use. The use of copper or plastic pipe for residential piping is encouraged, as is the use of low-lead solder and glass lined water heaters.

Factors other than cost that influenced the decision to use this process included the potential impact on industrial water use. Area firms are adjusted to the relatively mineral free water presently supplied. Use of the low carbonate process should only slightly affect ion-exchange units or carbon adsorption beds, and few industries are expected to need extensive adjustments. The waste water discharged into Puget Sound should have lower metals concentrations, as will the waste water sludge. The corrosion abatement process is presently being further tested and fine tuned, and a 21 month plan, design, permit acquisition, and construction time period is anticipated.

Carroll County, Maryland

Carroll County, Maryland is a formerly rural area located about 30 miles northwest of Baltimore that has experienced a large population increase in the last two decades. The county currently has about 13,000 residences that rely on individual wells for domestic water, and being built since the early 70's, many of these houses contain copper piping joined with 50% lead solder. The major geologic formations supplying the well water include the Triassic Sandstone, the Wakefield Marble, and the Pretty Boy Schist. The sandstone waters are of variable pH, the marble waters are neutral or higher, but the bedrock acidity of the schist formation produces low pH waters.

In November, 1975, a 45 unit apartment complex served by two wells was found to have a 0.34 ppm lead concentration in a daytime water sample. At the same time, several children living in the complex were found to have moderately elevated blood lead levels. The complex consisted of relatively new buildings (two to 14 years old), but the water system neutralzer was poorly maintained and consequently ineffective. The neutralzer problem was corrected, lead concentrations dropped to below the federal standard of 0.05 ppm, and the children's blood levels returned to normal over a period of a few months.

A year later, in another part of the county, the physician of a family experiencing a gastro-intestinal illness thought their drinking water might be contaminated. A bacteriological analysis was negative, but the water's pH of 5.5 coupled with an inspection of the type of plumbing used in the house, suggested heavy metal contamination. Sampling indicated lead levels of 1.8 ppm and copper levels of 7.9 ppm in the tap water. Subsequent sampling of 14 new houses in this subdivision showed that 12 houses had lead concentrations greater than 0.05 ppm and 10 houses had copper concentrations in excess of 1.0 ppm in early morning samples. A second sampling, after water had run for five minutes, showed six houses with lead levels in excess of the federal standard. Water sampled directly from the well showed no lead or copper, and homes with a neutralizer that adjusted the water to pH 7.2-7.5 also showed no lead or copper. It was concluded that poor workmanship during the copper piping installation and the use of lead solder had precipitated the problem.

This subdivision incident prompted a random sampling of the tap water in homes in new subdivisions, and when 11 of 35 samples were found to contain lead levels above 0.05 ppm in samples where the water had run for five minutes, The Carroll County Health Department requested investigative assistance from the Center for Disease Control. A subsequent random sampling analyzed 350 homes in the county, determining pH, alkalinity, early morning lead and copper levels, and lead and copper levels after running the water for five minutes. Test results showed that low pH and low alkalinity were associated with excessive lead and copper levels in the tap waters.

Consequences of this testing included the immediate recommendation that the water, after standing overnight or for long periods, be allowed to run for two minutes prior to use. The plumbing code in the county was revised to permit the use of NSF-approved (grade 14) plastic tubing in new construction and repair work, and a neutralizer to raise water pH above 7.0 is presently recommended in existing homes with copper piping. The calcite filter type of neutralizer is suggested as best for this problem. (31)

Orange County, California

Between July, 1961 and March 1963, 206 houses were built by a land development firm in an area of southern Orange County, California. All plumbing was copper and water was supplied from wells by a local water utility district. A typical water analysis showed calcium, 225 ppm; magnesium, 40 ppm; sodium, 160 ppm; sulfate, 525 ppm; chloride, 170 ppm; bicarbonate, 360 ppm; TDS, 1500 ppm; hardness, 735 ppm; pH 7.1; and free carbon dioxide, 36 ppm.

In January, 1963, a one million gallon reservoir and hydropneumatic tank system began operation to provide better water service to the development's growing population. Some houses had been occupied since the spring of 1962 and had used water pumped directly from the wells. However the problem of perforation leaks in copper tubing throughout the development did not occur until three months after the additional system began operation. In April, 15 leaks were reported, and the rate of about 15 leaks reported per month continued through October.

By May 1963, the builder had decided that the leaks were not the result of faulty workmanship and sought the assistance of consultants, including two corrosion experts, a testing laboratory, and a civil engineering firm. The subsequent investigation determined that the corrosion was a form of severe occasional pitting occurring on the inside surface of the copper tubes. The pipe surface was covered with a thin green layer, assumed to be copper carbonates, and the pits formed conical penetrations into the tube material and were covered by nodules up to 1/8 inch in height. The pitting occurred in a random pattern and was not necessarily found near joints. The copper that had been used was both "Type K" and "Type L", had been supplied by three different manufacturers, and was found to conform to ASTM Specification B88-62 for that material. Copper staining of plumbing fixtures, that may result from copper solubilization in soft, acidic waters, was not found. Leaks appeared in occupied and unoccupied houses, and although electrical services were grounded to the water systems, the unoccupied houses did not have their power turned on. It was also noted that water from the same wells was used in other parts of the utility's service area with few reports of errosion failures.

The consultants decided that the probable cause of corrosion was the combined presence of free carbon dioxide and dissolved oxygen, possibly enhanced by the high TDS concentration. Although the well water had minimal dissolved oxygen, field measurements indicated that a dissolved oxygen concentration of about 3 ppm existed after the water passed through the reservoir and hydropneumatic system. The consequent remedial action taken by the water purveyor was to begin using water from two other wells. Because one of the additional wells had high iron and manganese levels, a package treatment unit consisting of potassium permanganate, caustic soda, and chlorine addition with pressure-type sand filtration was installed. Delivery of the treated water began in September, and the rate of pipe failures dropped to nearly zero in November.

In early 1964, copper tubing failures began to recur. The consultants found that while the well providing water with high iron and manganese concentrations had initially shown low levels of free carbon dioxide, the level had risen to about 25 ppm. Also, the addition of caustic soda that had neutralized the free carbon dioxide, had been discontinued after less than two months of operation. In April, the wells were taken out of service and the delivery of imported water to the affected area began. After about a month, the rate of tube failure again dropped to almost zero.

Damage resulting from the corrosion cost the builder substantial financial losses, bad publicity, inability to sell houses, and the loss of FHA financing commitments. He consequently sued the purveyor for negligence in not reducing the carbon dioxide content of the water and for breach of warranty that essentially says that in the absence of an explicit warranty, a seller warrants his goods to be suitable for the purpose for which they are sold.

The court found that the water only was responsible for the corrosion and that the addition of caustic soda was a simple and inexpensive remedy. However, the water supplier was not found negligent for his failure to reduce the free carbon dioxide concentration in the water. He was found guilty of breach of warranty and the judgment was for the builder's claimed loss. The court ruled that the water was defective in that the purveyor warranted that the water would be reasonably fit for transmission through copper pipes and would not damage or corrode the tubing during its normal life expectancy.

Additional Corrosion Control Practices

The Catskill-Delaware water system supplying New York City is treated with fluosilic acid for fluoridation that lowers the pH of this soft water from 6.9 to 6.3. Caustic soda is then added to raise the pH to the 6.9-7.2 range. The effectiveness of this system has been judged on the basis of consumer complaints, with an increase in the number of complaints of green (copper) staining occurring when insufficient caustic soda was added. Samples of water at consumer taps by the New York City Health Department indicated that of 500 samples analyzed for lead, copper, and iron, only a few cases of lead in excess of 0.05 ppm were found, contrary to the health department's expectations (40).

In New London, Connecticut, severe corrosion problems experienced prior to 1969 have been controlled by pH adjustment and corrosion inhibitor addition. The raw water pH of 6.8 is adjusted to 7.2-7.5 and zinc orthophosphate is added to produce a 0.5 mg/l zinc concentration in the water. Tap water sampling has shown this treatment to effectively control lead and copper concentrations.

Other large metropolitan areas with corrosion control methods are Salem-Beverly, Massachusetts (lime and zinc metaphosphate); Long Beach, California (see October, 1970 issue of JAWWA); Middlesex, New Jersey (see August, 1974 issue of JAWWA); Waterbury, Connecticut; and Philadelphia's Schuylkill River Plant.

REFERENCES

1. Atkins, G. R.,"Soft Water Corrosion and Calcium Carbonate Saturation", The South African Industrial Chemist, Vol. 8, June 1954, pp. 104-111.

2. Baylis, John R.,"Corrosion Studies,"J. New England Water Works Assoc., Vol. 67, 1953, pp. 38-73.

3. Baylis, John R.,"Treatment of Water to Prevent Corrosion," J. American Water Works Assoc., Vol. 27, No. 2, 1935, pp. 220-234.

4. Bard, Allen J., Chemical Equilibrium, Harper & Row, New York, 1966.

5. Bopp, C. D., and S. A. Reed, Stabilization of Product Water From Sea Water Distillation Plants, U.S. Office of Saline Water Research & Development Progress Report No. 709, Oak Ridge National Laboratory, Tennessee, July 1971.

6. Brighton, William D.,"Dissolved Copper from New Service Pipes," Water and Water Engineering, Vol. 59, July 1955, pp. 292-293.

7. Brown, Joseph E., Charles R. Pitts, Jr., Evaluation of Three Different Agents for Stabilizing Water, City of Richmond, Department of Public Utilities, Richmond, Virginia, April 1966.

8. California Department of Water Resources, Southern District, Consumer Costs of Water Quality in Domestic Water Use Lompoc Area, District Report, June 1978.

9. Committee Report,"The Value of Sodium Hexametaphosphate in the Control of Difficulties Due to Corrosion in Water Systems", JAWWA, Vol. 34, No. 12, pp. 1807-1830.

10. Cornwell, F. J., G. Wildsmith and P. T. Gilbert,"Pitting Corrosion in Copper Tubes in Cold Water Service,"Br. Corrosion J., Vol. 8, No. 5, Sept. 1973, pp. 202-209.

11. Cox, Charles R.,"Corrosion Control by Water Treatment," Water Works Engineering, December 4, 1934, pp. 1514-1517.

12. DeMartini, F. E.,"Corrosion and the Langelier Calcium Carbonate Saturation Index,"JAWWA, Vol. 30, No. 1, January 1938, pp. 85-111.

REFERENCES (continued)

13. Esty, Roger W.,"Cement Lining of Pipe Corrects Bad Water Troubles,"The American City, April 1941, pp. 60-62.

14. Esty, Roger W.,"When Can Cement Lining of Pipe be Used to Advantage,"Water Works Engineering, September 10, 1930, pp. 1363-4.

15. Feitler, Herbert,"Critical pH Scaling Indexes,"Materials Protection and Performance, August, 1975, pp. 33-35.

16. Flentje, Martin E.,"Control of Red Water Due to Pipeline Corrosion,"JAWWA, December 1961, pp. 1461-1465.

17. Frye, S. C.,"Epoxy Lining for Steel Water Pipe,"JAWWA, Vol. 66, No. 8, August 1974, pp. 498-501.

18. Gardels, M. and M. Schock, (EPA Cincinnati Laboratory), Personnal Communication, via P. Lassovszky, January 12, 1981.

19. Goldfarb, Alan S., James Konz, and Pamela Walker, "Coal Tar Based Materials and Their Alternatives," Interior Coatings in Potable Water Tanks and Pipelines, The Mitre Corp., Mitre Technical Report MTR-7803, U.S. EPA Contract No. 68-01-4635, January 1979.

20. Gumerman, Robert C., Russell L. Culp and Sigurd P. Hansen, Estimating Water Treatment Costs, Vol. 2, Cost Curves Applicable to 1 to 200 mgd Treatment Plants, EPA-600/2-79-K2b, prepared for the Municipal Environmental Research Laboratory Office of Research & Development, U.S. EPA by Culp/Wesner/Culp Consulting Engineers, Santa Anna, California, August 1979.

21. Hatch, G. B.,"Inhibition of Lead Corrosion with Sodium Hexametaphosphate,"JAWWA, Vol. 33, No. 7, pp. 1179-1187.

22. Heller, A., and K. C. Chang, and B. Miller,"Spectral Response and Efficiency Relations in Semiconductor Liquid Junction Solar Cells," J. Electrochem Soc., Vol. 124, No. 5, May 1977, pp. 697-700.

23. Hopkins, Edward S.,"Basic Principles of Corrosion Control by the Use of Lime," Paper Trade Journal, Vol. 127, No. 1, July 1, 1948, pp. 61-63.

REFERENCES (continued)

24. Hudson Jr., H. E. and F. W. Gilcreas,"Heath and Economic Aspects of Water Hardness and Corrosiveness", JAWWA, Vol. 68, 1976, pp. 201-204.

25. Internal Corrosion Study, Summary Report, prepared for the City of Seattle Water Department by Kennedy Engineers, 7708 Bridgeport Way W., Tacoma, Washington, 98467, February 17, 1978.

26. Karalekas, Peter C., Jr., C. R. Ryan, C. D. Larson, and F. B. Taylor,"Alternative Methods for Controlling the Corrosion of Lead Pipe,"J. New England Water Works Assoc., Vol. 92, No. 2, 1978, pp. 159-78.

27. Kelly, T. E., M. A. Kise, and F. B. Steketee,"Zinc/Phosphate Combinations Control Corrosion in Potable Water Distribution Systems," Materials Protection and Performance, Vol. 12, No. 4, April 1973, pp. 28-31.

28. Lane, R. W., T. E. Larson, S. W. Schelsky,"The Effect of pH on the Silicate Treatment of Hot Water in Galvanized Piping," JAWWA, August 1977, pp. 457-461.

29. Langelier, W. F., "The Analytical Control of Anti-Corrosion Water Treatment," JAWWA, Vol. 28, No. 10, 1936, pp. 1500-1521.

30. Lehrman, Leo, Henry L. Shuldener, "Action of Sodium Silicate as a Corrosion Inhibitor in Water Piping," Industrial and Engineering Chemistry, Vol. 44, No. 8, August 1952, pp. 1765-1769.

31. Lovell, John, Richard Isaac, Ruth Singer, Control of Lead and Copper in Private Water Supplies, Carroll County, Maryland, Carroll County Health Department, October, 1978.

32. McCauley, Robert F. and Mahmond Omer Abdullah,"Carbonate Deposits for Pipe Protection,"JAWWA, Vol. 50, 1958, pp. 1419-1428.

33. McCauley, Robert F.,"Use of Polyphosphates for Developing Protective Calcite Coatings,"JAWWA, January 1960, pp. 721-734.

34. McLaughlin, P. L.,"Eliminating the Guess from Anti-Corrosion Treatment", Water Works Engineering, 1937.

35. Merrill, Douglas T. and Robert L. Sanks,"Corrosion Control by Deposition of $CaCO_3$ Films: Part 1, A Practical Approach for Plant Operators,"JAWWA, Vol. 69, November 1977, pp. 592-599.

REFERENCES (continued)

36. Murray, W. Bruce,"A Corrosion Inhibitor Process for Domestic Water,"J. American Water Works Assoc., Vol. 62, No. 10, Oct. 1970, pp. 659-662.

37. Orange County Water District, California, Water Quality and Consumer Costs, Santa Ana, Calfornia, May 1972.

38. Patterson, James W. and Joseph E. O'Brien,"Control of Lead Corrosion", J. of the American Water Works Assoc., Vol. 71, No. 5, May 1979, p. 264-271.

39. Patterson, James W.,"Corrosion Inhibitors and Coatings," JAWWA Conference, June 1978.

40. Report on Corrosion Control Practices, EPA Region 1, Water Supply Branch, Division of Water Programs, Sept. 1975.

41. Sargent, Harold E.,"Asbestos in Drinking Water", J. New England Water Works Assoc., Vol. 88, No. 1, 1974, pp. 44-57.

42. Sawyer, Clair N. and Perry L. McCarty, Chemistry for Sanitary Engineers, 2nd Edition, McGraw-Hill Book Company, New York, 1967.

43. Scholefield, Ronald J., Metal Corrosion Products in Municipal Drinking Waters, Thesis, Environmental Engineering, Illinois Institute of Technology, August, 1979.

44. Shuldener, Henry L., Sidney Sussman,"Sodium Silicate - To Keep Piping Young,"Water Works Engineering, September 1960.

45. Shuldener, Henry L., Leo Lehrman," Influence of Bicarbonate Ion on Inhibition of Corrosion by Sodium Silicate in a Zinc-Iron System,"JAWWA, November 1957, pp. 1432-1440.

46. Simmonds, M. A.,"Effect of Aggressive Waters on Cement and Concrete, with Particular Reference to Cement-lined Mains," The J. of the Institution of Engineers, Australia, Vol. 26, January-February 1954, pp. 9-16.

47. Singley, J. Edward, A. W. Hoadley, H. E. Hudson, Jr., Edna T. Loehman, A Benefit/Cost Evaluation of Drinking Water Hygiene Programs, U.S. EPA contract #68-01-1838, Univ. of Florida, 1969.

48. Stericker, William,"Sodium Silicates in Water to Prevent Corrosion," Industrial and Engineering Chemistry, Vol. 30, #3, March 1938, pp. 348-351.

REFERENCES (continued)

49. Stumm, Werner and James J. Morgan, Aquatic Chemistry, An Introduction Emphasizing Chemical Equilibrium in Natural Waters, Wiley-Interscience, New York, 1970.

50. Weir, Paul,"Effects of Pipe & Tank Lining on Water Quality at Altanta", JAWWA, Vol. 49, No. 1, January 1957, pp. 1-14.

7. Considerations for Corrosion Control Regulations

Detailed information is presented in this study which presents the nature and magnitude of corrosion and corrosion control in the water works industry. From the results of previous studies as presented in the literature, it is obvious that corrosion control is quite complex and development of a responsible corrosion control strategy for the water works industry requires a comprehensive approach. The objective of this Section is to summarize the information presented in the preceeding Sections through the use of charts and tables to provide the necessary basis for this comprehensive approach. Information presented in these tables is developed directly from data provided and discussed in this study. In as much as results of studies presented in the literature are often in conflict and disagreement and the fact that factors affecting corrosion and corrosion rates are most commonly synergistic, the information and comments provided in these tables should be considered as guidelines only. Additionally, in the interest of brevity to present a comprehensive and readily usable overview, comments and statements have been taken out of context and could be misleading to users lacking sufficient technical understanding of the nature of corrosion in the water works industry. Therefore, it is suggested that references be made to the respective sections of this study, as needed, for effective use of the tables.

The tables and information presented in this Section were selected and organized to follow a logical order for considering and developing a corrosion control strategy for the water works industry. The various materials and their specific use within the water works industry are first presented. Second, the general extent of each material's use as well as the respective associated potential contaminants are presented. This information is presented to help assess the significance of corrosion or deterioration of the various materials and can be used to assign priority for corrosion control strategies if desired. Once materials and their specific potential contaminants have been identified, applicable corrosion monitoring and detection techniques can be selected to determine the extent, if any, of corrosion. The third table is perhaps the most comprehensive and provides a brief summary of the various water quality and conditions of service parameters' effects on corrosion of each of the materials. This table can be used most effectively to assess the potential for controlling all corrosion in a system by comparing the preferred water quality and conditions of service to minimize the corrosion of each material. Finally, a table is presented which identifies and summarizes the applicability of corrosion prevention technologies for each material. The following is a brief description of each of the tables included.

Table 45, Materials and Their Application in the Water Works Industry, identifies specific uses of materials and attempts to provide a relative quantification of occurrance. Specific uses include in-plant systems (piping and appurtenances), transmission lines, storage, distribution mains, service lines, and household plumbing. Information in this table can be used to assist in identifying the types of materials currently in place as well as their relative quantity. This table does not, however, identify current use patterns.

Table 46, Significance of Corrosion or Deterioration of Various Materials Used in the Water Works Industry, presents a brief discussion of the known extent of use of each material as well as the contaminants that have been found to be associated with the use of each material. Concentrations of contaminants released are provided where values are reported in the literature. The importance of this table is its use for assessing corrosion significance of each material. It should be noted here, that it appears from this table that the use of lead and asbestos-cement should be of paramount concern as these two materials occur extensively within the industry and have been associated with releasing significant quantities or concentration of contaminants, lead being potentially toxic and the effects of asbestos fibers in drinking water yet to be determined. Additionally, it should be noted that it is estimated that one-third of all water distribution pipe currently being sold in the U.S. is manufactured of asbestos-cement pipe. Alternatively, lead pipe is currently being used less extensively but, owing to its relatively long life, many of the lead lines installed remain in service. Lead based solders continue to be used extensively.

Table 47, Preferred Water Quality and Conditions of Service to Minimize Corrosion of Materials Used in the Water Works Industry, is the most comprehensive table and provides a brief overview of the factors influencing or affecting corrosion of each material. Water quality parameters included are pH, hardness, alkalinity, dissolved oxygen, carbon dioxide, total dissolved solids, metal ions, and organic acids. Conditions of service included are velocity and temperature. This table can be used to assess the potential of controlling various parameters to control the corrosion within a water system which contains a variety of materials. It should be noted that pH, hardness, and alkalinity are the most controllable water quality parameters using conventional water treatment practices. The other water quality parameters are more difficult to control and/or maintain throughout a distribution system but are included herein to emphasize their synergistic affects on corrosion of materials. The effects of water velocity can only be controlled through design, and possibly, operation practices.

In general, water pH should be maintained near neutral or in the slightly alkaline range. However, it is noted that localized corrosion has been observed to peak in plain iron in the pH range of 6-9.

For most materials, the effect of hardness concentration is synergistic with other parameters. In general, hard waters are considered less aggressive then soft water to materials with the exception of aluminum in which the reverse is true. For iron-based materials, the presence of calcium ions has been shown to control corrosion if sufficient alkalinity exists. However, hardness concentrations do not appear to influence corrosion characteristics of stainless steel. In the case of copper, it has been found that soft waters may not be corrosive if the carbon dioxide concentration is low while pitting corrosion has been observed in cold hard water indicating that hardness concentration does not act independently on the corrosion or pitting of copper.

Corrosion and pitting appear to be inhibited in waters containing higher alkalinity concentrations for most materials except copper. In copper pipes, the addition of bicarbonate has been shown to actually enhance corrosion under specific conditions. Desired alkalinity concentration ranges of 20 mg/l or higher for lead pipes (12) and 40 mg/l or higher for concrete pipes (25) have been reported. For asbestos-cement materials, the aggressiveness of water is defined as a function of the combined levels of pH, hardness, and alkalinity concentrations. Although good studies on the pH effect on asbestos-cement independent of the other factors is lacking (12), a less "aggressive" water should be produced by increasing any one of the three factors.

Dissolved oxygen concentrations cannot be controlled or maintained effectively throughout a water system. Nevertheless its presence or absence can significantly affect corrosion or corrosion rates on various materials and is therefore included in Table 47. In all cases with the metals, a low concentration of dissolved oxygen is desirable to minimize corrosion. However, a higher dissolved oxygen concentration is generally desirable for the formation of protective films. For the iron-based materials, corrosion appears to increase linearly with increased dissolved oxygen concentration. For copper, however, the presence of dissolved oxygen is known to enhance corrosion, but corrosion or the corrosion rate is not dependent on the dissolved oxygen concentration. Little information is provided in the literature identifying the effects of dissolved oxygen concentration on nonmetallic materials.

As with dissolved oxygen concentrations, low carbon dioxide concentrations are desirable to minimize corrosion. For lead materials, it has been stated that the presence of excess carbon dioxide concentration will tend to dissolve protective carbonate films and assist corrosion.

Results of studies reported in the literature indicate that the effects of total dissolved solids (TDS) concentrations on corrosion of the metals is complex and no general characterization can be established. For plain iron it has been reported that while the presence of TDS can decrease dissolved oxygen and carbon dioxide concentrations resulting in a reduction in corrosion, the increased conductivity can, in fact, increase the range of galvanic coupling or lead to the formation of less protective films. It is also stated in the literature that the presence of chloride (Cl^-) and sulfate ($SO_4^=$) ions can increase corrosiveness of water while elsewhere (11) it is stated that the presence of these two ions can improve the protectiveness of scale.

For galvanized iron, it is reported that chloride ion concentrations observed in most potable water supplies do not increase corrosion or corrosion rates while higher concentrations will tend to accelerate corrosion. For the case of stainless steel, the literature reports that the presence of sulfate ions will inhibit corrosion while the presence of chloride ions can cause severe corrosion.

The corrosive effects of metal ion concentrations also varies with the specific material as well as with the specific metal ion. The presence of copper ions has been reported to be a major factor affecting and increasing the corrosion of galvanized steels. Additionally, the presence of copper, tin, nickel, and mercury have been shown to be detrimental to the corrosion of aluminum. Conversely, low concentrations of iron (Fe^{++}) have been reported to inhibit corrosion in copper and to precipitate on asbestos-cement to form a protective coating inhibiting calcium leaching (3).

Little information is presented in the literature describing the corrosive effects of the presence of organic acids. For iron-based materials, it is reported that the presence of humic acids can improve protective deposit formation and lead to reducing corrosion or the corrosion rate (30, 40). Alternatively for lead, the literature recommends that the occurrance of organic acids whose lead salts are soluble should be minimized to prevent corrosion (32). It is also reported that the occurrence of an unidentified high molecular organic acid found in surface but not in ground water may inhibit corrosion in copper piping (5).

For most materials used in the water works industry, it is reported that both stagnant waters and waters of high velocity will promote corrosion. Where protective coatings may be formed, some flow is required. However, excessive velocities can cause impingement attack and accelerate corrosion. In general, but not in all cases, water velocities ranging between 2-7 fps are desirable.

Temperature effects on the corrosion of materials are of most concern at elevated temperatures above that observed from operations at a utility. With the exception of aluminum, increased water temperatures will generally increase corrosion of materials used in the water works industry. Water temperatures in excess of 40°C are considered preferable for minimizing corrosion on aluminum. For plain iron, it is reported that the aggressiveness of water increases with temperatures up to approximately 80°C and then decreases at higher temperatures.

Table 48, Application of Corrosion Control Mechanisms, provides a summary of the various corrosion control alternatives and their application to the various materials. Corrosion control alternatives include both coatings and linings and inhibitors. Coatings and linings are coal tar, cement mortar, epoxy, vinyl, and miscellaneous non-coal tar paints. Inhibitors include calcium carbonate, silicates, and phosphates. Cathodic protection is also included as it is applied for protection of the inside of steel tanks.

TABLE 45

MATERIALS AND THEIR APPLICATION IN THE WATER WORKS INDUSTRY

MATERIAL	IN PLANT SYSTEMS: PIPING	IN PLANT SYSTEMS: APPURTENANCES	TRANSMISSION LINES	STORAGE	DISTRIBUTION MAINS	SERVICE LINES	HOUSEHOLD
WROUGHT IRON	✓				✓	✓	✓
CAST/DUCTILE	✓✓	✓✓	✓✓		✓✓✓ (cast iron)	✓	✓
STEEL	✓		✓	✓✓	✓	✓	✓
GALVANIZED IRON	✓				✓	✓	✓
STAINLESS STEEL		✓✓					✓
COPPER	✓	(brass)				✓✓✓	✓✓
LEAD		✓ (gaskets)	✓		✓	✓	✓
ALUMINUM		✓					✓
ASBESTOS-CEMENT	✓		✓✓		✓✓	✓	
CONCRETE	✓	✓	✓✓	✓✓	✓		
PLASTIC	✓			✓	✓	✓✓	✓✓

KEY: ✓✓✓ Used >50% for the particular service.
✓✓ Frequently used for the particular service.
✓ Has been or is used for the particular service.

TABLE 46
SIGNIFICANCE OF CORROSION OR DETERIORATION OF VARIOUS MATERIALS USED IN THE WATER WORKS INDUSTRY

MATERIAL	EXTENT OF USE	ASSOCIATED CONTAMINANTS
IRON-BASED MATERIALS:		
PLAIN IRON	Cast iron is used in 75% of all major U.S. water supply distribution systems (17). Also used in water appurtenances and treatment plants. Over 1/2 million steel water storage tanks exist in the U.S.	Iron concentrations in excess of the 0.3 mg/l approval limit occur, resulting in ferric oxide (red water) complaints.
GALVANIZED IRON	Generally limited to service lines, in plant systems, and households. It requires threaded joints and gooseneck connections and is declining in usage.	Zinc concentrations will increase 5 to 10 mg/l after 8 to 40 hour exposure to new galvanized pipe. Small amounts of iron will enter solution. Cadmium and lead (impurities in galvanizing process) concentrations will rise.
STAINLESS STEEL	Seldom used for piping, but used where low maintenance and reliable, continuous service is desired, such as pumps, valves, meters, venturis, and pressure regulators.	Besides iron, other metals used to manufacture stainless steel that may enter the water through pitting or corrosion are chromium, nickel and molybdenum.
COPPER	Extensively used in household piping and service lines. From World War II to 1972, over 6 million miles of copper tubing was put into service. Bronze may be used for appurtenances.	Copper, as well as iron, zinc, tin, and lead from associated pipes and solder may be oxidized into solution. Cu concentrations do not raise above about 5 mg/l. Impurities in brasses, such as manganese, arsenic, antimony, phosphorus, bismuth, and tin may also leak out.
LEAD	Little documentation available; according to Donaldson in 1924 (9) approximately 50% of water distribution systems in the U.S. had lead lines, used primarily for service lines and solders for copper pipes; ≈60% of residences in Boston are serviced with lead lines.	Lead concentrations ranging up to ≈ 3.0 PPM from both lead lines and lead based solders have been reported.
ALUMINUM	Use of aluminum is relatively limited; currently used for weir gates, storage tands, reservoir roofs and supports, hot water systems, and pipe lines.	No information available which identifies or quantifies potential contaminants; could release traces of copper, magnesium, silicon, iron, manganese, chromium, zinc, or titanium as well as alumina ions.
ASBESTOS-CEMENT	Approximately 1/3 of all water distribution pipe currently being sold in the U.S. is manufactured of asbestos-cement pipe; approximately 200,000 miles has been placed into service.	Asbestos-cement fibers counts in excess of 4.5 million fibers per liter have been observed; tetrachlorethylene concentrations as high as 2500 μg/l (3) have been observed from lined asbestos-cement pipes (22).
CONCRETE PIPE	Extensively used in water distribution (and storage) systems with service life of over 50 years in some locations. About 15% of new water tanks are concrete	Contaminant release is greatest when the pipe is first used and decreases thereafter. Water hardness and pH initially increase. Oxides of silicon, aluminum, iron, magnesium, and sulfur may hydrolyze, releasing these elements.
PLASTIC PIPE	Currently growing in use for service lines and household piping except in hot water systems; 1978 use was about 1/3 of all piping on a footage basis. Recent development has produced larger pipes being used in distribution mains.	Lead stabilizing compounds may be leached from PVC pipes. Other contaminants arise from the solvents used and include 2-butanone (MEK) and tetrahydrofuran (THF).

TABLE 47

PREFERRED WATER QUALITY AND CONDITIONS OF SERVICE TO MINIMIZE CORROSION OF MATERIALS USED IN THE WATER WORKS INDUSTRY

MATERIAL	pH	HARDNESS	ALKALINITY
IRON-BASED MATERIALS:			
PLAIN IRON	Long term: Little effect for pH 4-10 except localized corrosion may peak at pH 6-9 range (39). Short term: pH effects are a function of flow rate and time (10).	Calcium (Ca^{++}) inhibits corrosion in the presence of sufficient alkalinity.	1. Greater alkalinity produces less aggressive water (34). 2. Anodic dissolution of iron is accelerated by bicarbonate HCO_3^- through the localized formation of $Fe(CO_3)^=$ (7).
GALVANIZED IRON	1. Corrosion rate increases inversely with pH. 2. Optimum pH range is 7-12.	Hard waters are less aggressive than soft waters (21, 37).	Greater alkalinity produces less corrosive waters.
STAINLESS STEEL	Little effect within range of water systems	Not necessary for protection.	Increased bicarbonate alkalinity will inhibit pitting.
COPPER	1. A pH >7 will minimize uniform corrosion; also uniform corrosion will decrease with increasing pH (6). 2. Pitting corrosion will proceed at pH levels above 7.	1. Soft waters are not corrosive if CO_2 is low (18). 2. Pitting corrosion can occur in hard waters which are cold (5).	Addition of bicarbonate may increase corrosion (35, 36).
LEAD	1. A pH of 6-9 is preferred to minimize corrosion (13, 20). 2. A pH of 6.5-7.0 is preferred to minimize corrosion (26).	1. A hardness of 10-100 PPM as $CaCO_3$ is preferred (26). 2. A hardness of 125 PPM as $CaCO_3$ is desirable (32).	An alkalinity of 20 PPM is desireable to form a protective film (12).
ALUMINUM	Optimum pH is 7.0-7.5.	1. In general, soft waters are preferred. 2. The preferred concentration is dependent on the period of immersion. 3. $CaCO_3$ concentration should be approximately equal to the chloride concentration (2).	-
ASBESTOS-CEMENT	pH + Log (Hardness x Alkalinity) should be ≥12.0 (3).		
CONCRETE PIPE	pH levels of 7.0 and greater are preferred to inhibit leaching.	Hardness in excess of 16 PPM Ca^{++} is preferred to inhibit leaching.	Alkalinity in excess of 40 PPM a $CaCO_3$ is preferred to inhibit leaching.
PLASTIC PIPE	-	-	-

TABLE 47
PREFERRED WATER QUALITY AND CONDITIONS OF SERVICE TO MINIMIZE CORROSION OF MATERIALS USED IN THE WATER WORKS INDUSTRY (Cont'd)

MATERIAL	DISSOLVE OXYGEN	CO_2	TDS
IRON-BASED MATERIALS:			
PLAIN IRON	Corrosivity increases linearly with dissolved oxygen concentrations, however, better protective films are formed at higher dissolved oxygen concentrations.	Carbonic acid is aggressive to iron (8).	1. Presence of TDS can decrease O_2 and CO_2 content and, therefore, reduce corrosion. 2. An increase in conductivity can increase the range of galvanic coupling or lead to the formation of a less protective $Fe(OH)_2$ film. 3. The presence of Cl^- and $SO_4^=$ can increase corrosiveness (23). 4. The presence of Cl^- and $SO_4^=$ may improve protectiveness of scale (11).
GALVANIZED IRON	Corrosiveness of water increases directly with increased dissolved oxygen concentrations.	Corrosiveness of water increases with CO_2 concentration (38).	1. At levels required for water treatment, Cl^- does not increase corrosion (1). 2. At levels above that required for water treatment, Cl^- will accelerate corrosion. 3. Corrosion is enhanced with higher neutral salt concentrations (39).
STAINLESS STEEL	1. The presence of dissolved oxygen is necessary for the formation of a protective film. 2. The presence of dissolved oxygen is aggressive and encourages corrosion (14, 29, 33).	-	1. The presence of $SO_4^=$ will inhibit corrosion 2. The presence of Cl^- can cause severe corrosion (severity depends on type of stainless steel), (14, 29, 33).
COPPER	1. Corrosion is negligible in the absence of dissolved oxygen (31). 2. The presence of dissolved oxygen will enhance corrosion, but corrosion or the corrosion rate is not dependent on the dissolved oxygen concentration.	Dissolved CO_2 appears to enhance corrosion.	Specific effects are difficult to identify.
LEAD	Lower concentrations favor inhibition of corrosion.	Excess CO_2 may dissolve protective carbonate films and assist corrosion (32).	1. Chlorides should be minimized (31). 2. Ions that form soluble lead salts should be minimized (32).
ALUMINUM	Minimal dissolved oxygen is optimal (28).	-	1. Chloride concentration should approximate the $CaCO_3$ concentration (2). 2. Preferred TDS concentrations are dependent on the period of immersion.
ASBESTOS-CEMENT	-	Low CO_2 concentrations are preferred.	Possible inhibitory effect of dissolved solids.
CONCRETE PIPE	-	Low CO_2 concentrations are preferred	-
PLASTIC PIPE	-	-	-

TABLE 47.
PREFERRED WATER QUALITY AND CONDITIONS OF SERVICE TO MINIMIZE CORROSION OF MATERIALS USED IN THE WATER WORKS INDUSTRY (Cont'd)

MATERIAL	METAL IONS	ORGANIC ACIDS	VELOCITY
IRON-BASED MATERIALS:			
PLAIN IRON	-	Presence of humic acids will inhibit corrosion (30).	Stagnant waters can cause pitting and localized corrosion; velocities >15 fps accelerates corrosion; optimum velocity occurs between 1 fps and 8-15 fps.
GALVANIZED IRON	Copper even at low concentrations will increase zinc corrosion; the presence of copper is often a main factor in the corrosion of galvanized iron.	Presence of humic acids improves protective deposit formation and reduces corrosion (40).	Velocity differences produce little effect (4).
STAINLESS STEEL	-	-	1. Stagnant waters are corrosive. 2. High velocity can be tolerated.
COPPER	A low concentration of Fe^{++} (0.05-0.5 PPM) may inhibit corrosion.	An unknown organic corrosion inhibitor exists in surface water but not groundwater; this natural inhibitor is probably of high molecular weight and may be an organic acid (5).	1. Some flow is required to form protective Cu_2O film. 2. Velocity in excess of 5 fps can cause impingement attack (19).
LEAD	-	Minimize the occurance of organic acids whose lead salts are soluble (32).	Water should be running, not standing (27).
ALUMINUM	Metal ions, especially copper, tin, nickel, and mercury should be minimized (15,28).	-	Prefer a minimum velocity of 8 fpm (15).
ASBESTOS-CEMENT	Fe^{++} can precipitate to form a protective coating to inhibit calcium leaching (3).	-	Water should be running, not standing (24).
CONCRETE PIPE	Fe^{++} can precipitate to form a protective coating to inhibit calcium leaching (3).	-	Water should be running, not standing (24).
PLASTIC PIPE	-	-	-

TABLE 47.
PREFERRED WATER QUALITY AND CONDITIONS OF SERVICE TO MINIMIZE CORROSION OF MATERIALS USED IN THE WATER WORKS INDUSTRY (Cont'd)

MATERIAL	TEMPERATURE	COMMENTS	
IRON-BASED MATERIALS:			
PLAIN IRON	Aggressiveness of water increases with temperature up to ≈ 80°C; at higher temperatures the aggressiveness decreases (37).	Effects of any single variable are influenced by other parameters, especially the interrelation between pH, temperature, dissolve oxygen, alkalinity, hardness, TDS, and velocity.	
GALVANIZED IRON	Increasing temperature will increase corrosion (16).	Interrelation exists between pH, hardness, temperature, alkalinity, TDS, plus organic acids or other stabilizing agents like phosphates or silicates.	
STAINLESS STEEL	Increased water temperature above 25°C results in a significant increase in pitting susceptibility.	Different types of Stainless Steel have different corrosive tendencies. Cl^- and Dissolve Oxygen are two most important chemical factors in stainless steel corrosion.	
COPPER	Temperature effects are complex but usually not a major factor.	Copper concentration generally does not exceed 5 PPM--may be limited by solubility of reaction product.	
LEAD	Temperatures 20°C and less are preferred for corrosion control (26).		
ALUMINUM	Prefer higher temperatures of 40°C and up (15).	Aluminum corrosion is highly dependent on the period of immersion.	
ASBESTOS-CEMENT	-	-	
CONCRETE PIPE	-	Corrosion control is practiced by minimizing the dissolution of Ca^{++}, often by regulating $CaCO_3$ stability components.	
PLASTIC PIPE	-	1. Corrosion products that have been found are thought to leach from solvents used for joints. 2. No variable cause-effect testing results are available	

TABLE 48. APPLICATIONS OF CORROSION CONTROL MECHANISMS

	LININGS						INHIBITORS					
Surface Material	Coal Tar	Cement Mortar	Epoxy	Vinyl	Misc. Non-Coal Tar Paints		Cathodic Protection	$CaCO_3$	Silicates	Phosphates		
Tanks:												
Concrete												
Steel	✓	✓	✓	✓	✓		✓					
Pipes:												
Iron		✓						✓	✓	✓		
Steel	✓	✓	✓					✓	✓	✓		
Asbestos-Cement				✓				✓				
Reinforced Concrete								✓				
Lead								✓	✓	✓		
Copper								✓	✓	✓		
Plastic												
Galvanized								✓	✓	✓		
Aluminum										✓		

REFERENCES

1. Anderson, E. A., C. E. Reinhard, and W. D. Hammel, "The Corrosion of Zinc in Various Waters," JAWWA, Vol. 26, No. 2, 1934, pp. 49-60.

2. Bell, Winifred, "Effect of Calcium Carbonate on Corrosion of Aluminum in Waters Containing Chloride and Copper," Journal of Applied Chemistry, 12, February, 1962, pp. 53-55.

3. Buelow, R. W., J. R. Millette, E. F. McFarren, and J. M. Symons, "The Behavior of Asbestos-Cement Pipe Under Various Water Quality Conditions," Presentation-1979 AWWA Conference, San Francicso, June 27, 1979.

4. Burgmann, G., W. Frieke, and W. S. Schwenk, "Chemical Corrosion and Hygienic Aspects of The Use of Hot-Galvanized Threaded Pipes in Domestic Plumbing for Drinking Water," Pipes & Pipelines Int., Vol. 23, No. 2, 1978, pp. 11-15.

5. Campbell, Hector S., "A Natural Inhibition of Pitting Corrosion of Copper in Tap-Waters," J. Appl. Chem., Vol. 4, 1954, pp. 633-647.

6. "Cold-Water Corrosion of Copper Tubing," Task Group Report, JAWWA, Vol. 52, August, 1960, pp. 1033-1040.

7. Davies, D. H., and G. T. Burstein, "The Effects of Bicarbonate on The Corrosion and Passivation of Iron," Corrosion-NACE, Vol. 36, No. 8, August 1980, pp. 416-422.

8. DeWaard, C. and D. E. Milliams, "Carbonic Acid Corrosion of Steel", Corrosion-NACE, Vol. 31. No. 5, May, 1975, pp. 177-181.

9. Donaldson, W., "The Action of Water on Service Pipes," JAWWA, Vol. 11, No. 3, 1924, p. 649.

10. Eliassen, R., C. Pereda, A. J. Romeo and R. T. Skrinde, "Effects of pH and Velocity on Corrosion of Steel Water Pipes," JAWWA, Vol. 48, August, 1965, pp. 1005-1018.

11. Fergenbaum, C. L. Gabor and J. Yahalom, "Scale Protection Criteria in Natural Waters," Corrosion (Houston), Vol. 34, No. 4, 1978, pp. 133-137.

12. Gardels, M. and Schock, E.P.A. Cincinnati Laboratory, Personnel Communication via P. Lassovszky, Jan. 12, 1981.

13. Garrels, R. M., M. E. Thompson and R. Siever, "Control of Carbonate Solubility by Carbonate Complexes," American Journal of Science, Vol. 259, January, 1961, pp. 24-45.

14. Geld Isidore, and Colin McCaul, "Corrosion in Potable Water," JAWWA, Vol. 67, No. 10, October, 1975, pp. 549-552.

15. Godard, H.P.,"The Corrosion Behavior of Aluminum in Natural Waters," The Canadian Journal of Chemical Engineering, Vol. 38, No. 5, October, 1960, pp. 167-173.

16. Goetchins, D.R.,"Porcelain Enamel as a Protective Coating for Hot Water Tanks," J.Am. Ceramic Society, Vol. 25, 1942, pp. 164-168.

17. Goldfarb, A.S., J. Konz, and P. Walker,"Interior Coatings in Potable Water Tanks and Pipelines," MITRE Corporation, Technical Report MTR-7803, U.S. EPA Contract No. 68-01-4635, January, 1979.

18. Hale, F. E.,"Relation of Copper and Brass Pipe to Health," Water Works Eng., Vol. 95, 1942.

19. Hatch, G.B.,"Unused Cases of Copper Corrosion,"JAWWA, Vol. 53, 1961, pp. 1417-1429.

20. Karalekas, P. C., G. F. Craun, A. F. Hammonds, C. R. Ryan, and D. J. Worth,"Lead and Other Trace Metals in Drinking Water in The Boston Metropolitan Area,"J. New England Water Works Association, Vol. 90, No. 2, pp. 150-172, 1976.

21. Lane, R. W., and C. H. Neff,"Materials Selection for Piping in Chemically Treated Water Systems,"Materials Protection, Vol. 8, No. 2., February, 1969, pp. 27-30.

22. Larson, C. D., Chief Technical Support Section, EPA Region I, letter to John Hagopian, Rhode Island Dept. of Health, November 14, 1979.

23. Larson, T. E., and R. V. Skold,"Laboratory Studies Relating Mineral Quality of Water to Corrosion of Steel and Cast Iron," Corrosion, Vol. 14, June, 1958, pp. 43-46.

24. McCauley, R. F., and M. O. Abdullah,"Carbonate Deposits for Pipe Protection,"JAWWA, Vol. 50, 1958, pp. 1419-1428.

25. Merrell, D. T. and R. L. Sanks,"Corrosion Control by Deposition of $CaCO_3$ Films. Part 1, A Practiced Approach for Plant Operators,"JAWWA, Vol. 69, November, 1977, pp. 592-599.

26. Moore, M. R.,"Plumbosolvency of Waters,"Nature, Vol. 243, May 25, 1973, pp. 222-223.

27. O'Brien, J. E."Lead in Boston Water: Its Cause and Prevention," Journal of The New England Water Works Association, Vol. 90, No. 1, January, 1976, pp. 173-180.

28. Porter, F. C. and S. E. Hadden,"Corrosion of Aluminum Alloys in Supply Waters,"J. Applied Chemistry, Vol. 3, September, 1953, pp. 385-409.

29. Reedy, D. R.,"Corrosion in The Water Works Industry,"Materials Protection, Vol. 5, No. 9, September, 1966, pp. 55-59.

30. Rudek, R., Blankenhorn, R., and H. Sontheimer,"Verzögerung der Eisenoxidation Durch Natürliche Organische Wasserinhaltsstoffe und Duren Auswirkung auf die Korrosion Van Schwarzen Stahlrohren,"Von Wasser, Vol. 53, 1979, pp. 133-146.

31. Schafer, G. T.,"Corrosion of Copper and Copper Alloys in New Zealand Potable Waters", New Zealand Journal of Science, Vol. 5, Dec. 1962, pp. 475-484.

32. Slunder, C. J., and W. K. Boyd, Summary Report on Lead - Its Corrosion Behavior to ILZRO, Battelle Memorial Institute, Columbus, Ohio.

33. Streicher, Lee,"Effects of Water Quality on Various Metals,"JAWWA, Vol. 48, No. 3, March, 1956, pp. 219-238.

34. Stumm, W.,"Investigation on The Corrosive Behavior of Waters," Proceedings of The American Society of Civil Engineers, Vol. 86, No. 5A-6, November, 1960, pp. 27-45.

35. Tronstad, L., and R. Veimo,"The Action of Water on Copper Pipes", Water and Water Eng., Vol. 42, May, 1940, pp. 189-191.

36. Tronstad, L., and R. Veimo,"The Action of Water on Copper Pipes,"Water and Water Eng., Vol. 42, June, 1940, pp. 225-228.

37. Uhleg, H. H., Corrosion and Corrosion Control as Introduction to Corrosion Science, John Wiley & Sons, Inc., New York, 1963.

38. Uhleg, H. H., The Corrosion Handbook, John Wiley and Sons, Inc., New York, 1948.

39. Wagner, Ivo,"Influence of Water Quality and Water Treatment on Corrosion and Coatings in Steel and Galvanized Steel Tubes,"EUROCOR 77, 6th Euorpean Congress on Metallic Corrosion, (MET. A., 7807-72 0184), 1977, pp. 413-419.

40. Waring, F. H.,"Prevention of Corrosion by The Application of Inhibitors,"JAWWA, Vol. 30, No. 5, 1938, pp. 736-745.

8. Recommendations

The U.S. Environmental Protection Agency was mandated by the Safe Drinking Water Act of 1974 (PL 93-523) to safeguard public drinking water supplies and to protect the public health. Under the act, EPA is required to establish and enforce National Drinking Water Regulations which include corrosion by products in the water that pose a threat to human health. Information to form a basis for developing responsible and implementable corrosion control regulations is partially available from results of historical laboratory and field studies reported in the literature since the 1920's. Many of these investigations were conducted to identify and quantify water quality and conditions of service characteristics which influence corrosion of the various materials used in the water works industry. To supplement this data base, the EPA has initiated additional studies. These studies have focused on more clearly defining and quantifying the various aspects of potential corrosion control strategies which can assist in developing a reasonable corrosion control program. Recent primary emphasis has been to determine the magnitude of the problems and to search for both monitoring and corrective actions which can be enforced. Results of these studies have overwhelmingly concluded that the nature of corrosion and possible corrosion control alternatives are extremely complex. Limiting the corrosiveness of water by the use of a universal corrosion index or parameter is not feasible at this time. Instead it appears that corrosion control can only be accomplished through a comprehensively applied program on a community water system case by case basis.

An example of the complexity and effort involved in administering a responsible corrosion control program is provided by the Seattle Water Department. Their approach includes an extensive monitoring program coupled with an attempt to provide a wide range of treatment techniques including the addition of various corrosion inhibitors. Monitoring is continued simultaneously with the application of treatment alternatives to evaluate performance. In some cases, the results of the monitoring programs are not in agreement with, or at least do not reflect the expectations of the laboratory results. From this experience and from the results presented in the literature it is evident that a complicated and potentially expensive program is necessary to insure the success of corrosion control at each community water system. The following procedures outline a comprehensive program for corrosion control for specific water utilities.

1) Identify and quantify the materials used in the respective water works industry.

2) Characterize the delivered water quality with respect to pH, and concentrations of alkalinity, hardness, carbon dioxide, metal ions, total dissolved solids, organic acids, and temperature as a minimum.

3) Identify potential contaminants which would be indicative of corrosion with respect to materials used and water quality characteristics.

4) Develop and implement corrosion control treatment alternatives which may be feasible for the particular system.

5) Continue monitoring to determine effectiveness of corrosion control strategies.

6) Develop guidelines for future construction practices and local plumbing codes.

With an estimated 60,000 public water suppliers, it is evident that to administer or enforce a program, with these requirements and recommendations, on a nationwide scale would be both technically and economically prohibitive. In many cases the treatment plant operators do not have the technical skills nor do the small municipalities have the financial resources to implement an extensive corrosion control program. Because some potential control methods may enhance corrosion in specific instances, corrective action by unskilled personnel may further aggravate an existing problem. It is also important to note that relatively few water suppliers recognize a corrosion problem or consider corrosion control a high priority.

It is recommended that corrosion control regulations be developed which are both technically and economically reasonable and which can be enforced effectively. This recommendation implies that regulations should be developed which would first screen water suppliers to assess the potential for corrosion in their respective systems. Only those systems suspected of having corrosion problems should be required to initiate further investigations and corrective actions.

Specifically, water suppliers should be required to identify and quantify materials and practices used in their respective water systems. For smaller utilities this inventory can perhaps be easily compiled. For larger utilities, an accurate inventory may be impossible to compile. In these cases, an inventory estimate through use of recent records and historical plumbing codes can be made. This approach implies that developed areas within a large municipality can be sectioned with respect to historical growth, and applicable plumbing codes, which were in affect during that development period, can be superimposed on the respective area to provide some estimate. This procedure will require that large municipalities provide an historical review of respective local and state plumbing codes.

Water suppliers should also be required to conduct water quality investigations throughout their systems to assess corrosivity potential. As addressed in this study, water quality parameters which have historically been investigated to evaluate corrosion characteristics of specific materials used in the water works industry are pH, and concentrations of hardness, alkalinity, dissolved oxygen, carbon dioxide, transition metal ions, total dissolved solids, organic acids, and temperature. It is recommended that these water quality parameters be included as a minimum for a water quality characterization portion of a corrosion control program.

The water quality sampling and analysis program must be designed to effectively characterize the water quality conditions within the system. Samples should be taken from both the water source and at locations following treatment before it enters the distribution system. Samples should also be taken at various representative points along the distribution system as well as at consumer taps. It should be recognized, however, that water quality changes may occur as the water passes through the distribution system and these changes may produce erroneous results. For example, as previously discussed, corrosive water passing through asbestos-cement pipe will tend to leach calcium from the pipe and become less aggressive. Corrosive water samples taken from points long distances downstream of asbestos-cement pipe sections may appear non-corrosive.

The water quality data collected, specifically pH, alkalinity, and hardness can be used to develop corrosion indicators such as the Langelier Index and/or the Aggressive Index. Although limited in use, these two indices are apparently the most widely accepted indicators of corrosiveness of water and should be used as appropriate. However, their limitations should be recognized as previously discussed and corrosivity should not be assessed exclusively by these parameters.

Water suppliers which are identified as having conditions susceptible to corrosion should be encouraged to initiate monitoring programs designed in accordance with the results of the materials and practice inventory and the water quality survey. Those utilities which are found to use lead or materials which are sources of contaminants that adversely affect health extensively should be given priority consideration.

Water systems suspect of having corrosion problems should be encouraged to initiate a monitoring and detection program to determine the nature and extent, if any, of corrosion. The design and extent of this monitoring program is contingent on the potential contaminants which can be expected as dictated by the specific materials used. Potential contaminants which are associated with each material are listed in Table 49. It is suggested that non-subjective (i.e. analytical as opposed to visual) monitoring techniques be used and that samples be taken from consumer's taps and distribution mains. Coupon testing and electrochemical techniques should currently be considered only as indicators of corrosion. It is not necessary to monitor for every associated potential contaminant but rather to monitor for only a few. The potential magnitude of the corrosion problem and the staffing and financial resources of specific water suppliers should dictate the extent of the monitoring programs.

After identification of the specific corrosion related contaminants, an applicable treatment technique relating to the material involved and water quality parameters must be devised. Historically, attempts at corrosion control by $CaCO_3$ deposition have been used most often, but corrosion inhibitors or modifications in calcium carbonate control as noted in chapter 6 should also be considered.

Local jurisdictions should respond to corrosion problems by adjusting the local plumbing code to recommend some pipe materials and not allow others to be used. Lead and unlined asbestos cement piping should be reviewed. Plastic pipe is becoming more accepted and appears to resist corrosion, but the possibility of small amounts of potentially harmful organics entering the water should be further researched. Similarly, pipe coatings should be reviewed in relation to the local waters. The low flow resistance of epoxy coatings may be recommended pursuant to research on trace contaminant release.

TABLE 49. MATERIALS AND THEIR ASSOCIATED CORROSION PRODUCTS

Material	Potential Corrosion Products
Iron-based materials	Fe, Cd, Pb, Zn
Copper	Cu, Fe, Zn, Sn, Pb, Mn, As, Sb, P, Bi
Lead	Pb
Aluminum	Cu, Mg, Si, Fe, Mn, Cr,
Asbestos-Cement	Asbestos fibers, and tetrachlorethylene
Concrete	Si, Al, Fe, Mg, S
Plastic	Pb, components of various solvents including 2-butanone (MEK) and tetrahydrofuran

If several pipe materials are corroding, and the products are identified, it may not be possible to engineer a single program to alleviate all of the contaminants. In this case, where it is possible that the solution to one facet of the problem may aggravate another, the primary concerns should be control of those products deemed most detrimental in terms of health effects. Specifically, this would include lead and asbestos fibers. These may not be the easiest contaminants to justify in terms of economics or aesthetics, but their control should be paramount.

Irrespective of corrosion control regulations, the EPA should review local, state, and national plumbing codes and begin assessing the potential for discouraging the use of potentially dangerous materials and practices. Lead and lead-based solders should be seriously considered for discontinuing their extensive use. Although a large quantity of lead pipe is currently in service, its extensive use has begun to decline, but lead-based solders are currently the most widely used solders for joining copper pipers.

Other Noyes Publications

TRIHALOMETHANE REDUCTION IN DRINKING WATER
Technologies, Costs, Effectiveness, Monitoring, Compliance

Edited by

Gordon Culp

Pollution Technology Review No. 114

This book evaluates technologies for the effective reduction of trihalomethanes (THMs) in drinking water. It is based on studies by *Culp/Wesner/Culp; Temple, Barker & Sloane, Inc.; Malcolm Pirnie, Inc.,* and the *U.S. Environmental Protection Agency.* In addition to the treatment technologies described, their costs and effectiveness are evaluated. Monitoring methods and compliance with federal drinking water regulations are also covered.

The first part of the book will serve as a guidance manual for those planning changes in water treatment systems for THM control. It discusses THM formation, necessary steps which must be followed for compliance with THM maximum contaminant levels (MCLs), and procedures to ensure preservation of the finished water.

The second part of the book describes best available treatment methods as well as potentially available treatment methods. Cost analyses with ranges of possible costs are included.

Part III covers monitoring and compliance as they pertain to promulgated federal drinking water regulations. Suggestions and recommendations for implementation of the TTHM (total THM) amendment, issued in February 1983, are provided.

A condensed table of contents listing **parts and chapter titles and selected subtitles** is given below.

ISBN 0-8155-1002-0 (1984) **251 pages**

CONTAMINANT REMOVAL FROM PUBLIC WATER SYSTEMS

by

Daniel C. Houck et al

Pollution Technology Review No. 120

Based on studies by *D.H. Houck Associates; Rip G. Rice, Inc.; Wade Miller Associates; Purdue University and the City of Indianapolis, Indiana;* and *Environmental Science and Engineering, Inc.;* this book describes the rationale as well as methods for contaminant removal from public water systems. The book is for use by owners and operators, municipal managers and consulting engineers involved with the safe and efficient operation of public water systems. Its purpose is to assist personnel in understanding the importance of contaminant control and to explain design concepts, cost estimating techniques, and operational considerations associated with current technological approaches for maintaining such control.

The book is divided into four parts, each dealing with a distinct type of contaminant removal. Problem areas covered are microorganisms, nitrates, radionuclides, and turbidity.

The information provided is intended to explain:

1) why contaminant control is important,
2) theories of control,
3) process options for control,
4) design procedures for control,
5) process control methods,
6) operation and maintenance procedures,
7) cost estimation methods.

A condensed table of contents listing **part titles and selected subtitles** is given below.

ISBN 0-8155-1022-5 (1985) **524 pages**